W0232475

THE RIFLE MUSKET IN CIVIL WAR COMBAT

Earl J. Hess

The Rifle Musket in Civil War Combat

REALITY AND MYTH

University Press of Kansas

Published by the University Press of Kansas (Lawrence, Kansas 66045), which was organized by the Kansas Board of Regents and is operated and funded by Emporia State University, Fort Hays State University, Kansas State University, Pittsburg State University, the University of Kansas, and Wichita State University

ISBN 978-0-7006-1607-7

Printed in the United States of America

FOR PRATIBHA AND JULIE

with love

Contents

Acknowledgments

I wish to thank Wayne Wei-Siang Hsieh for contacting me in the spring of 2004, encouraged by his dissertation adviser, Gary W. Gallagher at the University of Virginia. Gary had told Wayne of my work on Civil War fortifications, which seemed relevant to the issues Wayne was exploring in his dissertation. An e-mail correspondence with Wayne inspired me to undertake this study after eighteen years of waiting for someone to respond to Paddy Griffith's call for a reevaluation of the standard interpretation of the rifle musket's impact on Civil War combat. David Perry also forwarded information about this issue to me from a Web site.

Most important, my gratitude to my loving wife, Pratibha, for everything.

Introduction

This book is about the rifle musket and its use in the Civil War. Ever since the end of the conflict, the prevailing view of this weapon has been that it revolutionized warfare because of its increased range. Participants and latter-day historians alike have assumed that because the rifle deepened the killing zone so much—from roughly 100 yards to about 500 yards—it produced significant results. The rifle musket has been blamed not only for the unusually heavy casualties of the war, but for prolonging the conflict by rendering engagements indecisive. Four years of increasingly brutal fighting, with battles that resulted in no clear-cut victory for either army, created a national tragedy. Those historians who accept the prevailing interpretation highlight the increased firepower of soldiers who defended a position with rifles, arguing that defensive measures ruled Civil War tactics. The widespread use of the rifle led to a dramatic increase in the employment of sophisticated field fortifications, so the traditional interpretation goes, further strengthening the use of defensive over offensive action. The rifle also reduced the ability of field artillery to support infantry attacks because the cannon could not advance close to the enemy without losing too many gunners, and it negated the ability of cavalry to attack and break up infantry formations. In short, those who believe that the rifle had a revolutionary impact on combat strive to account for every characteristic they see in Civil War battles by reference to the use of the new weapon.

The Standard Interpretation

Even before the firing on Fort Sumter, observers in Europe and the United States assumed that the rifle musket would revolutionize warfare. They wrote books and articles expressing their wonder at the possibilities inherent in the new technology. One can hardly blame them. Still largely untested in the hands of ordinary soldiers on a real battlefield, the new weapon drew speculation that was as yet unburdened by the realities of actual use in war.

But these writers issued an important caveat. They recognized that the new rifle's trajectory traced an arc—the bullet curved dramatically upward before descending to the ground, rather than sailing on a relatively flat line toward its intended victim. That meant that a wide space of the "killing zone" was safe passage for soldiers on the opposing side. The parabolic trajectory was so high that bullets would fly over the heads of many opponents, creating two killing zones. If a man adjusted the sights of his rifle musket for a range of 300 yards, the bullet ascended so that the first killing zone was about 100 yards long (almost exactly the same distance as the effective range of a smoothbore musket). The second killing zone lay at the far end of the bullet's trajectory and was only about 75 yards long, because it descended at a sharper angle than it had ascended. For nearly half the 300-yard range, enemy troops would be untouched by the balls. Pre–Civil War rifle enthusiasts tried to point out that it was absolutely necessary to train recruits how to use the weapon to compensate for this unique phenomenon. Soldiers had to be taught how to estimate distances and then to adjust the sights of their rifles to account for the complex problem of hitting a human target at a range greater than that of the smoothbore musket. In contrast, the old smoothbore had a much flatter trajectory and was easy to aim at the comparatively short distance that marked its range. It demanded little in the way of specialized skills. Without proper training, however, a soldier would be unlikely to maximize the full capabilities of the new rifle musket.[1]

The prewar rifle enthusiasts, including Lt. Cadmus M. Wilcox of the Seventh U.S. Infantry, noted that sophisticated musketry schools began to operate in Europe by the late 1850s to teach soldiers the needed rifle skills. They assumed it would happen in the United States as well, but that did not come to pass until years after the Civil War ended. Nevertheless, nearly all Americans who wrote about the subject continued to assume the rifle was the wave of the military future and was already in the process of revolutionizing warfare. A reviewer of William H. Morris's *Field Tactics for Infantry* proclaimed the old smoothbore musket obsolete in 1864. The new rifle reigned supreme on the battlefields of that year, making massed frontal attacks prohibitively costly, or so most observers believed.[2]

Twentieth-century historians have accepted the assumptions of the pre–Civil War rifle enthusiasts with little alteration or questioning, except to ignore their plea for rigorous training of all soldiers to use the rifle. They share the earlier generation's fascination with the increased range of the rifle, which they place at anywhere from 250 to 1,000 yards, and tend to assume that all their conclusions about the nature of Civil War combat stem from that increased range of fire. Historians readily take at face value comments by contemporaries that denigrated the effectiveness of the smoothbore musket, such as the observation of Brig. Gen. Stephen Vincent Benet, post–Civil War chief of ordnance, that it "was not much more efficient as a weapon of accuracy and range than a piece of gas-pipe closed at one end." Modern historians have readily employed phrases such as "the devastating impact of rifled musket fire" in characterizing the nature of Civil War combat. They stress the newness of the weapon against the old technology of the smoothbore, arguing without evidence that it "could be more reliably and accurately aimed." There is also no evidence that the new weapon could be fired faster than a smoothbore, yet many historians have assumed that it had a faster rate of fire.[3]

The tragedy of the war, in the view of traditionalist historians, lies in the fact that the Civil War generation continued to fight with old-fashioned tactics against devastating new weapons. "The tactical predominance of the defense helps explain why the Civil War was so long and bloody," writes James M. McPherson. "The rifle and trench ruled Civil War battlefields as thoroughly as the machine-gun and trench ruled those of World War I." Russell Weigley has written that the Civil War battlefield was "a killing ground par excellence," illustrative of "the devastating effect of the rifled musket in the hands of steady and well-positioned defenders against a frontal attack." In another study, Weigley argued that "Civil War battles soon demonstrated that rifles tended to tear any frontal attack to shreds before it could close." Many historians believe that "the advantage had swung sharply to the tactical defensive during the Civil War" because of the new weapon, especially when coupled with the digging of field fortifications. In fact, most historians accept the idea that fieldworks were a natural outcome of the use of rifles. "Because of the destructive power of the rifle," Weigley has written, "soldiers increasingly looked for the shelter of stone walls or dug rifle pits or trenches as soon as they halted anywhere."[4]

Believers in the standard interpretation are certain that both field artillery and cavalry had significant offensive capabilities in the era of the smoothbore. By moving forward with attacking infantry and firing at ranges in which the smoothbore could not reach them, artillerists helped to break up defending infantry formations; and the cavalry did the same by conducting mounted charges against enemy foot soldiers. Neither arm could do so anymore, their offensive capabilities neutralized by the long range of the rifle musket.[5]

Alternatives to the Standard Interpretation

It is time to reevaluate the standard interpretation of the rifle musket's impact on the Civil War. This book proposes an alternate view of the rifle as having only an incremental, limited effect on Civil War combat, mostly in the area of skirmishing and sniping. The operations of the battle line, the main body of Civil War infantry, were affected comparatively little by the use of the rifle, mostly because Civil War soldiers never utilized the long-range capabilities of their new weapon, for several reasons that will be discussed. Moreover, it becomes apparent that the average soldier was incapable for many reasons of realizing those long-range firing capabilities, and therefore the rifle could not revolutionize the operations of the main battle formations of Civil War armies. Sometimes on the skirmish line, and mostly in the highly specialized craft of sniping, some soldiers realized the long-range capabilities of the rifle musket in impressive ways. They tended to be men who had either a natural aptitude for firearms or had received some type of specialized training in gauging distances. Because the Civil War rifle was still a musket—a single-shot weapon loaded from the muzzle—its rate of fire remained the same as that of pre–Civil War smoothbores. Other than the possibility that it was a bit more accurate, a claim that has never been proven, the only advantage offered by the rifle musket was increased range. Given the difficulties of seeing man-sized targets 200 to 500 yards away, especially in wooded or hilly terrain, there were many obstacles to the soldier's ability to effectively use his weapon at that range. The most important difficulty, even on a level, open field of 500 yards, was that of accounting for the parabolic arc of the bullet and the consequently short killing zone afforded by the ball's trajectory. The vast majority of soldiers in the Union and Confederate armies received no training in estimating distances and therefore were not prepared to effectively fire their muskets at long-range targets.

Civil War commanders and soldiers alike preferred to wait until the enemy came quite close before opening fire. Several studies have indicated that most Civil War fire was delivered at distances well below the effective range of the rifle musket. In fact, most Union and Confederate soldiers opened fire at ranges quite consistent with, or a bit longer than, the range of the smoothbore musket.

If that is true, and there is ample evidence to support the argument that Civil War soldiers tended to fire at short range, that fact alone justifies a reevaluation of the rifle musket's impact on Civil War military operations. One problem with the standard interpretation has always been that it is too myopically focused on the theoretical range of the new weapon, without an attempt to verify that soldiers actually fired the gun at the new range. Those who have accepted the standard interpretation of the rifle musket's role in the Civil War also tend to ignore important characteristics of combat in wars that occurred before 1861 and after 1865. Heavy losses were not unique to the Civil War, and field artillery and cavalry never were as effective against infantry in eighteenth-century warfare or in the Napoleonic era as historians have assumed. Looking at other battles in other wars, one gets the impression that the Civil War was but one episode in the grand sweep of Western military history. It shared more characteristics with previous conflicts than the holders of the standard interpretation assume. At best, the rifle musket had an incremental effect on changing the nature of combat for a few selected functions on the battlefield, such as skirmishing and sniping. It did not revolutionize warfare.

The rifle musket had only a brief showing in world history. Within five years of Appomattox, opponents in the Franco-Prussian War fought with breech-loading rifles far superior to the rifle musket of the Civil War. Twenty years after that, most modern armies adopted bolt-action rifles that were superior to the breechloader. Increasing the rate of fire with magazine-fed weapons that had improved reloading capabilities represented the truly revolutionary aspect of small-arms development in the nineteenth century, not the mere rifling of a single-shot, muzzle-loading weapon whose reloading and firing mechanism was unchanged from that of a smoothbore. Ironically, despite the introduction of powerful small arms in the twentieth century, the evidence indicates that infantry fire continued to take place at ranges of roughly 100 yards in the major wars of that century. In short, the sophistication and power of long-range small arms had little effect on the range at which most infantry combat took place. Soldiers in World War II, Korea, and Vietnam continued to fire at ranges consistent with that of men in the War of the Spanish Succession and the American Civil War.

Ironically, historians have consistently denigrated the use of linear tactics as an outdated tactical formation, given the longer range of the rifle musket. If firing took place at ranges consistent with the smoothbore musket, then linear tactics, which were developed over centuries of European experience to conform to the smoothbore, were the proper tactics to use in the Civil War as well. Moreover, a careful study of battle reports supports the argument that unit commanders and soldiers alike learned those tactics well and executed them effectively to take their regiments forward into battle or extricate them from dangerous situations. Far from obsolete, linear tactics were highly successful in the Civil War. Gory and failed assaults like that which the Federals conducted against the Stone Wall at Fredericksburg, or like Lee's men conducted against Cemetery Ridge at Gettysburg, were not the fault of the tactical formations employed. Those are case studies of failures in command decisions—the question is whether those attacks should have been made at all, regardless of the tactical formations used. Judging tactics by the results of a handful of bad decisions on the grand tactical level is unfair. Linear tactics should rightly be judged by the hundreds of individual case studies on the regimental level in a variety of engagements. The irony is that Civil War soldiers learned linear tactics well and effectively used them on the battlefield, even though they failed to make the most of the long-range firing capability of their rifle muskets.

A handful of historians have suggested that the standard interpretation is in need of revision, and they have contributed significant ideas to that effort. The English historian Paddy Griffith, who has written on the art of war in several eras, became the first historian to question the validity of the standard interpretation when he published *Battle Tactics of the Civil War* in 1986. Griffith based his points, among other things, on a survey of the primary literature, which revealed that typical ranges at which Civil War soldiers delivered fire were much shorter than the effective range of the rifle musket. He concluded that the average range in battles fought in 1863 amounted to only 127 yards, while those of the following year averaged 141 yards. Griffith also found evidence in the primary literature that the range at which fire was most decisive in stopping attacks was much shorter than that, only about 33 yards in many cases. He rightly concluded that "the existence of a wide field of fire may be regarded as irrelevant to the effect of musketry in the Civil War." The rifle musket failed to bring down large numbers of opposing troops at long distance and therefore its arrival "actually made very little practical difference." At most, Griffith allowed that "minor improvements" in warfare followed the introduction of the rifle, while the truly important break with the past occurred only with the post–Civil War introduction of the breechloader.[6]

In Griffith's view, the length of the conflict was, in part, due to the lack of skill among higher-level commanders in handling their units. Griffith does not believe that the advantages shifted to the defense in the Civil War. All armies acting on the defensive usually had some advantages over the attacker; if those advantages seemed emphasized in the American conflict, it was due to other factors. Griffith asserts that linear tactical formations were not outdated by the weapon but were as relevant to the single-shot rifle as to the single-shot smoothbore. As in the days of Napoleon, linear tactics effectively brought infantry forces within short range of the enemy, where a heavy firefight decided the victor.[7]

Griffith chided Civil War historians for viewing their subject within a modernist framework, focusing on the introduction of new military technology to define the character of Civil War combat. Because those historians overlooked the fact that the long-distance killing potential of the new weapon was not realized, the result was a standard interpretation that was "anachronistic and exaggerated." It had "rather more to do with late twentieth-century habits of thought than with the military realities of the 1860s." Griffith convincingly portrays the Civil War as the last Napoleonic conflict, while the true shift toward modern war took place in the Franco-Prussian conflict of 1870–1871.[8]

Griffith issued a clarion call for revision of the standard interpretation that was ignored by most Civil War historians and vehemently rejected by the main exponents of the standard interpretation. I was struck by his evidence regarding the short range of most Civil War firing and believed it justified revision of the standard interpretation. Most of his other conclusions also are convincing, except Griffith's odd attack on the importance of field fortifications, a topic I have covered in previous work. It took time for me to give further thought to these issues. Indeed, I retained a knee-jerk adherence to the standard interpretation for some years until an opportunity to support Griffith's conclusion in print occurred with the publication of *The Union Soldier in Battle* in 1997. In that study, while attempting to delineate the experience of combat in the Civil War, I briefly dealt with the typically short range of firing and suggested further work was needed along the lines that Griffith had established.[9]

Mark Grimsley was influenced by Griffith's work and conducted his own brief exploration of the firing ranges typical of Civil War combat. Published in 2001, Grimsley's results indicate an average range of 116 yards. He concluded that it was "an improvement over the 80–100 yards characteristic of smoothbore warfare, but at best an incremental improvement."[10]

Brent Nosworthy, author of an important book on the development of linear tactics in Europe, published a study of Civil War battle tactics in 2003. He agreed with Griffith's findings regarding the short range of firing and conducted his own survey of Civil War battles to support them. Although Nosworthy believes the average range of firing during the Civil War was "more than a 50% improvement" over that of the Napoleonic era, he nevertheless concludes that the rifle enthusiasts who had predicted a revolutionary effect on warfare were proven wrong. Civil War soldiers would have had to fire at much longer ranges than they did, reaching out accurately 300 or more yards, to have fulfilled the promise of the rifle musket, and they clearly did not do so.[11]

In 2005 and 2007, I had an opportunity to approach the question of whether the rifle musket fostered the increased use of fieldworks in the 1864–1865 campaigns in Virginia. In the first two volumes of a projected series of studies on the use of field fortifications in the Civil War, I argued that continued close contact with the enemy was the most significant factor that led to the rapid development of trench warfare in mobile campaigning in Virginia during the last year of the war.[12]

The testimonies of Grimsley and Nosworthy are encouraging signs that at least a few Civil War historians are willing to reconsider the standard interpretation, although no one to date has conducted a thorough reevaluation of it. This study is an attempt to look at all aspects of the rifle musket's role in the Civil War, including fundamental issues such as the typical range of infantry firing, in order to understand the subject from the fresh perspective made possible by Griffith's seminal ideas. While it is important to understand in detail what happened on the Civil War firing line, it is just as important to understand what happened on the battlefields of other wars as well. When one realizes that casualties in European battles fought with smoothbore muskets tended to be just as high, if not higher, than Civil War losses, and that infantrymen in World War II, Korea, and Vietnam typically fired their small arms at ranges that were very close to those of smoothbore battles, the old idea that the rifle musket fostered revolutionary changes in warfare seems wrong.

Some definitions are in order. The term ***rifle musket*** (sometimes with a hyphen linking the two words) refers to a weapon that had spiral grooves (the rifling) inserted during initial manufacture. In contrast, a ***rifled musket*** was a smoothbore weapon with spiral grooves added sometime after its manufacture. I use the former term because the majority of rifles used during the course of the Civil War were originally made as such.[13]

THE SMOOTHBORE AND RIFLE HERITAGE

The Civil War was the last major conflict in which both sides were armed with a single-shot, muzzle-loading firearm. Civil War soldiers represented the end of a long, rich heritage of fighting men who relied on a musket for defense and offense on hundreds of battlefields. The Civil War was also the first large conflict in which that type of firearm was rifled, but it was still a musket, capable of only one firing at a time. That limitation was a more important influence on the nature of combat in the Civil War than was the rifling in the barrel.

Smoothbores, Tactics, and Firing Systems before 1756

An effective smoothbore musket had been developed at least 150 years before the firing on Fort Sumter. The matchlock, an early version, was in the hands of about two-thirds of all infantrymen in European armies by 1600. It was a heavy weapon, requiring the use of a forked stick to hold up the muzzle. The matchlock had a misfire rate approaching 50 percent. It was powerful enough to send a ball about 250 yards, but soldiers could hardly expect to hit their target beyond 60 yards. The wheel lock musket, with an improved detonation system, had a similar range.[1]

The Thirty Years' War (1618–1648) witnessed a great deal of improvement in all aspects of war making, much of it due to the innovations of Gustavus Adolphus of Sweden. Prepared ammunition for infantry, with a cartridge that increased the rate of fire by several times, was introduced. The wheel lock was lightened enough to enable soldiers to throw away their forked sticks. Soldiers continued to use pikes, but more and more troops were armed with muskets as well. Gustavus Adolphus tended to mass his musketeers into larger formations to more effectively deliver their fire.[2]

The development of a flintlock represented the next generation of musket technology and spelled the end of the pike. Initially called a fusil, it caused a shard of flint to strike a metal plate, creating a spark to detonate the charge. This made it safer to use around artillery and the grenades carried by grenadiers because the weapon involved no match. The flintlock was developed in Italy about 1630, but the armies of Europe did not adopt it until the period ranging from 1690 through 1710. The flintlock musket had a range of no more than 100 yards, but its rate of fire could be as much as two rounds per minute. The most important innovation was its flint-based mechanism, the most reliable detonation system yet developed for a firearm. Nearly as important, a socket bayonet became available by the 1690s. It was designed to fit around the muzzle without impeding the loading or firing of the musket. The socket design was a great improvement over the plug bayonet, which literally was stuck in the bore and therefore could be used only when firing was unnecessary. While initially used only as a defensive weapon, the socket bayonet soon became the centerpiece of infantry attacks that involved no firing at all. Many national armies honed their skills at bayonet attacks, which often turned the tide of battle. Another important improvement in the musket was the replacement of wooden ramrods by iron. They broke less often and allowed for an increased rate of fire. Prussia set the trend by adopting the iron ramrod in 1718, but the French waited until 1750 to follow suit.[3]

Rifles had been developed by the middle of the sixteenth century, but their range was not appreciably greater than that of the smoothbore musket. The rate of fire and the level of dependability of the rifle were much below the standards of the smoothbore. The difficulty of loading the tight-fitting balls so they could engage the spiral grooves upon discharge made the rifle a specialized weapon demanding skills beyond those of most peasant recruits. Expensive as well, rifles were not suited for general distribution to the armies until much later. Meanwhile, they were mostly used by elites for hunting.[4]

As the smoothbore musket became generally available by the start of the eighteenth century, Europeans were already developing a sophisticated system of linear tactics, which became an ideal way to organize troops armed with the weapon. The purpose of linear tactics was to mass troops in tightly controllable, compact formations so their fire could be directed for maximum effect. Officers indicated when and how to load, selected targets, and controlled at what distance their men opened fire. Field armies remained small and the entire line was considered to be one unit, rigidly operating under the orders of the highest-ranking officer on the field. This was the ultimate expression of the philosophy that underlay linear tactics, with no latitude of decision making accorded subordinate unit commanders along the army's battle line.[5]

European armies experimented with a variety of firing systems or methods of delivering fire in unison. Commentators widely believed that the psychological impact of punishing the enemy with one well-disciplined volley was great enough to decide the battle, but they differed in recommending how large and widespread the firing should be. The Swedes experimented with firing by two ranks at a time, the forerunner of volley fire, in the Thirty Years' War. They also experimented with firing by one rank at a time, a system used by the French and Germans well into the eighteenth century.

The French also tried firing by files near the end of the seventeenth century. They practiced it differently than was common in the Civil War, with two files rather than one firing at a time as the soldiers stepped forward to shoot and then returned to the formation. This type of file firing occurred more often when the unit was behind a fortification or a natural obstacle. The French also developed the practice of firing by divisions, which involved five or six files at a time. The term *division* referred to any fraction of a unit, a definition it retained until the time of the Civil War. Individual fire, or firing at will, rarely occurred because it was anathema to fire discipline. It was useful, however, in an emergency.[6]

Most European armies resisted volley fire by entire units because of the need to have some men ready to shoot at almost any moment, given the relatively slow process of reloading. Firing by single rank, division, or file avoided this problem. The word *volley* originates from the French *volle,* or flight, inspired by the mass flight of bullets through the air.[7]

Another fire system that became very popular with all European armies was platoon firing, which allowed all ranks to fire at once but in small groups, so that each unit would have some of its manpower ready to shoot at any time. The Dutch developed this concept in the late 1600s; the British and French adopted it soon thereafter. The British battalion had four grand divisions with four platoons in each one. The sixteen platoons, plus two platoons of grenadiers, were divided into three firings of six platoons each, which alternated from one part of the line to the other. When used by a well-trained battalion, platoon firing could be devastating. The Royal Irish Regiment employed it to good effect against a French position at the battle of Malplaquet (September 11, 1709), stopping at 100 paces from the target and delivering fire by six platoons. The defender (which happened to be another Irish unit serving the French) fired by ranks and began to waver before retreating. The British unit was able to concentrate its fire more effectively on a specific point while the defenders dispersed their fire across a wider area. Based on the opposing casualty lists in this fight, historian Brent Nosworthy has concluded that "British platoon firing was 400 percent more effective than firing by ranks."[8]

The Prussians used platoon firing by the 1740s, starting with a platoon on the left of the battalion to be followed by another on the right in alternating sequence. This was called "flickering fire" because of the spurts of flame issuing from different points of the battle line. The French initially rolled their fire from the left of the unit to the right by platoons, but later started with the center platoon and alternated between left and right. An interval of only two seconds separated the different firings. This allowed about 25 percent of the battalion's manpower to fire at once. Most European armies adopted platoon firing, but none were better at it than the British. It was arguably the most effective method of directing the fire of smoothbore muskets ever developed, but it had its limitations. Platoon firing tended to be effective for the first few rounds when soldiers were fresh, but with each repetition it degraded as muskets became fouled and flints changed shape. The soldiers became more excited, reloaded hastily, and aimed poorly. They tended to lose their sense of timing and shot out of sequence, with a natural tendency toward individual firing. Because of these limitations, the platoon method of delivering fire coexisted with the other methods well into the early nineteenth century.[9]

Eighteenth-century Europeans generally distrusted individual firing as "disorderly and leading to the irreversible confusion of the entire battalion." Yet the French regulations of the 1750s authorized firing by individual, rank, platoon, and file. Ironically, the evidence seems to indicate that voluntary fire was more common than the other practices during the Seven Years' War, at least in the French army. American Civil War units generally preferred volleys and individual firing, although they sometimes used platoon firing, firing by rank, and firing by file as well.[10]

From what distance the smoothbore musket was able to fell a human target has been the source of debate for a long time. Theorists of the day argued that the ball carried beyond the effective range of the musket and could damage dense formations of enemy troops as far as 250 yards away. B. P. Hughes, a modern student of fire effectiveness in the smoothbore era, has concluded that "there was a belt of a depth of between 30 yards and perhaps as much as 100 yards within which the musket could develop its maximum effect, and . . . from a range of 100 yards to 200 yards it was capable only of rapidly diminishing results." This is perhaps an optimistic estimate. Brent Nosworthy has concluded that eighteenth-century commanders usually liked to open fire at about 50 yards, and often did so at 20 yards, although the range seems to have increased a bit from the turn of the eighteenth century to its midpoint.[11]

When used correctly, smoothbore muskets were impressive weapons. Detailed data for a British attack against the extreme French right at the battle of Blenheim (August 13, 1704) shows that one volley by the defender at 30 paces brought down one-third of the 4,000-man attacking force. Another British brigade later attacked the same target and lost 1,370 men to two or three French volleys. Casualty rates such as these would be considered very high even by the standards of the American Civil War. At the battle of Fontenoy (May 11, 1745), 2,500 British troops engaged in a firefight with French forces at 30 paces and made more than 25 percent of their rounds count in French casualties. Hughes admits that effective rates such as those seen at Blenheim and Fontenoy were probably exceptional, yet he also thinks the effective rate of smoothbore musketry increased as time passed. Early in the eighteenth century, he estimates, between 2 and 5 percent of rounds fired by smoothbore-armed infantry hit a man. By the middle of the century, that figure rose to between 10 and 20 percent. The misfire rate of the smoothbore, however, was quite high, about 33 percent, making the fire effectiveness of the weapon even more impressive.[12]

Skirmishing before 1756

For most of the eighteenth century, the battle line operated without any skirmishers. The French began as early as the 1720s to experiment with the use of what were termed light troops to operate in front of a battalion to screen its advance or to cover its retreat. They often referred to these troops as "pickets" (*piquets*), employing fifty to serve each battalion. They were the precursors of Civil War skirmishers. By 1732, Marshal Saxe suggested that seventy skirmishers be placed from 100 to 200 paces before each battalion, and that they open individual fire at the enemy at a range of 50 to 300 paces.[13]

Skirmishing steadily grew in importance among European armies as the eighteenth century progressed, but light troops came to be used in a quite different capacity as well. During the War of the Austrian Succession (1744–1748), light troops were widely used as raiders to harass enemy communications and lines of supply, or as scouts and patrols to gather information. Rarely were they used as true skirmishers to full effect. One exception was a fierce fight within the battle of Fontenoy, when French light troops held at bay a larger British-Hanoverian force in the Wood of Barry. The French skirmishers fought in open order, each soldier about a pace from his neighbor, taking advantage of natural cover and firing at will. The French referred to this type of combat as "helter-skelter," a decentralized kind of warfare dependent on the individual soldier's initiative and skill that was anathema to the current concepts of the linear system. No wonder that commanders of the mid-eighteenth century hesitated to embrace skirmishing. Yet the French had organized a company of pickets for every battalion by 1755.[14]

The Austrians and French used light troops for reconnaissance, harassment of enemy supply lines, and to operate in rugged, broken terrain during the Seven Years' War (1756–1763). Croats, from the southern part of the empire, mostly performed this duty for the Austrian army. The British and Prussians tended to ignore light infantry or misuse them as line troops. The French widely used skirmishers for the first time in their history, employing them more than any other combatant of that war. All French battalions were instructed to organize and train piquet companies to fight in open order, using the individual firing system and developing some degree of tactical flexibility and initiative.[15]

Many armies saw the utility of light infantry, if not skirmishers, following the Seven Years' War. The French created whole units of light troops, and Prussia distributed rifles to ten men in each newly created fusilier battalion (the rest received a lighter version of a smoothbore musket). These fusilier units received some training in open-order formations as opposed to the tight formations required of the battle line. When the comte de Guibert published *Essai general de tactique* in 1772, proposing significant changes in the French drill manual, he called for eliminating the distinction between light infantry and line troops altogether. Guibert advocated the training of all infantry so they could perform both functions as needed. While the practice of organizing and using light infantry for various jobs off the battlefield had reached its apogee by the 1770s, the concept of skirmishing as a support of the battle line was still relatively new, awaiting the next series of wars to be more fully developed.[16]

French Revolution, 1789–1800

Skirmishing came into its own during the wars of the French Revolution. While the army regulations adopted by the revolutionary government in 1791 hardly mentioned skirmishing, the government's Military Committee issued a separate regulation in 1792 to rectify that mistake. Skirmishing developed as "an integral part of battlefield tactics." The French impressed their opponents by deploying a "skirmish cloud," an ill-organized swarm of skirmishers barely under the control of officers that covered the front of the main battle formation. This cloud was the result of inexperience rather than sophistication among revolutionary commanders and their men, and it soon gave way to more organized methods of locating and harassing enemy formations on the battlefield. The standard skirmish formation was two ranks, with the men operating in pairs several feet or several yards from each other. The French at times came to deploy whole brigades on the skirmish line. They preferred, in some theaters of operations, to develop a "semi-permanent skirmish company in each battalion." During the Napoleonic period, the French took the next step and indeed created designated skirmish companies, as the Confederates would do in the Civil War.[17]

Napoleonic Wars, 1800–1815

The French and the British continued to support light infantry as well as to develop skirmishing, and it seems that the previously separate roles of skirmishers and light infantrymen were now merging somewhat. Among the French, the regulations stipulated that a quarter of the troops should be trained as light infantrymen by the turn of the nineteenth century, but it is uncertain whether this goal was actually met or whether the training was significantly different from that accorded line infantry. Some light troops were armed with rifles, but mostly they used a lighter version of the standard smoothbore. The British concentrated more seriously on special training for many battalions and regiments that they designated as light infantry. About 20 percent of the entire British army was so designated by the height of the Napoleonic Wars, and they could operate either as skirmishers before a battle line or as independent forces traversing hostile countryside while scouting or harassing enemy logistics and communications.[18]

One element that remained constant from revolutionary to Napoleonic warfare was the standard smoothbore musket. It still had a maximum effective range of about 100 yards when fired by an individual soldier. Hughes has estimated, however, that when fired in a volley it could reasonably damage a mass of troops up to 200 yards away, but beyond that distance the bullet lost so much velocity that it could no longer be considered lethal. If a soldier worked hard at firing the smoothbore on a practice field, he might get in five rounds per minute. In large formations and under control of his officers, the best one could hope might be two rounds per minute. Rifles were used in very limited ways during this era, mostly by some light troops. The Baker rifle of Britain had an effective range of about 200 yards in the hands of an ordinary soldier, possibly up to 300 yards if used by a trained marksman. This was a relatively modest increase of range over the smoothbore.[19]

Napoleon's enemies gradually improved many aspects of their war-making capabilities until the French emperor was finally defeated at Waterloo in 1815. The British heavily emphasized light infantry, and the Duke of Wellington became adept at using concentrated musketry at close ranges to disrupt advancing French columns before they could deploy into lines. The Prussians also expanded their light infantry force. Their new drill manual called for more target practice to improve volley and individual fire and more skirmish drill for a wider percentage of the line infantry.[20]

The armies that fought the Napoleonic Wars used an array of firing systems on the battlefield. Their manuals explained how to fire by platoons, in volleys of varied magnitude, by file, and by individual. The evidence indicates that, no matter which firing system was used at the start of an engagement, it usually degenerated into individual firing after a while. Officers maintained greater control over the firing in better-trained battalions fighting in shorter engagements.[21]

Commanders supplied units with more ammunition during extended firefights with ad hoc arrangements that varied from unit to unit. They sent staff officers, or subordinate officers in line units, to bring more rounds forward from supply trains in the rear. Sometimes commanders pulled units out of the line to go to the rear and refill their cartridge boxes. These methods were among those typically used in the American Civil War as well.[22]

The rate of fire achieved by soldiers in the Napoleonic era was about one to two rounds per minute, under the best of circumstances. In a rare bit of evidence, a British soldier of the Seventy-first Light Infantry recalled firing 107 rounds during a full day of fighting at the battle of Fuentes de Onoro (May 5, 1811) and 108 rounds during another day at the battle of Vitoria (June 21, 1813). If we take it as eight hours of firing, that amounts to thirteen rounds per hour, or one round for every 4.6 minutes. If our assumption is accurate, then actual battlefield conditions greatly reduced the ability of a soldier to pour fire at his enemy compared to practice firing.[23]

The technology of the smoothbore musket, among other factors, restricted the range at which soldiers of the Napoleonic Wars could deliver fire. Paddy Griffith sampled the data from nineteen engagements to conclude that British troops opened at an average of 75 yards. Rory Muir believes this was comparatively shorter than other armies due to the degree of fire discipline that prevailed among British units. He thinks average ranges were between 75 and 100 yards.[24]

Historians vary in their estimates of smoothbore fire effectiveness in the Napoleonic era. B. P. Hughes intently studied the results of two trials conducted in the early nineteenth century, concluding that smoothbores had the potential to be more deadly than most modern historians admit. A French trial conducted in 1800 involved a target 1.75 meters (5.5 feet) by 3 meters (almost 10 feet). The results were that 60 percent of the rounds hit it at a range of 82 yards, 40 percent at 164 yards, and 25 percent at 246 yards. A much later trial conducted with smoothbores that used percussion rather than flint detonation achieved 75 percent hits on a target 6 feet by 20 feet at 100 yards, 42 percent at 260 yards, and 16 percent at 300 yards.[25]

Hughes concluded that, theoretically, a Napoleonic battalion of 500 men could have gotten 500 to 600 hits out of two volleys against an attacker whose formation was well within a killing zone about 150 yards wide and 100 yards deep. We should take this estimate as the most one could hope to achieve under the right conditions.[26]

The fact that this high rate was never achieved on the battlefield is not an indictment of the weapon but yet another proof that battlefield conditions were seldom ideal. At the battle of Albuera (May 16, 1811), a fight between Werle's Brigade and the British Fusilier Brigade at ranges of 30 to 40 yards lasted about twenty minutes. Approximately 540 Frenchmen fired twenty rounds (10,800 bullets) and inflicted 600 British losses. The English troops, numbering about 1,500, also fired twenty rounds (30,000 bullets) and inflicted 1,700 French losses. Both sides achieved a hit ratio of about 5.5 percent. At the battle of Talavera (June 28, 1809), a French force of 4,050 troops suffered initially from artillery fire before closing in on a British position atop the Cerro de Medellin, a hill that anchored Wellington's line. They were repulsed with a loss of 1,500 men, probably three-fourths of them due to musketry. Smoothbore fire was about 3 to 5 percent effective in this action.[27]

The smoothbore musket dominated the Napoleonic battlefield as modern artillery came to dominate the western front in World War I. It caused the great majority of casualties in Napoleonic battles, with estimates ranging from more than half to 75 percent of the losses due to musket fire. Statistics for the battle of Malplaquet indicate that two-thirds of those wounded and treated by surgeons had been hit by smoothbore muskets, and only 2 percent had suffered from bayonet wounds. The smoothbore infantryman achieved this dominance despite the technical deficiencies of his weapon. Its misfire rate, even in dry weather, has been estimated at 15 percent (and 25 percent in wet weather). Its accuracy was limited by what Hughes has called "the eccentricity of the bullet." His final conclusion is that the maximum effectiveness rate of the smoothbore musket was not more than 20 percent, with up to 25 percent of the rounds misfiring. That meant that at least 50 percent of rounds missed their target. With the increasing use of conscripts by the French, Hughes is certain that the effectiveness rating dropped much lower, probably to 5 percent or less, during the latter stages of the Napoleonic Wars. As we have seen, even veteran troops at Albuera and Talavera could hardly exceed an effectiveness rate of 5 percent.[28]

But this seemingly low effectiveness rate should not mislead us. It is safe, even necessary, to assume that the effectiveness of any weapon will prove higher in tests than in actual battle. Many factors, ranging from nervous troops to difficult terrain to smoke and confusion, come into play to limit the men's ability to make the most of their weapons.

Napoleonic armies refined the art of skirmishing to a level commensurate with that of the American Civil War. Commanders routinely sent out a skirmish line when near the enemy. The Prussians recommended using only a portion of their light troops while keeping the rest about 100 to 200 paces to the rear as a reserve. They preferred to place their main battle line 300 to 400 paces to the rear of this reserve. Troops dedicated to skirmish duty, often wearing distinctive dark green uniforms and armed with rifles, or line troops given some degree of training in loose order formations, typically manned the skirmish lines.[29]

Some line units gained so much experience as skirmishers that the men came to enjoy and take pride in their distinctive work. The Ninety-fifth Regiment in the British army was widely noted for its expertise on the skirmish line. Armed with Baker rifles, the men "possessed an individual boldness, a mutual understanding, and a quickness of eye, in taking advantage of the ground," according to an observer. The British were willing to shove large numbers of men onto the skirmish line in Peninsula battles. At Bussaca (September 27, 1810), 1,500 men from the British Light Division (which brought 2,300 troops onto the battlefield) were deployed as skirmishers. At Salamanca (July 22, 1812), the British Fifth Division deployed about 1,000 of its 6,700 men on the skirmish line, or one-sixth of the division strength. Rory Muir believes this "was probably a rather higher proportion than was normal."[30]

The Prussians devoted more attention to skirmish drill during the Napoleonic Wars than ever before. Recruits practiced loading while lying down, and officers promoted a greater sense of individual initiative. They also aimed to make all soldiers proficient at both skirmish drill and the evolutions of the line so they could do both tasks as needed. The Austrians, who had a long history in the use of light infantry for harassing enemy logistics and communications, had rarely used skirmishers before the battle of Austerlitz (December 2, 1805). The results of that crushing defeat led them to pay more attention to the art of skirmishing, but the Austrians and the Russians failed to achieve what the French and British accomplished in this regard. The Prussians also had limited success in skirmishing under actual battlefield conditions before the end of the Napoleonic Wars.[31]

After 1815

All of Europe accepted the Napoleonic model of war making after 1815 and tried with varying degrees of success to maintain it. The conservative Austrians increased light infantry training and distributed more rifles to their troops. Yet they were not as backward as the Russians, who virtually ignored skirmishing and preferred battalion volleys over all other firing systems. The Prussians, more firmly adhering to tactical changes instituted by the end of the Napoleonic Wars, called for the consistent use of skirmishers in their regulations.[32]

The French, once again, took the lead in developing new ideas to supplement the Napoleonic system after 1815. By the 1830s, they began to offer specialized training to meet the threat posed by Algerian fighters who resisted French occupation of North Africa. The enemy was highly mobile, taking good cover in rugged terrain and able to damage French units even with relatively primitive weapons. The French created *tirailleur* ("sharpshooter," but used loosely to refer to skirmishers) units, gave them intense target practice and physical training, and encouraged them to fight as line infantry or skirmishers when needed. Their members wore distinctive baggy pants and colorful uniforms to instill a higher esprit de corps. The tirailleurs did well against Algerian fighters when sent to North Africa in 1838. Zouave units, composed of native troops, adopted their training regimen. The tirailleurs were later called Chasseurs à pied (literally, "hunters on foot," but loosely refers to "light infantry") because they developed a faster way to deploy on the field from one formation to another as well as a quickened pace when advancing toward the enemy. In skirmish drill, the Chasseurs à pied developed an improved formation. They still deployed in two ranks, each man 4 to 8 feet apart from his neighbor, but now they operated in four-man sections.[33]

The Algerian experience taught the French that aimed individual fire was better than volleys. The Chasseur à pied training regimen became so popular it was adopted for the entire French army in 1852, but such a rigorous program was impractical as a general rule. While the majority of French recruits failed to learn the drill effectively, the army simply increased its emphasis on aimed fire and skirmish drill.[34]

Prussia became the first nation to fully arm its military force with a rifle, the famous Dreyse or needle gun, beginning in 1842. It would not be used in combat for some thirty years, yet the Prussian army adopted new regulations in 1847 to account for its employment. The regulations allowed for the use of entire companies rather than just the third rank as skirmishers, hardly a revolutionary change in doctrine. Other European nations would not arm their soldiers with rifles until the 1850s and later.[35]

The American Experience

The Americans largely shared in the European military heritage, even though some aspects of their own development as a society called for adjustments in the inherited military culture. The rifle initially came to North America about 1700 in the hands of Swiss immigrants to Pennsylvania. American gunsmiths greatly improved it, but use of the rifle never became widespread, although frontiersmen in Pennsylvania and colonies to the south favored the weapon for hunting. The rifle already had a considerable reputation by the time the Revolutionary War broke out, and many Americans wanted to arm their regiments with it to gain some advantage over the British army.[36]

Lewis Nicola, a French-born officer in the British army before he settled in Philadelphia, embraced the revolutionary cause and wrote a tactical manual for the five militia companies of the city. He prefaced his book with a plea for the Americans to recruit a company of riflemen for every regiment, as they were "the most useful body of troops we have." Nicola wanted to recruit rifle companies from among farmers rather than city dwellers, training them to endure long marches and using them to harass the enemy as European armies had been doing for decades. "Not having any game-acts in America, to restrain its inhabitants from the use of fire-arms, they are in general good marksmen, and it would be difficult for any of them to put a gun to his shoulder without covering some object." Nicola admitted that "using rifles in war is certainly savage and cruel," apparently because it smacked too much of hunting for game, but he considered it justified by "absolute necessity."[37]

The Second Continental Congress, which directed the American war effort, authorized the creation of ten companies of riflemen on June 14, 1775. They were the first units raised directly by the government for the newly created Continental army and were intended to act as Europeans used light infantry. The rush to organize them was so great that additional companies were authorized, nine from Pennsylvania and two each from Virginia and Maryland. The companies joined the other units of the army around Boston in August 1775. In a demonstration of skill, the members of one company reportedly hit targets only 7 inches wide at a distance of 250 yards. In another demonstration, a rifleman managed to place eight shots in a row on a target "the size of a dollar" 60 yards away.[38]

George Washington was impressed enough with the rifle to organize larger units armed with the weapon. He drew more than 100 men from each existing brigade to form a corps of light infantry in August 1777. The next year he recommended that a light infantry company be included in each battalion, which could be detached from its parent unit and brigaded into a larger light infantry organization when needed. This system worked well. Anthony Wayne used the light infantry brigade to take the fortified British garrison at Stony Point on the Hudson River (July 15, 1779) in one of the more brilliant American operations of the war. The light infantry brigade was disbanded each winter and reconstituted at the approach of the spring campaign season. Enough companies were available for Washington to create a two-brigade division of light infantry in 1780 and 1781. The marquis de Lafayette commanded the division at the siege of Yorktown.[39]

The rifle never replaced the smoothbore musket as a standard infantry weapon due to the increased cost of purchasing and maintaining it. Also, commanders correctly viewed it as a specialty weapon, demanding intensive training beyond the aptitude of the average soldier. The rifle was best utilized in battle as a weapon for the light, agile, and quick man capable of some independent thinking and action. Several unit commanders in the American revolutionary army thought there were too many rifles in their units and obtained permission to exchange them for smoothbores.[40]

The British also fielded light infantry units, hiring some of them from various German states for the war in North America. Some of these units were armed with rifles but, in the eyes of observers, the British never excelled in their use of that weapon. European students of the Revolutionary War were so impressed with the American rifle that it reinforced their own desire to organize rifle-armed light infantry units. Among Americans, the success of the Revolution inspired wholly unrealistic images of independence won by marksmen aiming their trusty rifles from behind trees at redcoats who foolishly stood in closed ranks on open ground. To explain this, patriots liked to claim that all American men were natural marksmen due to their frontier environment. As William Duane, who published the first American treatise on military affairs in 1809, put it, "The use of the *rifle* was an indispensible [*sic*] qualification to every man who had occasion to defend himself, or a taste for the sports of the field and the forest." Duane went on to claim that in the era of the American Revolution, "there was not a man in the country who could not *hit* a space of a foot diameter, at one hundred and fifty yards, with a single ball." Of course, there is no support for the notion that rifle-firing troops won the American Revolution because the vast majority of battles were fought with smoothbore muskets.[41]

The Americans and British used the same firing systems during the Revolutionary War as the Europeans used in their conflicts. Lewis Nicola cautioned his readers not to accept the popular notion that firing fast was firing well. It tended to lead men to load improperly and aim poorly, thus lessening fire effectiveness. "Two fires leisurely and properly given," he concluded, "will do more execution than three in the usual way." Nicola described platoon firing by six platoons in rotation, stopping the process with the taps of a drum. He sanctioned the idea that each platoon could advance before firing, to be followed by the others, and even thought it was possible to do this during a retreat. Nicola also prescribed how to fire by companies, with the first and second firings performed by four companies each, and the third firing done by the grenadiers or light infantrymen. He also included firing by wings and ranks in his tactical manual.[42]

Nicola thought soldiers preferred to fire straight ahead, and therefore recommended that they be trained to face a bit to the right or left while standing in place to facilitate oblique firing. Nicola prescribed how to fire from behind an earthwork by having the first rank step up to the parapet, fire, and retire through the intervals of the second rank. While the first rank reloaded, the second rank stepped up to the parapet to fire. Nicola also detailed how entire ranks could retire to the right or left of a column formation after firing, go to the rear of the column to reload, and wait their turn to advance again to the parapet. Each rank in the meanwhile had been duplicating this maneuver to present the enemy with wave after wave of volleys.[43]

Many of Nicola's recommended firing systems sound too complex for militia, perhaps even too complex for Continental troops. Frederick William Augustus von Steuben, with long service as an officer in the Prussian army, described firing systems that were much simpler. He drilled a model company at Valley Forge and drafted a tactics manual that the U.S. Army adopted in 1779. Steuben recommended platoon firing in four stages and taught the troops to fire while retreating as well as advancing. The Americans were aware of platoon firing even before Steuben wrote his manual, but it demanded quite a lot of practice. Joseph Plumb Martin remembered that in the summer of 1776 his regiment endured a rainstorm and was ordered to parade and discharge its guns prior to cleaning them. "We attempted to fire by platoons for improvement, but we made blundering work of it; it was more like a running fire than firing by divisions." Martin reminds us that no matter how simple or complex the tactical system, using it effectively on the battlefield depended on rigorous training mixed with combat experience.[44]

Developing and Using an Effective Rifle Musket

During the 1830s and 1840s, several individuals worked toward improving the loading and firing mechanism of the rifle to create the modern rifle musket, suitable for distribution to all infantrymen of a national army. The introduction of the improved rifle before 1861 brought with it a flurry of books and articles by military writers who tried to predict what the new weapon could do on the battlefield and to suggest how armies should adjust to its technical capabilities. While some European nations tried to institute changes in training to encourage their soldiers to take full advantage of these possibilities, this was not done in the United States. Moreover, there is little evidence to indicate that training to use the new weapon resulted in anything like a revolution in tactical performance even in Europe.

The French began the process of trying to find a faster, easier way to load the rifle, which had been the chief impediment to its widespread use. Inspired by the difficulties of confronting native troops in Algeria, Capt. Gustave Henri Delvigne changed the configuration of the bore to ensure faster loading by 1834. He created a chamber at the base of the tube that was smaller in caliber than the rest of the bore, so the powder could snugly fill it. Then the soldier rammed the ball onto the shoulder of the chamber to create a tight fit with the rifling. The weapon had an effective range of 400 yards, but its success was limited due to a number of technical problems, many of them associated with the climate of North Africa.[45]

The year before Delvigne perfected his improvement, two French artillery officers developed the cylindrical-conical bullet. It was the basic shape of most Civil War rifle ordnance, a lead plug with an elongated side and a flat base. Another officer, Capt. Thouvenin, developed a "stem rifle" that used this type of ball by 1842. Thouvenin's device consisted of a steel pin 1 1/2 inches long at the base of the weapon's tube. The loose powder settled around it, and the bullet had a wooden sabot at its base that stuck on the pin to hold it in place until firing. The weapon had an effective range of 600 yards and was used for more than ten years. But the basic problem of Thouvenin's and Delvigne's improvements was that the soldier had to ram the charge home in a careful, uniform way to seed the bullet properly, or else it failed to fully catch the rifling and lost much of its range and accuracy. Capt. Claude Etienne Minié solved that problem by creating a hollow on the bottom of the cylindrical-conical bullet and letting the gases produced by the powder explosion expand the sides to fit the rifling. This worked much better because it substituted a mechanical process for human skill. The Minié rifle, as it came to be known informally (actually a term that denotes the loading system rather than the weapon itself), began to be used in 1846. It was widely successful during the Crimean War of 1854–1856. Essentially, all rifle muskets in the Civil War used this loading system.[46]

The breechloader was a more innovative and important development in rifle technology. The Americans designed the "first truly functional" breechloader with the Hall rifle, first developed in the middle of the 1820s. The Norwegians followed a few years later, and the German Johann Nicholaus von Dreyse invented his "needle-igniting musket" by 1838. Dreyse created a bolt-action weapon with a cylindrical-conical ball. Prussia adopted it in 1842 to become the first army in history to issue a breechloader as a standard infantry weapon, but the needle gun did not have an opportunity to prove its worth until more than twenty years later.[47]

Concurrent with the improved loading systems, a new ignition system also improved the effectiveness of the rifle. Fulminates had been developed in the 1790s but it was not until 1820 that a Philadelphian named Joshua Shaw made a "copper percussion cap containing mercurie fulminate." It initially was used in hunting weapons, but the British army began to experiment with it in 1831, distributing percussion muskets to its Guards battalions five years later. The British began to generally distribute weapons using the new firing mechanism by 1839, soon followed by the French and most other European nations. The Americans lagged a bit behind the continent in this regard.[48]

The decade of the 1850s saw the dominance of the Minié system combined with the percussion method of detonating the charge. Britain's whole army was armed with the Enfield Model 1853 rifle musket, perhaps the best in the world, with a reported range of 900 yards. Austria used the Lorenz rifle, with a cylindrical-conical ball, beginning in 1854. Russia began to adopt the Minié system in 1857, after its defeat in the Crimea, the same year that France ensured the universal distribution of the weapon. With Jefferson Davis, a progressive secretary of war, the United States adopted a Minié system, percussion-fired weapon in the Springfield .58-caliber rifle musket, Model 1855. The Americans considered a multiplicity of newly developed breech-loading weapons in the 1850s, but they could not agree on how to use them. Worries about cost, wasting ammunition, and the propensity for technical breakdowns prevented the U.S. Army from being armed with the latest weapon of the day.[49]

As the technical details of the improved rifle were ironed out, the tactical debate began. Capt. Wittich opened the discussion among continental armies with the publication of a pamphlet in 1849, arguing that the weapon could help skirmishers hold back large formations of infantry and cavalry and devastate batteries of field artillery at long range. Wittich believed that the main battle formation would still operate in shoulder-to-shoulder lines, but skirmishers would see the dawn of a new day. A Belgian artillery officer, Capt. Gilluim, argued on the other hand that the rifle musket would make linear formations obsolete as commanders would be forced to deploy all their men in several skirmish lines. Few other theorists agreed with Gilluim's radical hypothesis.[50]

British theorists were more conservative in their estimates of the rifle musket. Some argued it would demand no changes in deployment, while others sought ways to speed up maneuvers performed within range of the weapon. The latter writers admitted that the battlefield would become a more dangerous place, with a deeper field of fire, which would reduce the role formerly played by cavalry and artillery when confronting infantry formations. But they assumed the basics of the linear tactical system would remain the same. The British theorists rightly pointed out that the long-range fire capabilities could easily be limited by the realities of the battlefield. Every field of conflict had a greater or lesser number of obstacles, including forests, buildings, and hills, that reduced visibility and therefore reduced accuracy of long-range fire. The British were aware of psychological factors that could affect how infantrymen used, or failed to use, their weapons. Fear, excitement, and disorientation impeded a soldier's ability to gauge distance and properly reload his weapon.[51]

Some of the more conservative British authors even argued that the old smoothbore musket would continue to be as effective in battle as the new rifle musket. Their argument was not just an expression of old-fashioned thinking, for they pointed out that the cluttered nature of the battlefield was not the only impediment to accurate, long-range musketry. The curved, or parabolic, trajectory of the new weapon was a major factor to consider. The problem was lessened within short range, about 100 yards, which was the effective range of the old smoothbore musket. The theorists recognized that it was not impossible to train infantrymen to deal with trajectory problems, but it demanded a great deal of practice and skill on the part of the rifleman to compensate for them.[52]

The continental writers were enthusiastic about the apparently revolutionary aspects of the rifle musket, and the British generally were subdued and careful—the Americans tended to side with the continental writers. Jefferson Davis reported to Congress that the rifle musket would change tactics by forcing an increased reliance on skirmishers. This thinking led to a slight change in the tactical manual used by the American army, instituted by William J. Hardee in 1855. Hardee's book replaced Winfield Scott's twenty-year-old manual. Like Scott's, it was almost an exact copy of existing French manuals, but Hardee brought to America new French developments in light infantry training as a way for line troops to compensate for the enhanced fire capabilities of the rifle musket. All that meant, in essence, was that Hardee wanted American soldiers to move faster when they maneuvered within range of enemy rifles. Despite their enthusiasm for the new weapon, the Americans prepared very little for dealing with its potential use by an enemy of their army.[53]

The Mexican War of 1846–1848 did not give the Americans an opportunity to test the rifle in action. Most American troops still used smoothbore muskets, and most regular regiments did not even have percussion caps. This new detonation system was more widely used by volunteer regiments. American units displayed a commendable ability to articulate their strength and to grapple with the enemy on the battlefield, and thus no cries for tactical reform were raised during or after the war. But the conflict with Mexico was not a good case study of American tactical proficiency because the opponent was not a well-trained European army. The Mexicans filled their ranks with peasant draftees, and their officer corps was weighted with aristocratic, often ill-trained and unscrupulous men unfit for their commands. The Americans outperformed the Mexican army, utilizing a corps of West Point–trained middle- and junior-level officers, displaying a vibrant sense of national pride, and fighting with an aggressive spirit of running over the opposition in quick order to win literally every battle fought in the war. Historians have given too much importance to the Mexican War in the thought processes of future Civil War generals, for the tactical experiences of these men in 1861–1865 did not consistently duplicate the method of military operations practiced by commanders in 1846–1848. Furthermore, in noting that tactical offensives almost always succeeded in Mexico against smoothbore weapons but usually failed against rifle muskets in the Civil War, historians have ignored the fact that there were a thousand other factors that also determined the success or failure of an attack. Not the least of which was the quality of the manpower using those defensive weapons. It is arguable that the Americans so consistently won battles in Mexico because of issues unrelated to the weaponry involved.[54]

The Crimean War of 1854–1856 was a better proving ground for the rifle musket than the conflict in Mexico, although the armies were unevenly armed with the new weapon. Most Russians used smoothbore muskets, while only "a tiny fraction" had a gun that used the Minié system. Perhaps one-third of French troops used a Minié rifle, as it was then generically called, while the rest were armed with smoothbore percussion muskets. The British were similarly armed. The British units that had been stationed overseas for long periods of time arrived at the Crimea with older weapons. Those units newly dispatched from home carried with them the new Enfield rifles, "a beautiful weapon," in the words of an officer in the Eighteenth Regiment of Infantry. The Turkish army had bought enough Minié rifles from Britain to arm about one-fourth of its men.[55]

The British and French pushed siege operations against the heavy field fortifications that protected the port city of Sebastopol for nearly a year. Given the long stalemate in fortified, close-range positions, both sides used the new rifle musket for sharpshooting. The British created sniping details consisting of a captain, a lieutenant, a sergeant, two corporals, and fifteen privates from each regiment. These men were "exempted from all trench and other duties" so they would be free to roam about and use their rifles to pick off Russians who exposed themselves. They dispersed behind rocks and undulations in the open ground between the lines or dug pits for protection. The Russians used their few rifles for the same purpose. In fact, the individual pits dug by Russian sharpshooters to harass allied personnel came to be called rifle pits, a term Civil War participants widely used to describe any trench designed to protect an infantryman.[56]

Capt. Nicholas Dunscombe of the Forty-sixth (South Devonshire) Foot led his unit's sharpshooting detail during much of the Sebastopol campaign. He placed twenty-five men in an "advanced trench" and others in adjoining communication trenches during the night of February 17, 1855. The Russian sharpshooters began firing at first light, and his men were ready to counter it. They fired all the rounds issued to them before dusk. Dunscombe participated in the firing, shooting sixty rounds that day.[57]

The use of the rifle for sharpshooting in the Crimea presaged developments to come in the Civil War, indicating that the new weapon would have its greatest impact in specialty uses rather than in the operations of the line. But the outcome at Sebastopol was not determined by sharpshooting. The allies captured Sebastopol simply by wearing out the Russian defender with weight of material, especially artillery, and by repeated frontal attacks on different sectors of the Russian line. Sustained heavy artillery bombardments, with as many as 75,000 rounds falling each day, followed by attacks that sometimes succeeded in limited ways, ate away at the defender's hold. By the summer of 1855, the Russians were outnumbered, nearly overwhelmed by a disparity of one to four in artillery, and were losing up to 400 men each day in the trenches. They lost 7,500 men in the three-day artillery bombardment preceding a French attack that captured Fort Malakhov on September 8, 1855. The Russians evacuated Sebastopol that night.[58]

By the late 1850s, the European and American armies used a variety of rifle muskets, most of them of the Minié pattern. Rifle enthusiasts were fully aware of the problems involved in the general distribution of these guns to average soldiers; the greater range and the parabolic trajectory meant that it was no easy matter to aim the weapon with expectations of hitting the target. Some observers thought that their contemporaries were just poor shots, regardless of the demands of the new weapon. Horace William Shaler Cleveland, a zealous advocate of rifles, believed the myth about American patriots at Bunker Hill and New Orleans hitting a British soldier with nearly every round they fired. By the mid-nineteenth century, Cleveland thought, only one out of 600 rounds fired on a battlefield hit a human target. Cleveland's solution was to call for able-bodied men in America to take on the responsibility of learning how to shoot a rifle. He advocated the formation of rifle clubs whose members would use the latest scientific methods of musketry training, so all men of military age could learn the "act which constitutes the vital spirit of military efficiency." Arthur Walker, a staff member of the English army's School of Musketry at Hythe, was Cleveland's British echo in the call for widespread training in how to fire rifles.[59]

Lt. Cadmus M. Wilcox of the Seventh U.S. Infantry was the American army's salient voice in this campaign. He visited the Hythe school and others on the continent while touring Europe and came back full of enthusiasm for scientific rifle training. Wilcox's book, *Rifles and Rifle Practice,* was an impressive study that mixed a thorough understanding of mathematics, geometry, and physics with a wide-ranging examination of the rifles currently in use throughout the world. Wilcox put together the most persuasive call for rifle training in America, noting that the new weapon offered the soldier an opportunity to fire at a range that was "equal, or greater, than the limit of distinct vision, and greater even than the extent offered by fields of battle in general." This demanded specialized training in gauging distance and fixing the movable sights of the rifle musket.[60]

Wilcox explained the trajectory problem better than anyone else. Ironically, the rifle musket had a muzzle velocity that was 40 percent slower than that of the smoothbore it replaced. This caused the curved trajectory that so worried rifle advocates. If a soldier correctly sighted a rifle musket to hit a man at 300 yards, the bullet would pass over the heads of everyone else between 100 and 225 yards from the shooter. The first 100 yards of deadly space duplicated the effective range of the smoothbore musket with no appreciable increase in accuracy over the older weapon. But the deadly space at the end of the rifle's trajectory shrank in size as the range increased. At 300 yards, the farther deadly space was about 75 yards. At 600 yards, it was only 60 yards deep, and 40 yards deep at 800 yards. At 993 yards, that deadly space was only about 12 yards, due to the sharper angle at which the bullet descended from the curved trajectory and fell to earth. Ironically, the greater the range, the less likelihood of hitting a human target even if the shooter knew how to correctly estimate distance and sight his rifle.[61]

The average soldier had never dealt with this problem before. Even though the rifle musket had a hugely increased range over the smoothbore, it was extremely difficult for any soldier to take advantage of that capability. The rifle musket needed a sighting mechanism that could be flipped up for use and laid down when unnecessary. This mechanism had a movable gauge that gave the soldier the impression he was leveling his weapon at a distant target, but in reality it forced him to angle the barrel according to the distance. He had first to estimate the distance accurately with his eye, and then adjust the gauge accordingly before firing.[62]

A man who had a natural aptitude for this process could become an expert marksman. In every society, a certain percentage of young men have the eye, the quick apprehension, and most important the native interest in guns to become sharpshooters. Such men had the natural skills to make the most of the new rifle's potential. But the problem lay in universal distribution of the rifle to the average men who filled the ranks of all national armies. To reap the full benefit of the weapon, these men had to be given special instruction. The British and French took the lead by establishing schools of musketry in which soldiers were trained to estimate distance and adjust their weapon's sights. The English school at Hythe drew the most attention. It was an impressive course of instruction, based on a thorough, scientific analysis of the technical problems as well as the human limitations of the task. The staff at Hythe also was successful at publicizing its methods in numerous publications, and it hosted an array of foreign visitors, who tried to encourage their own governments to create versions of the Hythe method. This experiment in creating an institutional system of rifle training for an entire national army in peacetime worked far better in England than elsewhere, reinforcing the distinctive British emphasis on controlled and rapid small-arms fire in battle during the late nineteenth and early twentieth centuries. Whether British troops were capable of consistently hitting targets at long range is not clear, but they could deliver controlled volleys at an impressive rate.[63]

The United States failed to institute a course of rifle instruction of any kind before 1861. Brig. Gen. Stephen Vincent Benet, the army's chief of ordnance after the Civil War, could find no evidence that the army had any official policy on target practice before the adoption of the rifle musket. In 1854, the army sent out a printed circular recommending practice with the new rifle, but it set no guidelines nor created a system for doing so. Two years later, another circular instructed officers to report on whether their men were practicing. The response clearly showed that essentially nothing was being done. The army had adopted a short manual for rifle practice written by Capt. Henry Heth of the Tenth U.S. Infantry just before the onset of the Civil War, but it had never been used in training. Heth based his book entirely on the English School of Musketry at Hythe.[64]

As a result, target practice was all but ignored in the U.S. Army before 1861. As Benet aptly put it, "Officers . . . had been brought up under the use of the smooth-bore musket, and even after the new arm had been put in the hands of the troops, were slow to abandon the ideas imbibed from the old smooth-bore." More important, the army's administrative officers had failed to issue a set of regulations mandating and regulating rifle practice. Subordinate officers were "expected to take up the subject of rifle practice spontaneously, and without specific orders as to allowance of ammunition, time of practice, and method of making systematic report," as Benet explained.[65]

Even if the U.S. Army had instituted an effective system of rifle practice before the Civil War, it may not have had an effect on the way that war was fought.

By the end of the conflict in 1865, there were only about 50,000 troops in the regular army, while the number of Union volunteers numbered well over a million. The volunteers were organized by state governments before transfer to the Federal authorities. They were typically officered by citizen-soldiers who had little more understanding of the military art than the rank and file. Any degree of training of any kind, whether in tactical formations, camp cleanliness, or rifle practice, was dependent on thousands of individual regimental and brigade commanders with varied levels of intelligence, sincerity, and access to information. In short, the highly specialized art of rifle practice hardly existed among the volunteers. The Hythe system was designed for a peacetime army, when opportunity allowed for thorough training by a centralized authority. It did not suit the American system of manpower mobilization for an emergency like that of the Civil War.[66]

Ignoring the real possibility that the average soldier would never be adequately trained to take advantage of the rifle's capabilities, American and European observers contemplated the new weapon's impact on tactics. Cadmus Wilcox asserted that the rifle musket was "fourfold more destructive" than the smoothbore, and therefore it would create a far more deadly battlefield. The killing zone would be enormously deepened, and the problem faced by commanders would be to move their men close to the enemy without wasting lives. Almost all other writers agreed with Wilcox, assuming that even unaimed fire at ranges of several hundred yards would prove deadly to advancing infantry. Hans Busk was a lone dissenting voice in this discussion. He noted that troops typically did not halt at the extreme range of their weapons to fire but were more prone to close in on the enemy. Infantry fire at long range was wasteful, inaccurate, and tended "to produce unsteadiness" in the men. Busk warned that the average soldier was not capable of learning the complex process of sighting a rifle to fire at long-range targets, and that field artillery was better adapted to hit the enemy at these ranges.[67]

Ironically, the French, who served as America's role model in this debate, backed away from their initial enthusiasm for long-range musketry following the Crimean War. They concluded that battlefield conditions tended to make close-range firing more practical, agreeing with previous arguments by British commentators. The French even removed the long-range sighting mechanism from rifles they issued to the Zouaves and Chasseurs à pied, training them instead to use their thumb as a sighting guide for ranges up to 200 yards.[68]

A number of American commentators pondered the impact of the rifle on varied aspects of combat. Wilcox believed that field artillery would become less effective on the battlefield due to the infantry's ability to deliver fire over a greater distance. Gunners would have to protect themselves with fieldworks to survive on the modern battlefield. He also predicted a greatly reduced role for cavalry, which would have fewer opportunities to win fights with infantry formations armed with rifles. John Gibbon, however, argued for the continued importance of field artillery by noting the difficulties inherent in firing the rifle at long range. He believed that artillerymen thoroughly trained to deliver fire at this same range could outshoot a formation of foot soldiers. English observers also pointed out that rifle muskets would make battles shorter because troops armed with them could deliver fire more decisively. That is, of course, unless the enemy resorted to artificial protection by digging fieldworks. Other observers reinforced the prediction that field fortifications would blossom in future warfare.[69]

Observers also debated what type of firing system best suited the new weapon. Experiments in France seemed to suggest that individual fire was the most effective, with file fire and volleys by company the next best. Wilcox cited trial results from France that seemed to show that two men firing individually were as effective as three men firing by file, and as effective as four men firing by volley. How much we can trust estimates like this is an open question, but the results reinforced the American tendency to view the rifle as enhancing the potential of each soldier as a marksman.[70]

Wilcox even believed that manipulating the movable gauge of the sighting mechanism would force the soldier to remain calm under fire by occupying his mind with a mechanical process instead of allowing it to dwell on the dangers of the battlefield. He noted this phenomenon among field artillerists, who had a bigger and more complex piece of ordnance to work.[71]

Rifle enthusiasts were encouraged by reports of field trials in England showing that rifle fire delivered on large infantry formations at long distance could be quite effective. A column of troops was twice as likely to suffer losses as a battle line at a distance of 450 yards, and up to six times as likely at greater distances. Americans in particular could take even more pleasure in Hans Busk's conclusion that they were "more practiced in the use of firearms" than Europeans. He based this on statistics that indicated that one out of every 459 rounds fired by British troops at the battle of Vitoria hit a man. The ratio during a skirmish against native fighters in South Africa in August 1813, was only one out of 3,200 rounds. In contrast, during the Mexican War, American troops fired 125 rounds for every Mexican casualty. The Mexicans fired 800 rounds for every American casualty during the battle of Churubusco (August 19, 1847).[72]

There were no large-scale wars in the late 1850s to demonstrate whether expectations regarding the rifle's impact on combat were justified. The Indian Mutiny of 1857 saw limited use of rifles, mostly in sharpshooting during static campaigns like the siege of Lucknow (July 1, 1857–March 21, 1858). The Italian War of 1859, between France and the Kingdom of Sardinia on one side and Austria on the other, lasted only three months. French troops using Chasseurs à pied tactics conducted a spectacular attack on an artillery position at the battle of Magenta (June 4, 1859) that seemed to indicate the value of the new training. It also helped to reinforce the developing view among the French and British that close-range fighting was still relevant in combat. American observers had little time to absorb the tentative lessons displayed by this short war. They continued to believe in the potential for long-range fire while they retained a contradictory but natural affinity for closing in at short range and winning battles quickly and with panache.[73]

2

THE FIRST RIFLE WAR

The Civil War was the first major war in which both sides were fully armed with rifle muskets, although not until the second half of the conflict. With its national policy of minimal preparedness for war during peacetime, the U.S. government was unable to supply rifles to all its units until the early part of 1863. The newly created Confederate government had even more difficulties supplying weapons to its troops, but dealt with the problem through vigorous measures to seize and purchase arms. By the latter part of the war, the two primary weapons were the American-made Springfield rifle musket and the English-made Enfield rifle musket, both of which used the Minié system. The Civil War was the only significant war in which the rifle musket predominated on both sides, for rapid advances in weaponry soon turned Western armies toward breechloaders as their standard infantry arm after 1865.

At the start of the Civil War, 35,335 muskets (which had either been retrofitted with rifling or had been made with rifling from the start) were stored in warehouses throughout the United States. In addition, there were 422,325 older types of rifles and smoothbores. Of these, 160,000 guns of all kinds were located in states that would join the Confederacy. The store of modern rifles was completely inadequate for the initial mobilization of manpower. Lincoln issued a call for 75,000 troops to serve three months on April 15, 1861, and called for 100,000 volunteers for three years' service in May. The Confederate government called for 100,000 volunteers for one-year service following Fort Sumter. Obviously, many of these first volunteers would have to settle for smoothbores or antiquated rifles, even though most of them demanded the latest in firearms. When Brig. Gen. George H. Thomas urged Capt. William F. Patterson to hurry the organization of his Kentucky company in November, 1861, he wrote: "As you allude to the passion which the men have for the rifle I will inform you that I can see no prospect for getting any for them. They will have to take the [smoothbore] musket." Patterson's recruits were not alone in this attitude. An inspector reported that the men of the Third Kentucky "dislike greatly common muskets & as Kentuckians wish for good Sharp's rifles." The recruits who volunteered for the Fifty-fifth Illinois had been promised Colt's revolving rifles as an inducement for signing up, but they never received them. In fact, they waited a considerable time to receive any weapons at all, and meanwhile "had coaxed, grumbled, swore and howled after the manner of Western volunteers for deadly rifles and glittering bayonets."[1]

When a few of these early volunteers actually received good rifles, they were greatly impressed. Cyrus F. Boyd of Iowa called the Springfield rifle musket "a most beautiful weapon—they are bright and in fine condition." But some of the early units received odd weapons that never became widely popular, such as the Whitney Plymouth Navy rifle, boasting a saber bayonet and designed for service aboard naval vessels.[2]

The volunteers' fascination with rifles was expressed in the names they adopted for their units. Sixteen Confederate battalions and regiments were authorized to add the word "Rifles" to their numerical and state designations. Many of them were mounted units, especially those raised in the Indian Territory, Texas, and Arkansas. In contrast, only four units serving in the Federal army carried a rifle designation, two of which were mounted regiments.[3]

There were more units that unofficially adopted the word Rifles, such as the Thirty-seventh Illinois. Advertisements calling for volunteers in the northern part of that state promised "the best Minnie Rifles" for anyone who joined the regiment. Upon completing its organization, the Thirty-seventh received only enough Colt's revolving rifles to arm the color guard and the two flank companies. The rest received "Austrian and English muskets and some 'old worn out

Harpers Ferry muskets.'" The latter weapons were so bad that the men nearly mutinied, until a promise to find better guns averted a crisis. Eight companies received Belgian rifles two months later, and the men were generally satisfied with them. The Thirty-seventh Illinois nevertheless was called the "Fremont Rifles" by its members and well-wishers.[4]

Scarcity of rifles was only one problem affecting the issuance of guns to newly raised units. While Abraham Lincoln, also a westerner, was fully aware of the value of the rifle musket, he had to contend against a different viewpoint held by the chief of army ordnance. Given the shortage of available rifles, Col. James W. Ripley preferred to issue smoothbores to the untrained volunteers and reserve the newer weapons for the regulars.[5]

Historian Carl Davis has divided the issuance of weapons by the Federal government into three phases. During the first phase, the government sought to find weapons of any type to fill a gaping void, and that tended to mean purchasing whatever was available on the European market. These foreign guns were by no means the best, but they were usually serviceable. Northern volunteers generally hated them, in part because they were not as finished in their workmanship as more recently manufactured guns. They also complained that these old European rifles kicked too much or were inaccurate. Even a handful of such weapons could result in scathing comments or a general protest, followed by loud calls for better guns. During the second phase, the Federal government pushed to increase its own arms production. The armory at Springfield, Massachusetts, had made only 10,000 rifle muskets each year before the war. It expanded its output to 300,000 per year by 1864. Private industry did its part by producing more advanced weapons such as breechloaders and repeaters, and even contracting with the government to produce rifle muskets. During the third phase, which lasted from early 1863 to the end of the war, the emphasis was on supplying repeaters and breechloaders to at least a few regiments.[6]

Historians have tended to deal pretty harshly with Ripley and his handling of the issuance problem, although a handful of scholars have argued persuasively that he was right. The Ordnance Department determined that the best rifles should go to the regulars and to the skirmish companies of volunteer regiments, suggesting that breechloaders and repeaters be used by skirmishers as well. The war lasted long enough to enable the authorities on both sides to almost completely arm their men with good weapons. During the first two years of the conflict, the Federals purchased close to 1,165,000 weapons in Europe. Less than 10 percent of them were smoothbores, and 436,000 were modern Enfields. The Austrian rifle, Model 1854, accounted for nearly a quarter of a million of that total number. Carl Davis has estimated that, the smoothbores notwithstanding, 80 percent of these purchased weapons were "of good quality."[7]

Thanks to the research of Ken Baumann in quarterly ordnance reports, we have a good idea of what types of weapons were issued to Illinois volunteers during the war. The state raised 150 infantry regiments. Virtually all of the 125 regiments organized from April 1861 to November 1862 made do with a mixture of different weapons, most of them inferior in quality to the best available. The worst weapon was the smoothbore flintlock, converted to percussion before the war. Eighteen Illinois regiments out of 125 were given these guns early in the war.[8]

Volunteers were not shy about criticizing what they considered bad guns. The Second Wisconsin received "old Harper's Ferry smoothbore muskets" in May 1861. These employed the famous buck and ball cartridge, one bullet and three buckshot "with powder sufficient . . . to drive them out of the gun barrel and at the same time nearly knock the life out of the man who stood behind the gun." Leander Stillwell of the Sixty-first Illinois called the Austrian rifle a "wicked shooter" because of its hard kick. "Men would hold them very tight, shut their eyes, and brace themselves to prepare for the shock," as Ulysses Grant told a congressional investigating committee. Allen Morgan Geer of the Twentieth Illinois called the smoothbores "only fit to drill with." Whether a unit got a good Enfield or a bad smoothbore in the early part of the war often depended on the political influence of its commander.[9]

Postwar veterans of the Thirty-third Illinois recalled the qualities of the Belgian rifles, converted from smoothbore flintlocks, that were issued to them at the start of the conflict. "They might have been used in the Napoleon wars," remarked one veteran. "They were crude enough to be a relic of that date." It was difficult to pull the trigger and the weapon "kicked at both ends, humped up their backs and kicked your hand underneath the rifle." Once, when told to discharge their pieces after standing guard duty, some members of the regiment tied them to a rail fence and pulled the trigger with a gun strap to spare their shoulders. A member of Bissell's Engineer Regiment of the West also refused to injure himself in a similar circumstance. He tied his American-made smoothbore flintlock, converted to percussion, to the back of another man who crouched on his hands and knees. The gun blast nearly deafened the poor soldier and recoiled off his back, almost knocking down the gun's owner.[10]

It is easy enough to understand the frustration felt by these men. The Fifty-first Illinois received "heavy ungainly looking" Belgian muskets, and the men were "not at all pleased with them." On February 16, 1862, several members of the regiment engaged in an intense target practice to see if they could rely on the weapons in combat. The guns "kicked like the business end of a mule" and choked up with powder residue so badly that they were unusable. When the group reported its results, the entire regiment voted to reject the weapons. That evening at dress parade, the men refused to take their guns from the stacks. The officers initially threatened them but then thought again and promised to find better guns. The old Belgians were left in their stacks overnight and were ruined by a heavy rain. Ordnance officers issued newer Harpers Ferry rifles the next day, and they proved to be acceptable to the volunteers. Some companies of the Fifty-first Illinois received Enfields in March, and the rest were issued Austrian rifles in July 1862.[11]

The Fifty-fifth Illinois received Dresden and Suhl rifle muskets from the German states when it organized in the fall of 1861. "Language fails when attempting to describe the grotesque worthlessness of these so-called arms," recalled the regiment's historian. "They were of foreign make having scarcely the similitude of guns." Sgt. H. H. Kendrick of Company K pleaded with Gov. Richard Yates, "For God's sake remember us in mercy, for the men cannot and will not fight with such guns. It is impossible; they have no faith in them."[12]

Kendrick's prediction that his comrades could not fight with inferior weapons had some validity. Col. Philip B. Fouke reported that the ability of his Thirteenth Illinois to perform at the battle of Belmont was severely reduced by the mixture of Harpers Ferry rifles and the altered flintlocks it carried. One-fourth of them "either exploded or were rendered useless before the battle was half over," he wrote.[13]

Yet some volunteers were happy to make do with whatever guns they were given. Leander Stillwell reported that his comrades of the Sixty-first Illinois were "quite proud" of their hard-kicking Austrian muskets. Many soldiers paid less attention to the quality of their weapons than to the fundamental reason for their enlistment. F. Y. Hedly of the Thirty-second Illinois recalled that as soon as the recruit was given a gun, he "was at last a full-fledged soldier. It little mattered that the weapon was a clumsy old 'Belgian,' thrown away as useless by its petty crowned owner in Europe; or an old government musket altered from a flint-lock; it was a gun, and the soldier asked no questions." A member of Allen Morgan Geer's regiment even claimed that the American smoothbores that had been converted from flintlock to percussion "proved a very efficient and deadly weapon, and some of them were carried all through the service."[14]

Moreover, some Union regiments received the best rifles from the very beginning of the war, or soon after. Samuel Beardsley of the Twenty-fourth New York enjoyed taking potshots at Confederate pickets in Virginia with his Enfield rifle in September 1861, while the Sixth Kentucky received Enfields at its organization two months later. The Fourteenth Illinois began to switch from altered flintlock smoothbores to Enfields by December 1861. Modern Springfield rifles were given as a reward for the best showing in target practice to Company K, Thirty-third Illinois, at about the same time that the Fifty-first Pennsylvania received enough Enfields to replace all its Harpers Ferry muskets in early 1862.[15]

Confederate Problems

The Federal volunteers from the western states tended to complain more about inferior weapons than their comrades in the East, but it is possible that they were given worse guns because of their distance from the center of political power in Washington. On the Confederate side, the western volunteers also had a more difficult time obtaining good weapons than their eastern counterparts. The state of Tennessee, the bulwark of the western Confederacy, had particular difficulty finding guns. Many of those issued to its regiments had to be returned as unserviceable, and the authorities often collected weapons from civilians. They wound up with a bizarre assortment of hunting guns, shotguns, and antiquated ordnance, one-fourth of them unusable. Volunteers often brought guns from home upon mustering in, adding to the variety of weapons shouldered by many units. The Seventh Texas arrived in Tennessee with 750 men, but only 377 of them had any weapons. They consisted of 123 shotguns, 25 of which needed repair; an assortment of 150 miscellaneous rifles, 48 of which were in "poor condition"; and 104 muskets with percussion detonation systems that had been given to the Texans when they passed through Louisiana.[16]

For the time being, this was the best the Confederate government could do for its western volunteers. Pres. Jefferson Davis told St. John R. Liddell, when Gen. Albert S. Johnston sent the latter to Richmond to plead for good weapons, that Johnston had "plenty of men in Tennessee, and they must have arms of some kind. Shotguns, rifles, even pikes could be used. We commenced this war without preparation, and we must do the best we can with what we have at hand." By July 31, 1861, Tennessee had raised 17,541 infantrymen. In all, 69 percent of them were armed with flintlocks, almost 20 percent had smoothbore percussion muskets, and only 11 percent were armed with rifles.[17]

The issuance problems in Tennessee were far worse than anywhere else in the Confederacy, although other states did not find it easy to arm the volunteers. Officers had promised new Enfields to anyone who enlisted in the Thirty-first Georgia, but the men drilled "with the old shotguns and squirrel rifles with which they had left home" for some time before pikes arrived. This "came near causing great trouble, and the officers did not attempt to compel the men to take them," according to I. G. Bradwell. The regiment finally received "old smooth-bore muskets which had done service in previous wars," and the men had to be content for the time being. The Second Mississippi received "old army muskets" upon organization, but Samuel W. Hankins had little desire to shoot again after he first tested their characteristics. "Why such a weapon was ever dealt us with which to fight the enemy is a puzzle to me, as there is about equal danger at either end."[18]

Yet there were Confederate regiments that received good rifles from the beginning, or nearly so, of their service. The Sixth Tennessee was among those few regiments that were well armed in the state during the initial rush to organize troops. Col. Thompson B. Flourney refused to accept old guns that were seized at the Little Rock Arsenal and waited until his First Arkansas passed through Lynchburg, Virginia, where it received Springfield rifles. The Seventh Georgia Battalion organized in September 1861 and was issued the "best Enfield rifles," while at least one company of the Eighteenth Louisiana received Mississippi rifles at the same time. Often the quality of guns depended on officers like Col. Flourney. A wealthy planter named Plowden C. J. Weston joined Company A, Tenth South Carolina, as a private and used his own money to buy his comrades 155 Enfield rifles. Weston later was elected captain of the company.[19]

As was the case in the North, Southern authorities managed to find better guns as the war progressed, and units were able to exchange their flintlocks for rifles using the Minié system. The First North Carolina had used flintlocks altered to percussion for nearly a year before receiving Enfields. The Tar Heels were now "proud possessors" of the latest in firearms, recalled Finley P. Curtis. "We thought we could 'lick' the entire Federal army." The Twenty-sixth Louisiana organized in early 1862 and received its first weapons soon after the retreat from New Orleans. Eight companies took Enfields and two were given Belgian rifles, which were later exchanged for Enfields.[20]

Nevertheless, in many areas of the Confederacy, regiments still received inadequate weapons as the second year of the war began. A regiment raised in East Tennessee sported "some old flint-lock muskets, some squirrel rifles with saber-bayonets and some without, and some shot-guns, almost all out of fix and wholly unfit for service." The Fortieth Alabama also took the field with shotguns and a few muskets in May 1862. The Confederate government eventually accepted civilian weapons and gave volunteers an allowance for their value as a way to supplement government-issued guns.[21]

The hastily organized Army of the Mississippi, which was later renamed the Army of Tennessee, fought the battle of Shiloh with many units inadequately armed. Following the engagement, an inspection of Brig. Gen. Alexander P. Stewart's brigade revealed that 53 percent of the men were using smoothbore muskets, and there were still 103 flintlocks in the brigade. In another brigade, the Forty-seventh Tennessee possessed no fewer than ten different kinds of guns. The Federals saw clear evidence of this when the Army of the Mississippi evacuated Corinth in late May and retreated toward Tupelo. Those Yankees who followed up the retreat saw the road scattered with equipment, clothing, wagons, and out-of-date guns. Marcus Woodcock of the Ninth Kentucky (U.S.) described them as "old-shotguns, sporting rifles, *war guns* of Southern invention, and old flint lock muskets marked 1776."[22]

Federal Improvements

On July 2, 1862, Lincoln issued a call for 300,000 troops to serve three years or the duration of the war. This single piece of paper essentially doubled the size of the Federal army and inspired a new wave of patriotic enlistments all across the North. As a result, the same problems associated with the raising of the earlier regiments were seen. Volunteers spared no criticism when reporting on the quality of their old weapons. Austrian rifles did not amuse the men of the Ninetieth Illinois when they tested them in September 1862. "Many would not go off at all, and those that would got heated, and went off prematurely, often taking along a finger, a cap-visor, or a piece of an ear," recalled George P. Woodruff. "The stocks were of soft wood and easily broken; and the bayonets were also easily broken. In fact, the ramrod was the only reliable part of the Austrians." The 104th Ohio received Austrian rifles upon its organization that month. The guns were "wonderful and painful kickers," thought J. W. Gaskill, and "in fact either end of the musket is more or less dangerous."[23]

The Belgian rifles were considered little better in the eyes of the new volunteers. They were "about the poorest excuse of a gun I ever saw," remembered Theodore F. Upson of the 100th Indiana. "I don't believe one could hit the broadside of a barn with them." Even older-model Enfields seemed "ungainly pieces having the look of old age" to some men. John Henry Otto had experienced war in a border clash with Denmark in 1848 and in suppressing the revolution in the Rhineland in 1849 while serving in the Prussian army. He had deserted while serving as a reservist in 1850 and immigrated to America. Now, having joined the Twenty-first Wisconsin, Otto was disgusted with the Belgian muskets issued to the regiment. "Oh! how I longed for the light handsome needlegun, which we used in Prussia." It was lighter than the Belgians and accurate, Otto reported, at 1,000 yards, capable of firing five shots a minute. In contrast, "it took almost five minutes to load one of these Clumsy pieces and then good only for a distance of 150 yards." Officers rejected these weapons and later replaced them with Austrian rifles. At first the Austrians did not seem much of an improvement, but the men came to rely on them. "Those innocent, plain looking shooting irons prooved [*sic*] to be fully equal to the demands made of them," Otto remembered. "They were very dangerous weapons in cool and steady hands."[24]

Col. Dan McCook could not have agreed with Otto. The Illinois regiments in his brigade of the Army of the Cumberland were burdened with Austrian muskets. He sent some of them to Louisville for repairs. "The defect seems to be principally in the locks, which weakened by use, so that now not more than *One half* will snap a cap until after two or three trials." McCook reported to the army's chief of staff that his "men have no confidence in them and in the event of an action the bayonet must be my reliance." McCook's other regiment, the Fifty-second Ohio, was armed with Springfield rifles. The 130th Illinois, in another brigade, also had to send at least one-third of its Austrians back to repair shops at Memphis. French rifles issued to the 124th Illinois were no better. In target practice "the locks proved too weak, the range inconsiderable, the recoil tremendous, they heated rapidly and their aim was entirely unreliable—the guns were worthless," according to Richard L. Howard. The Seventy-seventh Illinois, which organized in September 1862, received "flintlock muskets of ancient times—some with locks and some without." Fortunately, the regiment was soon able to exchange them for Enfields.[25]

In contrast to these examples, many new regiments received fine weapons in the fall of 1862. The 100th Pennsylvania obtained Springfields—"the best gun made," bragged Frederick Pettit. The Nineteenth Iowa received Enfields said to have been confiscated from Southern vessels trying to run the blockade, while the Ninth Vermont felt proud to be issued Springfields. "This, I take it, means something," commented Edward Hastings Ripley. "I don't believe they will put their best muskets into the hands of men to remain long idle."[26]

Both the new regiments and the old were gradually being armed with the best rifle muskets available, but the process of conversion was slow. The Thirty-third Iowa used smoothbore muskets for many months before shifting to Enfield rifles. The Eighth Illinois, the first infantry regiment organized by the state, still had seven different kinds of small arms representing three different calibers by December 1862. At the same time, there were 113 American rifles, converted from flintlock to percussion, in the Twentieth Illinois, more than any of the other three types sported by the regiment.[27]

Records dated anytime in 1863, especially the early months of the year, show the steady gains made by Springfields and Enfields among Union regiments. The Thirty-seventh Illinois received the former to replace its Belgian rifles, except for Companies A and K, which retained Colt's revolving rifles. The Forty-first Ohio used "altered muskets" for nearly two years before taking up new Springfields, while the 152nd New York had first used "an old fashioned State musket," then Austrian rifles, then Enfields by the end of winter 1863. When Enfields were given to the Second Minnesota near Franklin, Tennessee, in March 1863, the men "felt that they were better armed than ever." Leander Stillwell, who was an experienced shooter, had been willing to make do with the old Austrian rifles but judged the Springfield Model 1863, which was issued to his regiment in June of that year, "a very efficient firearm." The state of Illinois raised fifteen regiments in 1864, and all of them received Enfields at the time of their mustering in. All ten Illinois regiments raised in 1865 received Springfields.[28]

At least for Northern regiments, the Springfield and Enfield dominated in the issuance of small arms during the last half of the Civil War. This was a state of preparedness that should have been the norm in April 1861. These two weapons are widely viewed as the best rifle muskets available in the world by 1863. The Springfield was made at the U.S. Arsenal at Springfield, Massachusetts, as well as by private contractors. The Model 1861 and 1863 had replaced the original Model 1855, and all of them were .58 caliber. The arsenal made almost 800,000 units of the Model 1861 and 1863, while private contractors made close to 600,000 by the end of the war. The weapon cost $9 each in 1863 and could be dismantled into forty-seven different pieces. The arsenal compound was huge, embracing 72 acres, fifteen buildings, and employing more than 2,600 workers. The Enfield was made solely in Britain. The Model 1853 and 1856 were the most common types used in the Civil War, and both were .577 caliber. Both the Union and Confederate governments imported a total of 800,000 Enfields during the course of the war.[29]

The Army of Northern Virginia seems to have been equally well armed as its opponent, the Army of the Potomac, by 1863, but this was not the case with the Confederacy's major army in the west. The Army of Tennessee lagged behind all other major field armies. By April 1863, it had a total strength of 37,232 men. Only 37 percent of them held Enfields, and 14 percent were armed with Springfields. Fully 44 percent of the troops still used smoothbore muskets, and 5 percent of them had an assortment of different weapons. Six months later, at the battle of Chickamauga, one-third of the guns used by Brig. Gen. William B. Bate's brigade were smoothbores. Yet the Army of Tennessee managed to address some of these deficiencies by the time the Atlanta campaign began in May 1864. Some 55 percent of its men had Springfields or Enfields, and 32 percent had Austrian rifles. But 11 percent still had to make do with smoothbores, and 359 men had no weapons of any kind.[30]

The process of trading old guns for new was sometimes difficult. The colonel of the Twenty-seventh Connecticut blamed poor-quality Austrian rifles for the high casualty rate suffered by his regiment at Fredericksburg. He called them "lamentably deficient" and "unfit for service." They were therefore "'inspected,' condemned, and finally exchanged for better ones," according to Pvt. Robert Goodyear. Tired of their Dresdens, which were "heavy, clumsy, and hard to keep clean," the men of the Twenty-second Wisconsin held an inspection and target practice to try to get them condemned.[31]

Some Federal and Confederate regiments were simply out of luck due to circumstances beyond their control and had difficulty acquiring adequate firearms even near the end of the war. Brig. Gen. John P. Hawkins complained about administrative delays in shifting from poor-quality Austrian rifles to a better-class weapon. "Our present arms are constantly breaking and bursting, and their fire is very inaccurate. I could not go into a fight with much confidence in their efficiency. The men fear the effect on themselves of their own fire." Hawkins commanded a black brigade on occupation duty in Louisiana and apparently received low priority. Another black unit, Lt. Col. John C. Chadwick's Ninety-second U.S.C.T., fought a skirmish with Rebel troops during the Red River campaign armed with Springfield smoothbore muskets, "many of them becoming useless at the first fire."[32]

Confederates in the Trans-Mississippi found themselves critically short of weapons during the last half of the war. They had to organize clandestine shipments of spare guns from Mississippi and eastern Louisiana to partially make up the deficiency. Brig. Gen. Thomas P. Dockery's Arkansas brigade was "wholly or nearly unarmed" in January 1864, but received 1,400 weapons in one shipment from east of the Mississippi River. A month later, Brig. Gen. William L. Cabell outlined his policy for distributing scarce rifles in his cavalry brigade for his division commander, Brig. Gen. John S. Marmaduke. Cabell thought it best to give "long-range guns" to both flank companies in each regiment rather than divide the rifles and smoothbores equally among the regiments. This followed the precedent set for decades in the U.S. Army and those of the European nations. "I am very anxious to get rifles for all my men," Cabell assured Marmaduke, "but if I cannot do that I would greatly prefer having my arms distributed as I have them, in order to drill and use [the flank companies] as skirmishers."[33]

Soldiers who had surrendered to the enemy and given up their arms went through a parole process, at least until early 1864, before going back to duty. They had to be reissued weapons at that time and often found themselves low on the priority list. Brig. Gen. John C. Moore led a mixed brigade of Texas, Mississippi, and Alabama regiments at Vicksburg. After its surrender, the brigade was broken up and reorganized into an all-Alabama unit, but it received "a lot of arms and accouterments that had been condemned as unfit for service and piled up in an outhouse near the railroad depot" at Demopolis, Alabama, upon its official exchange. Moore was told this was only a temporary expedient to allow his men to start drilling, but his hope that better guns would be forthcoming before his brigade took the field were not realized. Sent to the Army of Tennessee for the battle of Chattanooga, his men found to their dismay that the ammunition issued them did not fit the guns. Although it was a bit too small, the troops used it anyway, but fire effectiveness must have suffered. If the rounds were too big, they just jammed in the barrel when fired, "and some of those guns may remain loaded to this day," Moore sardonically wrote in 1898.[34]

Both Union and Confederate armies created institutional means of evaluating weapons and accoutrements in the hands of their men, replacing them if possible and charging the soldiers for damage due to neglect. The colonel of the Thirteenth Tennessee read orders to his men in November 1862 that spelled out exactly what they could expect if they lost or damaged anything without a good explanation. The charges included $50 for an Enfield, $35 for "Minnie Rifle or Musket," $25 for "Improved Muskets," and $15 "for old United States Muskets." Capt. Willis H. Claiborne, who served on the staff of Brig. Gen. Alexander W. Reynolds's brigade, inspected arms and accoutrements in April 1864, prior to the Atlanta campaign. He decided to condemn eight Enfields belonging to the Fifty-fourth Virginia and forty-four Springfields in the Sixty-third Virginia, plus a number of other pieces of equipment associated with weapons, including nearly a hundred damaged cartridges. Claiborne planned to send all the condemned ordnance stores to an arsenal for any feasible repairs.[35]

Harvesting Guns from the Battlefield

A surprisingly fruitful source of weapons in the Civil War was the battlefield itself, for it was a fact of military life that a certain percentage of soldiers lost their guns in the excitement of combat. How many of them did so, and whether purposely or accidentally, depended on a number of factors. Of course, the rate of dropped guns soared if a unit suffered a severe tactical defeat and evacuated the field in haste and panic, or if large numbers of men threw down their guns before surrendering. Whoever controlled the field, either temporarily while the fighting shifted to new ground or permanently after the battle ended, had the opportunity to harvest a crop of usable weapons.

Periodically commanders recognized this phenomenon in general orders, cautioning their men to take better care of their guns in the heat of battle. One such order circulated among Eighteenth Corps troops following their repulse in the battle of Drewry's Bluff on May 16, 1864. "Those who may have thrown away their arms on the field or march will be brought to trial if it be possible to find any guilty of such crime." Nineteen regiments in the Third Corps, Army of the Mississippi, lost a total of 172 guns at Shiloh. The causes, as listed by inspectors, ranged from "Neglect" to "Threw it away" to "Stolen," "unavoidable," "In action," and "Unfit for use." It is estimated that the same army, now named the Army of Tennessee, lost 4,000 guns because its dejected troops simply threw them away while retreating from the hard-fought battlefield of Stones River in early January 1863. The same army lost 6,175 small arms when severely defeated and sent in hurried retreat from Missionary Ridge the following November. At times troops made strenuous efforts to save guns abandoned by their comrades. Some Confederate gunners loaded muskets onto the caissons of their artillery pieces while retreating from the battlefield of Shiloh. Also, Union troops "broke, bent, and secreted a large number of their small-arms" before either giving up or retreating from the battlefield of Brice's Crossroads on June 10, 1864. Yet Maj. Gen. Nathan B. Forrest's Confederates managed to find and send off for repairs many of these damaged weapons.[36]

Of course, a soldier often had no control over whether his gun remained in a usable condition during combat. A comrade of John Henry Otto in the Twenty-first Wisconsin had his musket deformed by a bullet that converted the barrel into "the shape of an Elbow." Cpl. Dan Wells of the Fourteenth Illinois was driven to exasperation at Shiloh. He approached Col. William Camm with a musket that had been rendered useless by a bullet, its stock only barely attached by a single screw. "That is the fourth gun smashed in my hand since I have fired. What in hell shall I do?" Camm simply pointed to an abandoned weapon lying nearby, and Wells "pettishly" threw down his useless gun and soon was "blazing away."[37]

John Calvin Hartzell of the 105th Ohio admitted in his memoirs that he became excited and threw away his usable weapon at the battle of Perryville, but he regained the presence of mind to take another, for "guns lay loose everywhere." After the fighting ended at Stones River, Marcus Woodcock of the Ninth Kentucky wandered through the gathering dusk until he found a weapon to replace his damaged Enfield. He did not examine his new firearm until the next day, when he realized that it, too, was useless. Woodcock had to wait nearly a week before he was issued a better gun. He knew of many other soldiers who had deliberately thrown away their smoothbores and replaced them with rifles they found on the battlefield. A few months later at Chickamauga, Confederate troops in Brig. Gen. Arthur M. Manigault's brigade took extra ammunition from the dead and wounded, and if they could not find any to fit their weapons, they took the abandoned guns that matched the available ammunition.[38]

David Holt, a young soldier in the Sixteenth Mississippi, was very particular about the weapon he used. Right after the battle of Malvern Hill, he spent quite a bit of time surveying the field for the right gun, looking at many that were still clutched in the hands of dead Federals and others lying loose upon the ground until he found one that was brand new. Nearly two years later, Holt allowed the gun to slip out from under his tent during a heavy rainstorm and it was ruined in the muddy water, but he acquired a new one from a Federal he wounded in a skirmish a few days later.[39]

Individuals by the thousands picked up replacement weapons on many battlefields, but there were organized efforts to harvest guns as well. The entire Forty-seventh Tennessee equipped itself from weapons scavenged from the field of Shiloh on the night of April 6, while the Confederates garnered some 6,000 weapons from the Stones River battlefield, which they largely controlled for three days before retreating south. Chickamauga provided yet another rich haul of guns for the Rebels. The primary duty of ordnance officers was to resupply their units with more ammunition, but whenever they had time during a battle, they also told their men to pick up all the guns they could find. The ordnance train of Brig. Gen. Marcus J. Wright's Tennessee brigade collected 1,100 guns during the battle. Maj. John A. Cheatham, ordnance officer of Maj. Gen. Benjamin F. Cheatham's Tennessee division, used empty ordnance wagons to haul away piled-up guns and accoutrements on the second day at Chickamauga, but a shortage of wagons limited his collection to some 3,000 guns. In another division, the Thirty-sixth Alabama spent enough time after the battle to take in an impressive amount of weaponry and ordnance: "274 rifles, 69 muskets, 122 bayonets, and 274 cartridge boxes, with belts."[40]

As the biggest engagement of the western theater, and one of the worst Union defeats, Chickamauga represented the mother lode of Confederate weapon retrieval from a battlefield. Gen. Braxton Bragg's Army of Tennessee gathered a total of 23,281 small arms. In addition, the Confederates piled up 135,000 rounds of small-arms ammunition and 461 bayonets. They even found 237 blank cartridges among the captured Union ordnance stores. In addition to all this perfectly usable material, they also picked up 10,000 cartridge boxes, 2,000 cap boxes, and 1,800 ramrods that were damaged. Brig. Gen. William B. Bate's Tennessee brigade benefited from this haul in a very immediate way. One-third of Bate's men went into battle at Chickamauga armed with smoothbores, "but after the first charge Saturday evening every man was supplied with a good Enfield rifle and ammunition to suit, which was used with good effect on their original owners the next day." Bate's ordnance officer also picked up an additional 2,000 guns from the field after the fighting ended.[41]

Gen. Robert E. Lee's Army of Northern Virginia also had opportunities to reap a weapons harvest on fields from which it drove Union opponents. His chief of ordnance, Lt. Col. Briscoe G. Baldwin, reported collecting 11,091 small arms after the battle of Fredericksburg on December 13, 1862, but he estimated that perhaps 2,000 of them had been abandoned by Confederate troops. Baldwin also recovered 255,000 rounds of ammunition and 1,800 sets of accoutrements. All types of small arms were represented, including 13 flintlocks, but 1,406 of the number were also damaged. Baldwin kept 7,720 guns in his reserve train for future distribution while giving 2,166 to the First Corps, 513 to the Second Corps, and sending 692 to Richmond warehouses. More than a year later, the ordnance chief of Richard S. Ewell's Second Corps collected 4,000 small arms from the Wilderness battlefield.[42]

The Confederates secured Union weapons wholesale when Maj. Gen. Earl Van Dorn led a cavalry raid to Holly Springs, Mississippi, in December 1862. He easily captured the town with its rich depot, supplying Maj. Gen. Ulysses S. Grant's initial campaign against Vicksburg. Van Dorn had little time to haul away the loot, so his men destroyed up to 7,000 Springfield and Enfield rifles and took hundreds of revolvers and carbines before riding away.[43]

Union forces also recovered weapons from every battlefield they held after the fighting ended. After Pickett's Charge at Gettysburg, they found that the best way to deal with a loaded and cocked musket was to stick its attached bayonet into the ground so the muzzle would not point at anyone. An observer recalled that so many such weapons dotted the field they were "standing as thick as trees in a nursery." At least 3,900 Confederate guns were collected from a Confederate attack force of some 11,000 men. One wounded Rebel left behind at Gettysburg saw Union cavalrymen breaking abandoned weapons on the field and "a pile of old broken guns from six to eight feet high extending along the track for several yards" when he was transported to the railroad station. At Gettysburg, Federal authorities had to contend with avaricious civilians who apparently scavenged more than 15,000 small arms from the sprawling battlefield before the army could get to them. Guns were probably the most valuable article littering the field, for soldiers' personal belongings would hold little interest for a civilian wanting to make a dollar. Army officials estimated that from 3,000 to 5,000 civilians wandered across the battlefield for days. Seventy-six wagons were sighted leaving the area on July 7 alone, and thirty of them were stopped. Each was loaded with a variety of looted army property.[44]

The Army of the Potomac had more opportunities to acquire a rich haul of abandoned weaponry during the Overland campaign. During the bloody battle of May 12, 1864, at Spotsylvania, the Second Corps captured a large salient in Lee's fortified line, along with more than 3,000 Confederate prisoners. Maj.

Gen. Winfield S. Hancock reported gathering over 7,000 guns from the captured salient over the next two days, but he had no empty wagons with which to transport them. Ordered to move to the Union left, Hancock had no choice but to destroy or bury them.[45]

Like their Confederate counterparts, many Federal units took advantage of battlefield acquisitions without waiting for official approval. The Thirty-first Illinois used flintlocks that had been converted to percussion at Belmont, many of which "choked and burst while in battle," but Col. John A. Logan happily reported that his "boys soon had better [guns] in their hands." The Eighth Illinois picked up as many Enfields on the Shiloh battlefield as possible to replace old smoothbores, although it meant that the regiment would have seven different kinds of small arms representing three different calibers. Some of Sherman's troops captured fourteen "entirely new rifles" in a skirmish with the Rebels during the advance on Corinth. The guns had been issued to the Confederates only the day before, and some of them had never been fired. The surrender of Arkansas Post on January 11, 1862, included some 5,000 Rebel troops, and the Ninety-seventh Illinois quietly took enough Enfields from them to replace all their old Belgian flintlocks that had been converted to percussion. Regimental officers then turned in the Belgians as if they were captured enemy ordnance.[46]

The surrender of 30,000 Confederates at Vicksburg on July 4, 1863, revealed to Grant that their small arms "were far superior to the bulk of ours." Grant then authorized all regimental commanders to trade their inferior weapons for the enemy's. During the latter part of the war, the Union army often acquired large amounts of weapons through battlefield success or through the capture of depots. More than 3,000 small arms and 1,200 cartridge boxes fell into the Union's hands at the battle of Nashville. When Sherman's troops occupied Columbia, South Carolina, in February 1865, they found a large cache of "guns, large and small" stored at the South Carolina Military School. Division leader William B. Hazen took charge of it and issued a circular offering the loot to anyone who was interested. "The officer in charge is directed to recognize all requisitions, however informal," he announced. This pile was probably the source of 100 guns issued to the mayor of Columbia so he could attempt to maintain law and order, but he was made to sign an affidavit promising not to use the weapons against the interests of the U.S. government. Sherman's men also took possession of 20 tons of cartridge paper and 10,210 rifles and muskets at Columbia. They found 8,424 small arms, 363,500 cartridges, and half a million caps at Greensborough, North Carolina.[47]

Whenever an army held undisputed possession of a battlefield or a depot, there was no question that to the victor belonged the spoils. But what if there was no clear winner of an engagement? After the Union attack on Kennesaw Mountain was bloodily repulsed on June 27, 1864, the Federals maintained their forward position only a few yards from the entrenched Rebel line on Cheatham Hill. A couple of hundred Union dead lay in that narrow no-man's-land, and a truce was called to allow for their burial. An interesting debate occurred between Lt. Col. James W. Langley of the 125th Illinois and Col. Horace Rice of the Twenty-ninth Tennessee as to who had the right to the abandoned guns that lay between the lines. Rice seemed anxious to grab them, but Langley refused, saying "they should be regarded as neutral property, and not touched by either party until one or the other should occupy the ground." Rice felt compelled to agree or appear to admit that the Confederates had no chance to hold their line. After the Army of Tennessee evacuated the Kennesaw position on the night of July 2, the Federals took possession of the 250 Enfields.[48]

Repeaters

There were weapons available during the Civil War that were far more advanced than the rifle musket, yet they never became standard issue in either army. The cost involved and the relative newness of the repeater limited its issuance. The U.S. War Department made it clear that the government would not commit to putting the new weapon into the hands of troops even if recruiters promised it as an inducement for enlistment. If private individuals, or the soldiers themselves, chose to pay for the weapon, that was acceptable as far as the government was concerned.[49]

Many soldiers did spend their own money to purchase the latest in killing tools, and most of them seem to have been westerners. Theodore F. Upson of the 100th Indiana bought a Henry rifle, perhaps the most advanced repeater of the era, from a soldier in the Ninety-seventh Indiana who had been wounded and was about to go home. Upson had only $35 on hand, $10 less than the asking price, but the man sold it anyway. "They are good shooters," Upson commented, "and I like to think I have so many shots in reserve." After nearly a year of using the Henry, Upson told a correspondent, "I feel a good deal more confidence in myself with a 16 shooter in my hands than I used to with a single shot rifle." Many other western Yankees purchased the Henry, paying up to $65 for them. Several men in the Tenth and Sixty-fourth Illinois purchased the weapon, and in battle they fired them standing alongside men who still used the Springfield or the more obscure Whitney rifle.[50]

Whole regiments sometimes were armed with repeaters solely at the men's expense. This was the case with the Seventh Illinois by late summer 1864. "They are the nicest guns I ever seen & the handiest to load," reported Robert L. Mountjoy. "It can be loaded and fired in a second. All we have to do is jerk a lever forward & back & it is loaded & cocked. . . . They will shoot to kill half a mile." The men paid over three months' salary, about $50 apiece, for the guns. Altogether, three Illinois regiments were armed with the Henry and five used the Spencer repeater. All of them received these weapons in 1863 and 1864. Sherman's 60,000 men fired a total of 1,742, 338 rounds during the march through the Carolinas in 1865. Of that number, 38,654 were Henry rounds, 112,000 were Sharps, 213,448 were Spencer, and 1,223,636 were rifle musket rounds.[51]

Eastern regiments also acquired repeaters but apparently not in such numbers. Col. Edward Hastings Ripley's brigade of the Eighteenth Corps received enough repeaters to arm one regiment as it marched to take part in the Fifth Offensive at Petersburg on the night of September 28, 1864. Ripley was told to literally choose a regiment as his command approached one end of a pontoon bridge to cross the James River, and the guns were issued to that unit as it marched off the other end of the bridge. The officers tried to explain how to work the guns before the regiment entered combat.[52]

The Thirty-seventh Massachusetts received Spencer repeaters in mid-July 1864. The regiment was among Sixth Corps troops sent to deal with Lt. Gen. Jubal A. Early's raid on Washington, DC. The chief of ordnance of the U.S. Army happened to chat with brigade leader Col. Oliver Edwards as his command was marching through the capital, and asked him if he needed anything. Edwards told him to give Spencers to the Thirty-seventh Massachusetts, and it was done in a day or two. "This new rifle was undoubtedly at that time the most formidable weapon that could be placed in the hands of infantry," asserted the regimental historian. "It had wonderful range, shooting with accuracy and immense force." The Thirty-seventh was the first regiment in the Sixth Corps to receive the Spencer. The rate of fire was so much greater that the men had to carry 100 rounds instead of 40, leading to some waste as the weaker among them threw cartridges away to lighten their load.[53]

The most famous unit in the Union army to use repeaters was Wilder's Lightning Brigade, a western organization led by a successful entrepreneur named John T. Wilder. Born in upstate New York, the descendent of men who had fought in the Revolution and the War of 1812, he made his fortune as a foundry and millwrighting businessman in Indiana. Wilder became commander of the Seventeenth Indiana and was forced to surrender a small command at Munfordville, Kentucky, in September 1862, during Bragg's invasion of that state.[54]

After exchange, Wilder was given command of a brigade of Indiana and Illinois regiments attached to the Army of the Cumberland. He received permission to convert his unit into mounted infantry early in 1863 and spearheaded efforts to purchase Spencer repeaters for his entire brigade. The government refused to foot the bill, so Wilder raised a loan from bankers at his hometown of Greensburg, Indiana, in exchange for nothing more than his personal note as collateral. The men were to reimburse Wilder for the guns, but eventually the government decided to cover the cost after all.[55]

The Indianans and Illinoisans in the brigade were very proud of their new guns. "We felt ourselves to be well nigh invincible," one of them later recalled. B. F. McGee, a soldier in the Seventy-second Indiana, quipped that a single-shot muzzle loader was, "for a *fast* people like the Americans, . . . entirely too slow." McGee, like all of his comrades, praised the Spencer to the skies. It was easy to dismantle for cleaning and was light and balanced so that one could go through the manual of arms with ease. "The Spencer rifle demonstrated the superiority of breech-loading repeaters," concluded McGee.[56]

The chief of ordnance later reported that four out of every five repeaters issued to the Union army went to western troops, demonstrating the immense interest in the most advanced guns felt by young men from that region. On the Confederate side, there was interest in repeaters but very little opportunity to use them. The South made none of its own, and about the only time a Rebel saw a repeater was in the hands of a Union captor or lying on the battlefield, dropped by a Federal casualty. The Army of Tennessee reported having 49,193 guns in the middle of the Atlanta campaign. Only 58 of them were Spencer rifles, all in the hands of troops riding with the cavalry.[57]

Bureaucratic resistance to the repeater began to melt as soon as Federal soldiers demonstrated its value in combat. The battle of Hoover's Gap on June 24, 1863, was one of the first engagements to display the repeater in action. Wilder moved swiftly to occupy the gap as a line of advance for the Army of the Cumberland and defended it against a counterattack by a Confederate brigade. The Federals stopped the Rebel advance with the Spencers used by only two regiments, the Seventeenth Indiana and the Ninety-eighth Illinois. From Wilder down to the lowest private, all Federal veterans of this battle praised the usefulness of the weapon. "No human being could successfully face such an avalanche of destruction as our continuous fire swept through their lines," wrote Wilder. His men felt able to take on "ten times" their number of opponents. A story circulated among Union troops after the battle about a corporal in the Seventeenth Indiana who had vowed never to let his weapon fall into enemy hands. Wounded in the chest, the desperate man unscrewed a key component of the weapon and threw it away before he died so the Spencer could not be used if it was captured.[58]

Three months later, at Chickamauga, repeating rifles again made their mark. The Twenty-first Ohio was armed entirely with Colt's revolving rifles and fired at least ninety-five rounds per man on the second day of the battle, giving opposing Confederates the impression that they were fighting an entire Union division. Members of the 123rd Illinois thought their Spencers saved them at Chickamauga, "and our men adore them as the heathen do their idols."[59]

The Seventh Illinois made good use of its Henry rifles at the battle of Allatoona, where a badly outnumbered Union force was nearly overwhelmed by an entire division of Confederate troops. "Indeed, it is questionable whether the comparatively slow fire of muzzle-loading guns would have been able to cope with the dense masses of the enemy in their desperate assaults," wrote F. Y. Hedley. "The moral influence of this firearm was probably as great as its destructive power." Several Federal regiments had repeaters at the battle of Griswoldville on November 22, 1864, the only pitched battle of Sherman's March to the Sea. The attacking Confederates, most of whom were green troops, older men, and boys, were shattered. Theodore F. Upson called the battlefield "a terrible sight, . . . a harvest of death." A Confederate artilleryman, whose unit opposed a Federal brigade armed with Spencers at the battle of Franklin, called them "a terrible weapon compared with the muzzle loading Enfield rifle. It was like a Monitor compared with a wooden vessel."[60]

The Thirty-seventh Massachusetts first used its Spencers while skirmishing with Early's troops on August 21, 1864, near Charlestown in the lower Shenandoah valley. They found that they could fire at surprisingly long distance with a reasonable degree of accuracy. Col. Elisha Hunt Rhodes borrowed forty Spencers from the Massachusetts regiment to arm some of his men in the Second Rhode Island, and they also found the weapon to be especially helpful while on picket or skirmish duty. The Spencer was not perfect, as Daniel W. Sawtelle of the Eighth Maine found out in the attack on Confederate Fort Harrison during the Fifth Offensive at Petersburg. His weapon jammed when he jerked the lever too hard, forcing him to break off action, dismantle the slide with a screwdriver, and clean spilled powder from the mechanism. But mostly the Federals praised the Spencer highly. The historian of the Thirty-seventh Massachusetts described its use at the battle of Sailor's Creek: "Volley followed volley with almost the rapidity of thought, tearing the opposing line into demoralized fragments."[61]

By the latter months of the war, it seemed apparent that repeaters were far superior to muzzle-loaders on the battlefield, due primarily to their greatly increased rate of fire. The single-shot breechloader, like the Sharp's rifle, was an intermediate step between the rifle musket and the repeater, and its star also began to rise in the latter part of the war as a less risky alternative to the Spencer. The breechloader's rate of fire was higher than the musket but, of course, not as high as a repeater. The new chief of ordnance, A. B. Dyer, called for all Union troops to be armed with a breech-loading weapon in the fall of 1864, and he began to investigate whether the Springfield could be converted into one. On March 1, 1865, the secretary of war, Edwin M. Stanton, directed the issuance of breechloaders to all Federal soldiers, but of course nothing was done toward that goal before the war came to an end. Nevertheless, the U.S. government bought a total of 105,000 Spencers and 1,730 Henrys during the war.[62]

Assessing the Repeater

Not only in the United States but in all advanced nations of the world, government officials were thinking of converting to some type of breech-loading weapon by the mid-1860s. The rifle musket had just proven itself in a major war but was very soon to be superseded by a far more advanced gun. The Prussian needlegun already existed as a model; it eliminated "the two difficult and troublesome operations in loading,—the use of the ramrod to send the cartridge home, and putting the cap on the cone." Even though it was not a repeater, the needlegun sped up the rate of fire by this elimination. In addition, it and all other breechloaders and repeaters could easily be used while the soldier lay prone on the ground. In every way, they were the weapons of the future.[63]

But we need to be aware that the euphoric acceptance of these modern weapons by their supporters was not shared by everyone. Horace William Shaler Cleveland, the Northern rifle enthusiast, doubted that repeaters were suited for use by the battle line because he thought soldiers ought to take more time to aim and fire carefully rather than spew bullets in the general direction of the enemy. If a soldier fired without thinking, his fire discipline would inevitably break down, Cleveland thought, and so would the quality of his marksmanship. Col. Thomas W. Hyde, an officer on the Sixth Corps staff in 1864, saw a demonstration of this when he led a group of cavalrymen armed with Henry rifles in a skirmish. Even after the Confederates retreated, Hyde's men continued to pour fire into the woods as if on automatic pilot, despite all efforts to stop them. "It was quite a lesson on the improper use of rapid-firing arms," he later wrote.[64]

Moreover, there is only inconsistent evidence that units armed with repeaters or breechloaders actually shot down more of their enemy than those armed with rifle muskets. There was a very strong tendency for everyone who witnessed repeater fire to be hugely impressed with the volume of noise and smoke that resulted and to naturally assume that the targets would be knocked down accordingly. Even some Confederates eloquently described incoming repeater fire and the emotional effect it had on Rebel soldiers. Robert Stiles saw the repulse of Col. Lawrence M. Keitt's brigade (mostly Keitt's former regiment, the large and relatively inexperienced Twentieth South Carolina) during an impromptu attack on Federal cavalry armed with breechloaders on June 1, 1864, at Cold Harbor. "The regiment went to pieces in abject rout and threatened to overwhelm the rest of the brigade," Stiles later wrote. "I have never seen any body of troops in such a condition of utter demoralization; they actually groveled upon the ground and attempted to burrow under each other in holes and depressions."[65]

It is tempting to take Stiles's words for granted and assume, as did all observers, that the repeater and breechloader were powerfully effective weapons. But a recent historian has pointed out that Keitt's brigade suffered 150 casualties in this attack, which is a loss rate fully consistent with that of combat involving rifle muskets (and, for that matter, with battles involving smoothbore muskets, too). Confederate casualties at Griswoldville amounted to 650, which was quite high for the force engaged. But it must be pointed out that in both cases, the Rebel troops were relatively inexperienced in combat and were trying to attack veteran Union soldiers. We can certainly discount the claim by one of Wilder's men that, because the Spencer was more accurate than muzzle-loaders, it "expended less ammunition in battle . . . while doing thrice the execution." There is no evidence that breechloaders or repeaters were consistently more deadly than other arms. The occasions on which it can be proven that they mowed down the enemy were also occasions when other tactical or human factors also came into play to determine the rate of fire effectiveness. Otherwise, the new weapons seem to have produced a spectacular display of pyrotechnics without necessarily a proportionately higher rate of killed and wounded.[66]

In short, it is important for historians not to take for granted the awestruck descriptions of repeaters in action that fill Civil War literature. They are certainly an expression of what participants thought of the new weapon. But we need further study to determine exactly what tactical effect these weapons had on a variety of different battlefields.

Ironically, there were some Civil War contemporaries who did not accept even the rifle musket as a superior weapon. Admittedly they were in a tiny minority, but there was a debate in the 1850s among army officers as to whether a rifle musket was even necessary. Smoothbore defenders admitted that rifles would be useful for skirmishing, but they believed that the battle line could continue to do well with older weapons. They assumed that the range of fire would remain short, an assumption that actually was more or less fulfilled by the combat experience of the Civil War. And they also pointed out that buck and ball ammunition, consisting of a large-caliber bullet and several smaller-caliber buckshot, was very effective at short range because it spewed not one but several projectiles at every discharge. Buck and ball was not possible with rifles because the projectiles tended to damage the spiral grooves.[67]

Some Union officers continued this argument into the war years. George H. Thomas told a subordinate in the fall of 1861 that accepting smoothbores rather than rifles for his unit should not be seen as a problem, for the former weapon was "in reality . . . the best arm of the two." George L. Willard, a prewar army officer and colonel of the 125th New York, published a pamphlet entitled *Comparative Value of Rifles & Smooth-Bored Arms* in February 1863. His argument in favor of the smoothbore was based on numerous European studies and his own early war experience that most soldiers did not know how to accurately sight their rifles for long-range fire. Willard proposed that rifles be limited to skirmishers, one regiment for each brigade of volunteers and one battalion in each regular regiment, so the smaller number of men could be intensively trained in how to use them properly.[68]

A Confederate counterpart of Willard existed in Col. William E. Baldwin of the Fourteenth Mississippi. The Rebel attack at Fort Donelson on February 15, 1862, demonstrated, in Baldwin's view, "the efficiency of the smooth bore musket and ball and buck-shot cartridge." He wanted Confederate troops to be "impressed with the advantage of closing rapidly upon the enemy, when our rapid loading and firing proves immensely destructive and the long-range arms of the enemy lose their superiority." Baldwin expressed a common Northern and Southern desire to close with the enemy at short range to exact the greatest damage in the shortest space of time. The smoothbore musket was indeed an effective weapon for such tactics.[69]

Several Union regiments chose to retain their smoothbores for the duration of the war even though rifle muskets were available. The Sixty-third Pennsylvania, Sixty-ninth Pennsylvania, 116th Pennsylvania, and Eighty-eighth New York were among them. The Twelfth New Jersey carried .69-caliber smoothbores with buck and ball cartridges throughout the war. In preparing to meet Pickett's Charge, many Jerseymen adjusted the cartridge by taking the bullet out and putting in more buckshot, making their smoothbores even more effective at short ranges.[70]

3

THE GUN CULTURE OF CIVIL WAR SOLDIERS

A significant number of Union and Confederate soldiers were intimately familiar with firearms and extraordinarily interested in the quality and performance of their weapons. These men usually became the most effective killers in their units, because of their ability to adapt their native skills to whatever weapon was issued. They also were respected and admired by their less adept comrades. It is impossible to know how many soldiers fit this category; the percentage was well below half, probably well below a quarter of the total.

Brent Nosworthy has concluded that the Civil War soldier achieved a higher rate of fire effectiveness than that accomplished by any soldiers of previous wars. He estimates that the percentage of rounds Civil War soldiers fired that hit a human target denoted a ratio two to three times higher than that of British soldiers in the Peninsular campaigns of the Napoleonic Wars. While it may be tempting to conclude that the difference can be accounted for by the Civil War use of the rifle and the British use of the smoothbore, an even bigger difference in fire effectiveness can be seen when comparing the Civil War with the Austrian use of the new rifle musket in the Italian War of 1859. Moreover, American soldiers using the smoothbore in the Mexican War achieved a very high ratio of fire effectiveness, on the same level as they would later achieve in the Civil War.[1]

Nosworthy correctly concludes that the human factor rather than the quality of weaponry was the key in determining the differences of fire effectiveness. The greater American familiarity with weapons, and perhaps more effective aiming habits, could account for it. Nosworthy rightly points out that not all American soldiers in Mexico or in the Civil War practiced good marksmanship. He guesses that a minority of them who were "highly proficient marksmen" accounted "disproportionately for the casualties that were inflicted."[2]

There is interesting evidence in the personal accounts of Civil War soldiers to support this conclusion. John Calvin Hartzell of the 105th Ohio wrote pointedly about it in his memoirs. Born in 1837 and raised in Portage County, Ohio, he grew up in a family and a community that fostered an intense interest in firearms. "The guns of that day were very highly prized by the older men," Hartzell remembered. "They always spoke of them in the feminine gender." Hartzell's Uncle John used a flintlock to bring meat to the family dinner table. "That was long before my time," Hartzell admitted, "but as she hung in her hooks I admired and revered her, and I never knew her to be loaned to anyone."[3]

As a boy, Hartzell used an old flintlock, altered for percussion, which "was a cheap affair, to be sure, but I was proud of it and shot many a ground-hog and fish with it too." The local community held Saturday afternoon shooting matches on the river bottom. They nailed the lid of a cap box to a tree and stood 100 yards away. Sometimes cash was offered as a prize. Drinking was a common accompaniment to the shooting, although Hartzell argued that no one ever became so inebriated as to pose a danger to his competitors. He did admit, however, that late in the day it was possible to see the barrels "weave slowly up and down, yet, when the finger pressed the trigger the ball sped true." When Hartzell was yet too young to participate, he helped keep score at the target and greatly admired the old guns, some of which were "real works of art" with "bits of silver, bone and the like . . . curiously inlaid in the stocks." Hartzell could not wait to grow up so he could own a gun like these.[4]

Hartzell wrote his memoirs in 1896–1898, addressing them to his nephew Wilbur Johnson Hartzell. He felt it was necessary to explain to Wilbur the difference in temperament between himself and his brother, who was Wilbur's father, by blaming their own father. "Your grandfather never owned or fired a gun that I recollect, and just the same with your own dad; books ruined them both that way."[5]

Few soldiers were as descriptive in explaining the gun culture in which many Civil War participants grew up as was Hartzell. But there were many Northern and Southern soldiers who belonged to the gun culture. Richard W. Johnson, who later commanded a division in the Fourteenth Corps, grew up in Kentucky with the habitual use of guns. "I became a fine shot with the rifle for one of my age," he remembered, "and could 'bring down' a squirrel from the tallest tree, rarely ever missing my aim." Eugene F. Ware of the First Iowa, a three-months' regiment that fought at Wilson's Creek, was so interested in his gun that he began to test his weapon as soon as the regiment was issued its first firearms. It was an old smoothbore converted to percussion. "The guns had no rear sight, only a notch, and we had to tinker them and change the front sight so as to get an approximation." Try as they might, Ware and his comrades "could not make them shoot straight. But we understood guns and could get as close to it as anybody could, or as the gun itself could be made to go."[6]

The fascination with guns, even the reverence felt toward them, can be seen in a call by the adjutant general of the state of Tennessee to the colonel of a local defense regiment at Nashville. "Impress upon your soldiery that the Revolution of '76 was won by the Tennessee rifle, and that we fight in defense of our homes and all that we hold dear." It can also be seen in the fact that quite a few companies in the Confederate army adopted names that incorporated some reference to rifles or guns. Unlike regiments, the designations of which were officially approved by higher authorities, companies could adopt any nickname they chose. A survey of 3,035 Confederate infantry companies reveals that 570 of them, or more than 18 percent, used the word "Rifles" in their nicknames. Forty-six other companies used "Sharpshooters," "Marksmen," "Repeaters," or "Targeteers." A much smaller percentage of cavalry companies, 6 percent (31 companies out of 495) used Rifles as part of their nickname. Mississippi, South Carolina, Texas, and Virginia all had higher than average percentages of infantry companies that adopted words from the gun culture of America as part of their nicknames. In the Second Mississippi, nine out of eleven companies used the word Rifles, while seven out of ten companies in the Tenth Mississippi did so as well.[7]

Affection followed interest. Eugene Ware possessed an "old-fashioned, long, big, heavy, sturdy weapon." Nothing "could out-kick it but a Government mule," he asserted. The year 1829 was stamped on it, and Ware found that the old flintlock hole leading to the base of the tube had been filled in with brass. Despite the fact that most soldiers would have howled in protest at having to use such a weapon, Ware polished it up and gave it the affectionate nickname Silver Sue. "We all named our guns," he later reported. "The boys generally named them after their pet girls,—it was 'Hannah,' or 'Mary Jane,' or something else." Ironically, Ware was just as happy when he lost Silver Sue during Nathaniel Lyon's Missouri campaign. He and a regular soldier stopped to drink water at a springhouse and the regular accidentally took his gun and left his own behind. Ware adopted it, naming the weapon Orphan. An 1861 rifle musket, it was a great improvement over Silver Sue.[8]

Leander Stillwell of the Sixty-first Illinois had hunted squirrels and rabbits since he was old enough to fire a gun. Initially, he was so small that he had to rest the gun barrel on a bush. It was "a little old percussion lock rifle, with a long barrel," firing a bullet that weighed only 1/60 of a pound. Stillwell gathered, cracked, and sold hazelnuts to raise money for ammunition. It was a tedious chore, so he was careful not to waste a shot. After joining the army, Stillwell called his Springfield rifle musket Trimthicket.[9]

Many Confederate soldiers were no less enamored of their weapons. William A. Hughes of the First Tennessee saved the life of Sam R. Watkins, a comrade whom he'd known before the war. Hughes grabbed the muzzle of a Federal rifle just as the Yankee was about to shoot Watkins at close range during the terrible fighting at Cheatham Hill, along the Kennesaw Mountain line, on June 27, 1864. Hughes was mortally wounded, but he told stretcher bearers to give his gun to Watkins. Hughes had named it Florence Fleming, even embossing the name on the stock. He bequeathed his blanket and clothes to Watkins as well. "He gave his life for me," Watkins later remembered, "and everything that he had."[10]

It is unclear if Hughes was a gun aficionado before the war, but many soldiers found a new interest in weapons because they had to rely on their firearms for survival. As David Holt of the Sixteenth Mississippi put it, "Our guns were our best protection." Daniel W. Sawtelle of the Eighth Maine admitted that "a soldier always thinks of his gun first even before his blankets or clothing." The weapon assumed an importance beyond anything else. In fact, according to Robert J. Burdette of the Forty-seventh Illinois, the soldier "cares for it like a baby."[11]

Whether the gun-adept soldier was the product of a particular culture or simply a native genius with firearms will probably never be fully answered. To a degree, at least, he was influenced by the distinctive gun culture of America, which developed because of the frontier experience and the colonial heritage of the American people. The British failed to exact gun-control regulations on the colonials similar to those that historically had been imposed by virtually all European governments on their people. As a result, the colonial governments encouraged widespread gun ownership to support their militia forces, which were essential for frontier defense. The experience of the Revolution reinforced these ideals by implanting a fear of a standing army and a faith in a well-armed citizenry, buttressed by the passage of the Second Amendment as a prominent aspect of the Bill of Rights.[12]

As the frontier advanced, the importance of guns advanced with it, leaving the more settled areas, especially the cities, behind. Westerners especially wanted access to the latest developments in new weaponry. Arms manufacturing sharply increased in the United States by the 1850s as industry and commerce fed interest in firearms. In 1860, on the eve of the Civil War, some 50,000 guns were manufactured in the country, some by privately owned companies and some by the federal government.[13]

Crack Shots and Gun-Inept Soldiers

Right after the war's end, Theodore F. Upson of the 100th Indiana overheard some officers engaged in a discussion about sharpshooting. Upson had marched through Georgia and the Carolinas with Sherman and, like many of his comrades, was enormously self-confident. He told the doubting officers that he "could put half the shots in the magazine" of his Henry repeater into a man-sized target at 500 yards. The group went to a shooting range to see him try. Upson fired a dozen rounds and each one hit the target. "Those fellows opened their eyes; said they had no idea such shooting could be done. It really was no shooting; lots of men could beat me all hollow."[14]

Upson was a reliable soldier who had adjusted to combat conditions, but a number of good marksmen were not necessarily good killers. Samuel W. Hankins of the Second Mississippi recalled, "Some crack shots at home who always returned from the woods with a dozen squirrels, each shot in the head, when in battle could not hit a 'barn door' through excitement." Whether it was excitement or religious training that taught the soldier to preserve human life, there were many factors that could inhibit a recruit's ability to become a killing machine. These factors operated on the gun aficionado as well as on anyone else. A calculating effort to make every shot count inhibited Leander Stillwell's ability to fire during his first battle at Shiloh. Stillwell alone failed to shoot when the Sixty-first Illinois was ordered to blaze away at the enemy, who were mostly hidden by vegetation and a cloud of smoke. Lt. Bob Wildern excitedly urged him to pull the trigger even if he could not see a target. Based on his prewar experience of hunting squirrels and rabbits, Stillwell thought it was "ridiculous" to blaze away at an unseen target, but he did so anyway to appease the officer. For the rest of the war, Stillwell tried to carefully pick out individual targets whenever possible, but he rarely had an opportunity to do so.[15]

The recruit, whether proficient in the use of firearms or not, had to cross the gulf that separated civilian life from that of the soldier. He had to realize that taking human life was his raison d'être. As Robert J. Burdette wrote, it was the primary reason for his enlistment. "When the recruit is handed his Springfield rifle, and the corporal shows him how to load it, and teaches him the best methods of taking careful and accurate aim, and how to secure the most rapid firing with most effective results, the soldier is aware that he is not going to fire blank cartridges." Killing other men "is what he is shooting for. It is what he is paid for. That is his 'business.'"[16]

There were many soldiers, of both North and South, who were far from crack shots—many who actually were inept soldiers more dangerous to their comrades than to the enemy, though through little fault of their own. John G. Phillips of the Seventy-seventh Illinois was one such man. Described as "half-witted," he had no clue how to handle his musket safely. As the regiment advanced across ground entangled with vines and timber at the battle of Chickasaw Bluffs, Phillips trailed behind, overexcited and carrying a loaded gun. He tripped and the musket went off, the bullet narrowly missing his comrades in front. Phillips reloaded and continued, and the same thing happened. Now the men had had enough. They told their captain to send Phillips to the rear or they would shoot him, "for if they had to be shot they wanted it to be by the enemy and not by that d——d fool." The captain obliged by sending Phillips to the rear to help in the field hospital.[17]

During the battle of the Opequon on September 19, 1864, Capt. John William DeForest of the Twelfth Connecticut noticed a sixteen-year-old recruit who was not firing. This man, "a small and girlish youngster," could not be coaxed into it.[18]

Theobold Foltz of the Sixteenth Mississippi was certainly willing to fire but had no idea how to do it with effect. David Holt called him "an everlastingly good soldier and a sporting old numbskull that knew more about a yardstick than he did about a gun." The men often kidded him about his lack of aim. "Py tam, my gun goes off just like hell, and de pullet mus' go some phaeres. Phere he goes is de pisness of de Almighty. My pisness is to bull de drigger." One day early in the Petersburg campaign, he began to lose a duel with Union skirmishers. The company commander told him to go to the rear and let a better shot take his place. Foltz was irritated and refused to leave. Instead, he exposed himself to take better aim and was instantly shot in the forehead. His death infuriated his comrades. "Don't let that fellow live," they shouted, and eventually they shot the Yankee who had killed their beloved "numbskull."[19]

Levi S. Pogue of Company F, Fourth Texas, was a counterpart of Foltz. His story was preserved by a company comrade, Cpl. Joseph B. Polley, who enjoyed relating anecdotes about Pogue and Pogue's friend J. C. Jordan of Company I. The two men were opposite personalities. Jordan was a superb shot but "utterly and indescribably lazy." Pogue, whom Polley consistently called Pokue, was a huge man, 6 feet 4 inches tall, and almost 200 pounds. "While he keeps in line as long as the advance is continuous and artillery is not used against us, he never fires a gun. If a shell or round shot hurtles over or through the commands, he lets all holds go and drops broadcast to Mother Earth. If there is a halt, he is so fond of exercise that he runs. In short, Pokue is as much a noncombatant as any member of Stokes's Cavalry."[20]

Lolling in the hot trenches at Petersburg, the men often called on Jordan to show them how to shoot. He waited until the calls became widespread, then aimed across the parapet and waited for a target before showing off. The men next called on Pogue to do the same thing. He was willing and, "with a great show of alacrity and desire to please," barely poked the muzzle of his rifle out of the trench. This brought calls from his comrades: "Lower the muzzle of your gun, Pokue, for you will hit nothing but a quartermaster or commissary that way, and they ain't worth killing." Pogue shook so much that the barrel of his weapon wobbled before he pulled the trigger. Then, his task accomplished, he sank "back, exhausted, pale, and perspiring, into the arms of his friends, ready to receive their laughing congratulations."[21]

There were plenty of Pogues and Foltzs in both armies, perhaps as plentiful as the crack shots. The rest of the armies, the majority of soldiers, fell somewhere between those two extremes. J. W. Gaskill of the 104th Ohio went so far as to subtitle his book of reminiscences *Everyday Life of the Man under a Musket on the Firing Line and in the Trenches,* which indicates that he never felt completely comfortable using a firearm. In fact, Gaskill included a drawing entitled "Austrian Musket in Action," showing a distraught soldier right after firing the weapon, his hand to his right eye, apparently because the gun's kick had injured him.[22]

Target Practice

Serious and persistent target training could have made better shots out of many soldiers in both armies, but neither the Union nor Confederate armies instituted such a regimen. Target training was sporadic and uneven in quality; there is no evidence that it appreciably increased fire effectiveness. Native talent, more than training, accounted for the development of a small number of super-effective soldiers in the Civil War.[23]

This laxness in target practice mirrored prewar conditions, for the U.S. Army never instituted a true system of rifle training in the 1850s. Volunteers had little exposure to Minié rifles before 1861 because they were not widely used by civilian hunters. Stephen Vincent Benet, who served in the ordnance department during the war, later wrote that "but little attention, strange to say, was bestowed upon rifle instruction, and tens of thousands of men fired in battle their first shots with the Army rifle. Thus the value of the rifle as to accuracy was in a great degree lost for want of proper training to the soldier."[24]

Many Civil War commanders understood this problem and attempted to deal with it. Maj. Gen. George G. Meade tried to "familiarize the men in the use of their arms" in preparation for the spring campaign of 1864 by authorizing the use of ten rounds per man for practice. He sent a circular to all subunits in the Army of the Potomac: "It is believed there are men in this army who have been in numerous actions without ever firing their guns, and it is known that muskets taken on the battle-fields have been found filled nearly to the muzzle with cartridges." This situation existed despite the fact that the army had been campaigning ever since the early period of the war, more than two years before. Gen. Braxton Bragg authorized the publication of a special manual for his Army of Tennessee to walk the men through the basics of loading, gauging distances, and firing in the summer of 1863, after that army had fought major battles at Shiloh, Perryville, and Stones River.[25]

Hardee's drill manual offered instructions for how to conduct firing practice. He admonished the instructor to tell his soldiers to always aim at a particular spot. Company commanders were also advised to "make a short pause between the commands *aim* and *fire,* to give the men time to aim with accuracy." Hardee also emphasized the need for the soldier to remain calm while firing and to keep his feet in the prescribed place so the ranks and files could remain properly aligned, otherwise the men in the front rank might get in the way of those firing in the second rank.[26]

It appears that few if any of the practice firings conducted among volunteer units during the war were done by the book. Soldiers often mentioned the range of the firing and the type of target used, and sometimes the results. For Daniel E. Burbank's New Hampshire regiment, the range was 600 yards and the weapon was the Springfield rifle musket. He recorded that 360 men managed to hit an area the size of a man at that range. For Benjamin M. Seaton of the Tenth Texas, the targets were 3 feet in diameter and the distance was the width of the White River in Arkansas. His comrades put eighteen bullets into the mark—"a very good shoot," Seaton thought. A distance of 137 yards did not prevent Joseph K. Taylor of the Thirty-seventh Massachusetts from hitting a target 6 inches square.[27]

Col. Dan McCook set up a target practice for his brigade of the Fourteenth Corps in June 1863 by requesting blank cartridges and permission to fire live cartridges as well. He had his men firing an hour every day for several days and continued the practice in August. McCook also required his regimental commanders to report the names of their best marksmen. He continued the training in March 1864 but restricted it to only five rounds per man per day. "The object is to experience the men in firing," he explained, "and test the quality of the Ammunition now in use."[28]

The Fourth New Hampshire conducted target practice in March 1863 at its station near Beaufort, South Carolina. While the range was not recorded, the results of 790 men taking a shot at the target yielded the fact that only 10 percent of them (81) could place a ball within 15 inches of the bull's-eye. That was approximately the width of a man's shoulders. Only 22 of the 81 marksmen managed to place a ball within 2 1/2 inches of the bull's-eye, and 6 within 1 1/2 inches. If we were to judge fire effectiveness based solely on this minor evidence, we would have to conclude that only one out of ten men in the Fourth New Hampshire could take out an enemy soldier. Only one-third of those effective shooters (or 3.5 percent of the regimental total) could deliver bullets within a small enough target area to be termed marksmen. Although an observer called the members of the Fourth New Hampshire "Excellent target-shooters," the statistics do not seem so impressive.[29]

In the western Confederate army, there were at least sporadic efforts to instill a sense of marksmanship among the men. G. T. Beauregard issued a general order in the spring of 1862 extolling the American heritage of rifles. "It was the deliberate sharpshooting of our forefathers in the Revolution of 1776 and at New Orleans in 1815 which made them so formidable against the odds with which they were engaged," he asserted. Beauregard urged lower-level commanders to prevent "useless, aimless firing" among the men.[30]

Historian Andrew Haughton has surveyed the ups and downs of target practice in the Army of Tennessee, noting that it tended to take place in spurts between major campaigns. It also tended to occur more often in the latter half of the war than in the first two years. Braxton Bragg instituted target practice into the training regimen of the army in the late spring of 1863, using shooting competitions to increase interest among the men. Targets were set up at 200 yards, 400 yards, and 600 yards, but with less than impressive results. While one man hit within 1 1/4 inches of the target center at 500 yards, most missed the target altogether. Only seven men out of a company of thirty-one hit the target at 300 yards in another brigade. In still another unit, sixteen men out of seventy hit the target at 500 yards.[31]

Thomas Jefferson Newberry of the Twenty-ninth Mississippi was a moderately good shot. His brigade fired one round per man at a human-sized target set up at varied distances—200, 300, 400, and 600 yards. "Some of them hit it every time and some of them missed it every time," he reported. A few months later, in the winter of 1864, half the men of the Twenty-second Alabama missed a target that was 10 feet by 66 feet at a distance of 300 yards. But Companies E and I of the Thirteenth Tennessee improved their marksmanship. Only seven out of thirty-one men hit a target 300 yards away on April 29, 1863, while fifteen out of thirty-four hit it at 400 yards the next day.[32]

Joseph E. Johnston, who commanded the Army of Tennessee early in 1864, encouraged target practice. Given shortages in the Confederacy, he ordered that the targets should be placed in front of a bank of earth so the bullets could be recovered. Johnston had not always been a strong advocate of target practice. Early in the war, he argued that Southern soldiers needed "less instruction in shooting than in any other military exercise—It is one of our great advantages over the northern people." Yet Johnston admitted that his men had not shot well at First Manassas, blaming it on excitement rather than lack of skill. "It must be admitted, too, that since entering the service, they have had no systematic practice—for want of ammunition."[33]

Shooting competitions were organized sporadically in both Union and Confederate armies according to the level of initiative among regimental, brigade, and division commanders. A typical competition took place among African American troops along the South Carolina coast on Thanksgiving Day. Company commanders offered small prizes of $1 to $3 each for the three best shots in the regiment. Sometimes officers organized more impromptu shooting demonstrations.[34]

Col. William Camm of the Fourteenth Illinois had often told his men that "handling a gun by rule would make the fire more accurate." Some of the Illinoisans did not believe him, so Camm tried to prove it. He went to the flooded bottomlands near Fort Donelson and selected a group of "clean stemmed white oaks" about 100 yards away. Several men selected a tree as their target and used guns taken at random. Some of the best shots in the regiment participated, including Camm himself, and all were to fire at their own pace for five minutes. "I worked as nearly like an automaton or machine as I could in loading, aiming and firing; pulling the trigger almost the instant the butt came to shoulder, and got eight balls in my tree, to my nearest competitor's seven." What type of gun each man used was not recorded, but the Fourteenth Illinois possessed a mixture of Austrian and Enfield rifles. Camm's demonstration seemed to prove that loading and firing by the book was superior to instinctive firing by untrained men, although the difference was minor.[35]

Celebrating

The martial culture of Civil War soldiers sometimes led them to use their weapons to celebrate extraordinarily important news. When Twenty-fourth Corps troops captured Confederate Fort Gregg on the Petersburg line, the climax of one phase of Grant's April 2 attack, they spontaneously fired their weapons into the air to celebrate. Other Confederate troops nearby in a still-intact position apparently thought this meant the Yankees were killing Rebel prisoners in cold blood, and they opened artillery fire on them in retaliation. More than three weeks later, when news of Joseph E. Johnston's surrender in North Carolina spread among Sherman's troops, many of them spontaneously fired their weapons into the air. This was done "partly as an endorsement of the surrender, partly as the last rites at the burial of the defunct Confederacy, and partly as the inauguration of peace and prosperity." Other Federal units occupying East Tennessee also fired their weapons into the air on the news of Lee's surrender. One Federal claimed that several Yankees were killed by this wanton fall of bullets.[36]

Cleaning Guns

Taking care of their weapons was a high priority for most Civil War soldiers. Given repeated use and rough field conditions, these rifle muskets did not last forever. An ordnance officer estimated that they lasted on average about seven years among infantry troops. Typical causes of damage included "the clogging of the muzzle with mud, and thoughtlessly allowing the tompion to remain in, in firing." More important, rust and the buildup of powder residue inside the barrel deteriorated the performance of the weapon. Sherman issued general orders soon after the battle of Shiloh that the "muskets of the men must be kept clean and unloaded." The barrel was the most difficult part of the rifle to keep clean, inside and outside. William Gould of the 144th New York, who was stationed on Folly Island, South Carolina, explained how he managed this task by detaching the barrel from the rest of the weapon and placing it on sandy ground. Using a strap and sand, he scoured the barrel clean and then used a woolen cloth filled with sand to burnish the scratches. The barrel became "as bright & as clear that you can see your self in one quite plain. We have to handle them with gloves or mittens on our hands or we would stain them. Be the hands ever so clean." Most soldiers did not have a supply of sand available, so they used rags, "powdered dirt," corncobs, or pine twigs. For the inside of the barrel, they used "a greased wiper and plenty of hot water." Rainy weather created rust, so many soldiers greased the barrel with bacon to keep off most of the moisture.[37]

Gun aficionados invested the time and effort needed to keep their weapons spotless, while other soldiers only grudgingly did their chore. "Now if there is any kind of work that I dislike more than another it is the cleaning of a dirty musket," complained James M. Williams of the Twenty-first Alabama. After two and a half hours at it, Williams finally proclaimed, "Well! The greasy job is done." Williams had to do it or face disciplinary action; inspection by regimental officers was designed to make sure of that. The result of all this cleaning had an aesthetic effect as well as a utilitarian one. John S. Jackman of the Ninth Kentucky (C.S.) noticed on a bright June day in 1862 how strikingly the polished muskets caught the sun while the soldiers were marching. "There seemed a tide of melted silver flowing on by the green fields," he later recalled.[38]

Campaigns and battles dirtied muskets faster than living in camp, and the Civil War soldier had to learn to clean his weapon oftentimes under trying circumstances. Joseph K. Taylor of the Thirty-seventh Massachusetts stood picket one night in June 1863 while other troops crossed the Rappahannock River during a fierce rainstorm. After the picket line fell back, he quickly drew out the charge from his musket and wiped it clean before reloading, to be ready in case the Confederates tried to follow up the withdrawal. Col. John M. Stone told his Second Mississippi just before the opening of the fight on July 1 at Gettysburg, "Men, clean out your guns, load, and be ready. We are going to have it!"[39]

The rifle musket could fire about twenty to twenty-five rounds before the powder residue became bad enough to require attention. George D. Bowen fired thirty rounds at Gettysburg on July 3 before his smoothbore became so fouled that he literally could not load it anymore. Instead of cleaning it, Bowen threw the musket away and found an abandoned weapon instead. This problem was on the minds of many men as they blazed away at the enemy. Typical remarks heard in battle included "My gun's getting mighty dirty." Often regimental commanders took their entire units to the rear for gun cleaning. This happened with the Forty-first Ohio at Chickamauga, which was relieved on the line by another regiment. Robert McAllister's brigade in the Third Division of the Second Corps "sent men back by companies to wash [their guns] out, after which they would return and renew the fight" during the attack on the Mule Shoe Salient at Spotsylvania. On the Confederate side of that same engagement, William W. Smith of the Forty-ninth Virginia "had to stop in the midst of the battle and with my gun-screw take it [his Sharps carbine] to pieces and clean it."[40]

Whether he fired enough in any given battle to require emergency cleaning on the field, every soldier had to care for his weapon after the battle was over. In fact, "cleaning and scouring up the guns" was "the soldier's first duty after a battle is over," according to the historian of the Seventh Illinois. During the Union advance from Nashville in December 1862 and the subsequent battle of Stones River, the men had no time to keep their weapons clean other than "an occasional wiping with the sleeve of the blouse." But after that horrible conflict ended, Sgt. John Henry Otto of the Twenty-first Wisconsin set a good example for his men by taking his musket entirely apart and thoroughly scouring and oiling every piece of it. More than a year later, after two days of fierce fighting at the Wilderness, the men of the Sixth Wisconsin did little else the whole day of May 7 except clean guns and "wash the powder out of our faces."[41]

Those troops who failed to tend to their weapons after a fight found themselves in a quandary. Many members of the Twenty-sixth Alabama did not discharge their guns after the fighting ended on the first day at Shiloh, and water soaked into the tube and ruined the cartridges that were already in place as a torrential rain fell that night. Lt. Col. William D. Chadick had no tools with which to dismantle them. This caused his men to become "exceedingly dispirited" as the fighting resumed on the morning of April 7, yet "the most of them did the best they could." If soldiers had to get rid of charges in the tube that were no longer needed, the easiest way to do so was to fire the gun into the air. This often caused panic among other troops, who assumed it was a sign that the enemy were once again nearby. Complaints about this occurrence were common, and orders often went out specifying that the men should draw out the load with proper tools instead of firing it away.[42]

The cleanliness of guns was normally seen as an indication of the state of discipline and morale among troops, along with personal cleanliness and the order and neatness of their camps. That usually was a fair assumption, but it did not always hold true. When the Nineteenth Corps transferred to Virginia in the late summer of 1864, after two years of mostly occupation duty in Louisiana, the men were shocked at the dirty, disorderly nature of Sixth Corps camps. This corps had seen hard service since early May in the Overland campaign and at Petersburg. John William DeForest assumed the Sixth Corps was "badly demoralized" by its experiences. The men and weapons were dirty, and the camps were a mess. "These fellows lurk around our clean, orderly camps and steal our clean, bright rifles," he complained in a letter home. When he went looking for the stolen guns, a Sixth Corps sergeant told him, "Well, if you find a clean gun in this camp, you claim it. We hain't had one in our brigade since Cold Harbor." Despite this, no one could fault the battlefield performance of the Sixth Corps.[43]

On the Confederate side of the war, the Second Texas was a fine unit that had seen service in the defense of Vicksburg. Much of it constituted the garrison of Galveston after being exchanged. An inspection report revealed a great many problems with its proficiency at company and battalion drill, the quality of its clothing and blankets, and the level of vigor with which the men handled their weapons. But the guns themselves were another matter. "The condition of arms, accoutrements, etc, cannot be excelled by any regiment in the service," the inspector reported. "The guns, especially, are kept in excellent condition, and in some companies approach a standard of perfection."[44]

Ammunition

The Minié system used balls slightly smaller than the bore of the musket. For example, the Springfield was designed as a .58-caliber weapon, but the bullet actually was .577 of an inch in diameter. The balls were wrapped in a cartridge consisting of two layers of rough brown paper. One layer enclosed the ball itself, and this was enclosed in the other layer, which also contained the premeasured amount of powder. The soldier had to bite off a length of unfilled paper at the powder end of the cartridge to pour the charge into the barrel. He then dropped in the rest of the cartridge containing the bullet and rammed it home. Factories could produce 800 cartridges in every ten-hour shift. A coating of grease, consisting of "one part beeswax and three parts tallow," aided the Minié ball's passage through the barrel when fired.[45]

The cartridges were arranged in packets containing 10 rounds, and the packets were transported in boxes containing 1,000 rounds. The number of cartridges used in the Civil War was enormous. The U.S. Ordnance Department alone made or purchased 470,851,079 rounds of .58-caliber ammunition from January 1, 1861, through June 30, 1866. Its production of .69- and .54-caliber ammunition amounted to 230,338,378 rounds. Comparable data for the Confederacy does not exist, but records indicate that the C.S. Ordnance Bureau produced or purchased 36 million rounds between September 30, 1862, and September 30, 1863.[46]

Gun aficionados among the soldiers often paid a lot of attention to their ammunition. Eugene F. Ware remembered receiving his first issue of rounds as a soldier in the First Iowa. Forty cartridges and fifty percussion caps were put into his possession. "The cartridges were tough paper with big charges of coarse black powder," he recalled. Ironically, the ammunition was of the Minié variety, with a wooden plug inside the cartridge to help expand the ball when rammed, even though the regiment was using old smoothbores converted from flintlocks. Moreover, Ware remembered that the officers specifically told them "that the cartridges did not belong to us but to the Government, and that if we wasted any of them we would be charged 10 cents each. We were also told that we would be held responsible for their getting wet. And that a record would be kept of all we were ordered to fire, and that if we fired one without orders it would be charged to us. We were also told that ammunition was scarce, and that we must be careful, and that anyone willfully wasting any or stealing any would go under guard and suffer on the payroll besides." When fully loaded with forty rounds, the cartridge boxes weighed 4 pounds.[47]

Ware and his comrades immediately began to practice firing, presumably with their officer's permission. He recalled that "we always got a few grains of powder in our mouth, and as the taste was not unpleasantly peculiar, we chewed the paper which we had bitten off, and by the time we had fired a few times we had a good wad of paper in our mouths which we would chew as a school-girl would chew gum."[48]

The Second Rhode Island also used smoothbores converted from flintlocks, but the men were given buck and ball, a type of cartridge designed for such a weapon. It was the first time that Elisha Hunt Rhodes had ever seen "warlike ammunition," and he "examined [it] with much interest." Imagine the feelings of the men of the Fifteenth Iowa, who put their first cartridges into their muskets on the battlefield of Shiloh. The Iowa soldiers arrived by steamer on April 6, never having test-fired their Springfield rifle muskets.[49]

The rifle gave rise to new, experimental types of ammunition. Explosive bullets apparently were not feasible with smoothbores; at least no one paid much attention to that concept as connected with a smoothbore weapon. But rifles gave rise to the possibility of delivering some sort of miniversion of a shell because of its increased range. In fact, one historian has asserted that the rifle was "essential for its use." A British army captain named Norton devised an explosive bullet for use in the Baker rifle in 1824, and an East India Company officer named John Jacob devised another explosive bullet in the 1840s, claiming he could explode caissons at over 1,000 yards. The British used explosive bullets in the Indian Mutiny of 1857, and West Point textbooks discussed their theoretical employment in the United States.[50]

Samuel Gardiner Jr. of New York developed the only explosive bullet used in the Civil War. He called it a "musket shell" and approached Lincoln with his invention in June 1862. The round was similar to a Minié ball except that it contained a copper chamber filled with fulminate. A time fuse extended from the chamber out to the base of the ball. The time fuse allowed the bullet to be detonated from one to three seconds after firing. U.S. Army tests indicated that the shell did not fragment well, and the assumption was that if it hit a man it would inflict unnecessarily cruel injuries.[51]

Nevertheless, it could readily be used to explode caissons or set fire to wagons, and the U.S. Army issued 10,000 Gardiner rounds to the Eleventh Corps in the last half of 1862 on a trial basis. The experiment worked well enough for the War Department to order 100,000 Gardiner rounds in November of that year, three-fourths of them at .58 caliber and the rest at .54 caliber for cavalry arms. The Second New Hampshire, armed with Sharps rifles, was given 24,000 Gardiner rounds in June 1863. It later abandoned 10,600 of them, which were picked up by the Confederates. The regiment expended 4,000 Gardiner rounds during target practice and fired 10,000 in action from July 1 through October 1, 1863, becoming the only regiment in the Union army to intensively use the experimental ammunition. The regiment fired 14,000 rounds of it at Gettysburg. A Rebel artillery shell burst near enough to send hot fragments into one man's cartridge box, exploding all the Gardiner rounds in it. There is evidence that the Confederates used the captured Gardiner rounds at Gettysburg, too, for Col. E. L. Dana of the 143rd Pennsylvania claimed he was the target of a Rebel who was firing something other than regulation ammunition. All told, close to 35,000 Gardiner rounds were issued to troops in the field, and about 75,000 rounds were left in storage by war's end.[52]

Whether any other type of explosive bullet was used during the Civil War is questionable. Ordnance records indicate that the U.S. Army purchased as many as 200,000 rounds of a type of fragmentary bullet developed by Ira W. Shaler, but there is no indication that they were issued to troops. Another type of round, the Williams bore-cleaning cartridge, often was mistaken for an explosive bullet. The U.S. Army "routinely" used it from the beginning of the war to clear powder residue from the tube before it clogged the rifling. Some of the Williams bore-cleaning cartridges were placed in every packet of ten cartridges, wrapped in blue- or red-stained paper to distinguish them from regular rounds. The Williams cartridge included a zinc washer, a pin, and a plug, along with the Minié ball, to achieve its purpose. It is quite possible that these component parts, flying through the air in company with the ball, gave intended victims the idea that the enemy was using explosive bullets. At least one Confederate veteran swore that the black troops who participated in the Fifth Offensive at Petersburg used explosive rounds, and he picked one up as a souvenir. His description of it proves that it was a Williams bore-cleaning cartridge. The Federal army, however, stopped using the Williams cartridge in September 1864, due to unsubstantiated complaints that it damaged the rifling.[53]

It became a point of honor for Southern patriots to condemn the Yankees for using explosive bullets while claiming the Confederates never stooped to that low point. The truth was that the Southerners used them even before the Federals, and manufactured their own explosive bullets in the Richmond Arsenal. At least 100,000 rounds were manufactured, and a modern relic hunter has found one of the remaining rounds, which has the markings of the arsenal to prove it. Capt. William Farley, a member of J. E. B. Stuart's staff, owned an English target rifle and carried eighty rounds of explosive ammunition at the battle of Glendale in June 1862, roving about looking for targets of opportunity as if he were hunting game. Farley later claimed to have shot at least six Federal officers during the first half of the war.[54]

Grant claimed the Confederates used explosive bullets during the siege of Vicksburg, trying to explode them over the Federal trenches so the fragments could do injury like an artillery round. He could not recall a case where a man was injured in this way, but he condemned the use of explosive bullets on moral grounds. If one hit a man directly and then exploded, the effect was predictably gruesome, and Grant thought their use would "produce increased suffering without any corresponding advantage to those using them." The Confederate chief engineer at Vicksburg stoutly denied that any explosive bullets were used during the siege, but it is possible he was never made aware of their presence. Modern relic hunters have found six explosive bullets at Vicksburg and Champion's Hill. They were not Gardiner rounds but had marks indicating they were manufactured in the South.[55]

It would be bad enough for a soldier to be hit by an explosive bullet, but worse if he were struck by a poisoned bullet. Whether such rounds were prepared and used became controversial during and after the war, with many accusations flung about by both sides. A newspaper reported in March 1862 that Federal troops found poisoned ammunition in Nashville when they occupied the Tennessee capital. Benson J. Lossing visited Gettysburg right after the battle and claimed to have seen wounded Federal troops who were victims of both explosive and poisoned bullets. His description of the cartridges indicates that the former was probably a Williams bore-cleaning round and the latter had a small ball filled with poison inside a hollowed-out Minié bullet. It is possible that individuals might have improvised poisoned bullets, but there is no evidence that either government made them or authorized their use.[56]

It was difficult for the authorities to gauge the number of men wounded by explosive bullets, because regulation Minié balls could create similar damage to flesh and bones. One survey of 4,000 wounds to the scalp concluded that only two were caused by explosive bullets, while 2,612 were made by Minié balls and 384 by smoothbore ammunition. The rest, 1,002, were of undetermined origin. According to the compilers of the *Medical and Surgical History of the Civil War,* doctors reported 130 cases as having been caused by explosive bullets, but a postwar examination of the evidence indicated that only six of them seemed well documented and an additional sixteen were possible. Thirteen of the sixteen possible cases recovered from their injuries, while one was killed in battle and two died later of complications. The majority of the sixteen cases occurred in the last half of the war. All six of the authenticated cases recovered from their injuries, and all but two took place in 1864. The *Medical and Surgical History of the Civil War* includes an illustration of an exploded bullet taken from the thigh of a man in the 120th New York. The bullet obviously exploded from within and probably was a Gardiner round.[57]

There are precious few accounts by soldiers that clearly indicate that they were victims of explosive bullets. They sometimes came to this conclusion because the sound of the bullets whizzing overhead seemed unusual, and sometimes they even testified to seeing them explode near their faces. Some men claimed to see pieces of tin or zinc fly off the bullets, which would indicate a Williams bore-cleaning round rather than an explosive bullet. Thomas G. J. Doughman of the Eighty-ninth Ohio swore he was wounded by an explosive round at Chickamauga and retained the fragments after they were cut out of his hip by an army surgeon.[58]

Many commentators found the use of explosive bullets inhumane. The U.S. chief of ordnance concluded that their use was "inexcusable among any people above the grade of ignorant savages." The Russians called for a ban on their use, which resulted in an international conference at St. Petersburg in 1868. The delegates agreed to refrain from the use of explosive projectiles less than 400 grams in weight, but the U.S. government did not sign the agreement. The English government was a signatory but received a great deal of criticism from newspapers and military officers. In short, not everyone agreed that the use of explosive bullets was unthinkable.[59]

Bulletproof Vests

The widespread use of the rifle musket opened up the door to the introduction of more deadly ammunition than the simple lead slug. In general, most participants in the Civil War felt a certain sense of awe when contemplating the power of the rifle, and many of them quietly sought ways to protect themselves as they joined the army. A fad for bulletproof vests developed, spurred on by extensive newspaper advertisements placed by entrepreneurs who sought to capitalize on the natural anxieties felt by all recruits going to war. One such ad for "The Soldiers' Bullet Proof Vest" claimed it had been thoroughly tested with rifle fire at 220 yards. "It is simple, light, and is a true economy of life—it will save thousands." This particular product sold in different categories in New York City—privates could purchase one for $5 while officers had to pay $7.[60]

Body-armor marketers worked throughout the duration of the war to sell their products, often sending models to Washington to interest the army in adopting what they had to offer. Some of them went beyond the concept of a mere vest to develop a modern version of a "coat of mail" to cover much of the body. J. S. Smith of New York wanted to protect soldiers against bayonet and sword thrusts as well as against rifle bullets with a vest made of "thin spring steel." It weighed only 3 1/2 pounds.[61]

More soldiers actually bought and used body armor than is generally known, although it is impossible to estimate how many. J. Cabell Early saw a vest taken from a fallen Union soldier at First Manassas. It was made of steel strips 1 inch wide, connected in such a way as to allow for maximum flexibility. Early saw another dead Federal with a vest that covered both his back and front on the battlefield of Gettysburg. The Yankee had been killed by an artillery shell that tore his arm out of the socket. Other Confederates remembered that several vests had been thrown away by the troops of Maj. Gen. Nathaniel Banks's command as they retreated from Winchester, Virginia, in May 1862. Those members of the Fourteenth New Hampshire who had purchased bulletproof vests upon enlistment also dropped them in front of the White House on October 24, 1862, as the regiment moved out for the field. A Confederate lieutenant took the body armor of a fallen Union major at the battle of Dranesville, Virginia, and often shot at it for amusement long after the war ended.[62]

Purveyors of bulletproof vests seem to have been more active in the West than in the East. When the newly raised Fifteenth Iowa assembled at Benton Barracks near St. Louis, civilians sold several vests to regimental members at prices that ranged from $8 to $16. Tests indicated that rifle bullets easily penetrated half of them. "If the bullet did not go through it would knock a man into the middle of next week," thought Cyrus F. Boyd, "so that he might as well be killed first as last." Another soldier, Chesley A. Mosman of the Fifty-ninth Illinois, noticed several breastplates among the garrison troops of Cape Girardeau, Missouri, as his regiment passed through town in May 1862.[63]

Most men who purchased bulletproof vests were defensive about it. "I have a steel armor that covers me quite well," announced Capt. William Vermilion to his wife in January 1863. "It will resist any rifle shot. I intend to wear it, not through cowardice but because I consider it my duty to protect myself in every manner possible." However, Vermilion wanted his wife to keep his secret. "Don't speak of it to anyone, Dollie. The boys here don't know it." Maj. William T. Cunningham had been popular with the men of the Fifteenth Iowa, but for some reason they grew to dislike him after a few months of service. When Cunningham resigned, no one was sorry to see him leave. "He carried away with him his *'bullet proof vest,'"* as Cyrus F. Boyd sarcastically put it. Sgt. Hamlin B. Williams of the Twenty-first Wisconsin owned a "wirecloth" vest, but he was careful to keep it a secret from his comrades. John Henry Otto happened to see Williams trying to stuff it into a package, and he allowed Otto to examine the vest. It was "a mail waistcoat, made of fine steel wire and quite heavy. Only the front parts which cover the breast were made of wire." Otto was not impressed. "I suppose the manufacturers of the article took it for granted that in warfare the bullets only came square from the front." Williams transported it in the quartermaster wagon because it was too heavy for him to carry. At the battle of Perryville, the wagon was unable to keep up with the regiment and Williams was shot in the back, living as an invalid for the rest of his life.[64]

There is at least one documented case of a Federal officer whose life was saved by a steel vest that was hit by two balls, but most of the documented cases indicate that bulletproof armor was a complete failure. Several observers took armor off the bodies of men who had indulged in a futile hope. One such vest, taken from a Rebel captain at the battle of Williamsburg, was displayed at a store in Philadelphia with a puncture hole and two noticeable dents in it. Another vest, made of boiler iron, that was molded to fit the body of a different Confederate officer, was taken from his remains outside Fort Stevens in July 1864, with a hole over his heart. The best-documented case, however, relates to Col. William P. Rogers of the Second Texas, who was killed in the Confederate attack on Battery Robinette at Corinth, Mississippi, on October 4, 1862. A popular and impressive regimental commander, Rogers was determined to win a general's promotion in the attack or perish in the attempt. He was hit with from seven to eleven bullets at close range. One half of his vest has survived, probably taken home by a Union soldier—it may be seen today in the Wisconsin Veterans Museum in Madison with a large puncture hole through it.[65]

Victims of the Rifle Musket

As a hospital steward with the 130th Illinois, Charles Beneulyn Johnson saw with his own eyes that the rifle musket was the most deadly weapon of the Civil War. "I think wounds from bullets were five times as frequent as those from all other sources," he noted. Wounds from shell explosions and canister rounds came next. Johnson never saw a bayonet wound, and only once did he notice a sword cut. The medical director of the Army of the Potomac tabulated wounds treated from early May to the end of July 1864. At least 36 were caused by bayonets, 5 by swords, 2,200 by artillery, and 25,454 by bullets. The rifle musket caused a staggering 91.9 percent of the wounds treated in the Army of the Potomac during the Overland campaign and the early phase of the Petersburg campaign.[66]

The Minié ball invariably "produced large, ugly wounds." When it hit a bone, the ball often fractured into long sections, and it rarely left a round puncture hole on entrance. Joints were particularly vulnerable to extensive damage when hit, and head and abdomen wounds often proved fatal.[67]

A bizarre effect of the Minié ball was that it sometimes caused instantaneous death, freezing the victim in the act of doing something on the battlefield. Leander Stillwell recorded a classic case of instantaneous rigor mortis at Shiloh, where he saw a Confederate shot in the forehead. "He was in the act of biting a cartridge when struck, his teeth were still fastened on the paper extremity, while his right hand clutched the bullet end. His teeth were long and snaggy, and discolored by tobacco juice."[68]

On and off the battlefield, rifle muskets could produce unexpected casualties. Because the bullets often flew in weird ways, bouncing off trees and drastically changing their angle of flight, one could be hit by them even if not very close to the fighting. The extended range of the weapon meant that the bullet could "descend somewhere a long way off often killing some one over a hill or in a ravine entirely out of sight of the one firing the gun."[69]

Accidental killings were not uncommon, either. Charles Beneulyn Johnson reported that three men of the 130th Illinois were accidentally shot by their comrades. "Looking back," he wrote, "I can but regard our record in this direction as especially fortunate, when the handling of so many loaded guns through so long a period is taken into account." It is tempting to assume that the accidents were mostly caused by men who were inept at handling guns, but often it was simply because a loaded musket was stacked or leaned against a tree and fell to the ground, discharging into whoever happened to be in the way. In other cases, it was due to the negligence of the victims. Some soldiers of the 116th Pennsylvania tried to make firecrackers out of old cartridges. One of them was seriously injured in the eyes when a round suddenly exploded.[70]

The rifle musket also was a ready tool for anyone who decided to commit suicide while in the army. Sgt. William Vogel of Bissell's Engineer Regiment of the West did so when told he would have to return to the ranks due to the consolidation of his regiment with the Twenty-fifth Missouri. The weapon was also used when a court-martial sentenced a soldier to death for a serious military crime. For Pvt. Michael Nash's execution, a firing squad was made up of men from the Fifty-second Ohio. Officers issued live ammunition to eight of the twelve guns and gave blank but heavier charges to the other four "so that there will be no difference in the report or rebound of them."[71]

Disposing of Guns at the End of the War

For the majority of Union soldiers, saying good-bye to their weapons was easy and simple. They turned them in to ordnance officers and happily demobilized. A few men did something extraordinary with their weapons at the end of the war. A soldier in the Twenty-first Wisconsin smashed his rifle against a cottonwood tree upon hearing the news of Lee's surrender. His colonel charged him $14 for it. The man did not care, saying, "I have carried that rifle for nigh three years. I don't know how many Johnnies it has hit, or knocked over. It has done its part. I will gladly pay Uncle Sam for the rifle, for the good news is worth that to me."[72]

Based on Grant's understanding "that great numbers of soldiers going out of service are very desirous of retaining their arms," the Federal government allowed veterans to purchase their muskets for $6. One of them called this "a most generous act." His 152nd New York had been in service 1,000 days, 200 of them under fire. These veterans had grown to look upon their guns as trusted companions. "Now it was ours. To the veteran volunteer the value will be greater than the highest work of art America ever produced."[73]

Yet relatively few Union soldiers took up the government offer. Records indicate that veterans bought only 138,000 small arms and 20,000 pistols at muster-out, although there were well over 1 million men in Federal uniform by April 1865. Among the buyers was Thomas Goodman of the First Missouri Engineers, who purchased his Springfield rifle musket and all the accoutrements, including the bayonet, cap pouch, cartridge box, and gun sling. A sampling of 295 service records of this regiment indicated that only .02 percent of the men purchased their weapons, but that may be a low figure due to the fact that it was an engineer regiment.[74]

For most Confederate veterans, giving up their guns often involved a formal surrender agreement at war's end. Lee's army stacked arms on April 12, 1865, at Appomattox, with only officers allowed to retain their side arms. Johnston's army in North Carolina officially gave up but never held a formal surrender ceremony. Sherman agreed on April 26 that Johnston's men could keep one-seventh of their arms in order to protect their homes against marauders. The different units of Johnston's command were to deliver six-sevenths of their weapons to Federal authorities. Some unit commanders had to be urged more than once to comply, but most of them did so. Johnston's ordnance officers concluded that 177 guns were unaccounted for, probably due to desertions that took place before the due date for delivering the arms. Federal officers guessed that some Rebels even destroyed their weapons rather than turn them over, but they were satisfied that the number deposited in Greensboro was more or less commensurate with the agreement signed by Johnston and Sherman.[75]

The rifle musket was more than just a weapon: it became loaded with symbolism among those soldiers who treasured its potency. No wonder that thrusting the musket into the air, butt first, became a signal that a unit wanted to surrender. Anyone could readily understand this mute admission of defeat in the noise and confusion of combat.[76]

4

THE RIFLE MUSKET IN BATTLE

The experience of battle was intense, full of danger, and emotionally challenging, especially for the neophyte soldier. Those men who could remain calm focused their attention on their weapons, loading in the prescribed nine steps, aiming, pulling the trigger, and repeating the process despite the chaos around them. Of course, they hoped their comrades to right and left were doing the same thing, and that the regiment, their military household, would not dishonor itself on the day of battle.

"Every man was shooting as fast, on our side, as he could load, and yelling as loud as his breath would permit," recalled Eugene F. Ware of his experience at Wilson's Creek. "Most were on the ground, some on one knee. . . . The other side were yelling, and if any orders were given nobody heard them. Every man assumed the responsibility of doing as much shooting as he could." The "reiterated kick" of his Springfield rifle, nicknamed Orphan, made Ware's shoulder "feel as if I had the rheumatism." Confederate soldier Sam Watkins fired his rifle so much while defending Cheatham's Hill at Kennesaw Mountain on June

27, 1864, that his arm was "all battered and bruised and bloodshot from my wrist to my shoulder, and as sore as a blister." A fifteen-year-old recruit in the Fourteenth Illinois remembered that he had to hold his Enfield rifle tightly against his shoulder; otherwise, the weapon nearly knocked him down every time he fired.[1]

While some soldiers complained that it was difficult to bite off the loose end of the paper cartridge, most did not mind it. In fact, Eugene Ware remembered that many of his comrades chewed bits of powder-coated cartridge paper while firing away at Wilson's Creek, and streams of dissolved powder trickled down their chins.[2]

Most soldiers did not adapt so quickly to combat in their first battle as Ware and his comrades. Elisha Hunt Rhodes of the Second Rhode Island was astonished at the "peculiar whir of the bullets" that flew over his head upon entering the battle of First Bull Run. The regiment flopped to the ground when the first volley sailed over, but its colonel quickly ordered the men to get back on their feet and continue advancing. The regiment was armed with old smoothbore flintlocks converted to percussion, and Rhodes fired only a few rounds before the gun "became so foul" with powder residue that he had to force the ramrod into the tube by pressing it against a fence.[3]

Leander Stillwell's first battle experience at Shiloh produced an unforgettable view of the enemy. "Suddenly, obliquely to our right, there was a long, wavy flash of bright light, then another, and another! It was the sunlight shining on gun barrels and bayonets." A second later, Stillwell saw "a long brown line, with muskets at a right shoulder shift, in excellent order, right through the woods they came." As the Confederates approached, Stillwell's Sixty-first Illinois let loose its first volley of the war. "From one end of the regiment to the other leaped a sheet of red flame," as he later put it.[4]

On the other side of the field at Shiloh, Henry Morton Stanley of the Sixth Arkansas vividly recalled his first encounter with fire.

> "There they are!" was no sooner uttered, than we cracked into them with leveled muskets. "Aim low, men!" commanded Captain Smith. I tried hard to see some living thing to shoot at, for it appeared absurd to be blazing away at shadows. But, still advancing, firing as we moved, I, at last, saw a row of little globes of pearly smoke streaked with crimson, breaking-out, with sportive quickness, from a long line of bluey figures in front; and, simultaneously, there broke upon our ears an appalling crash of sound, the series of fusillades following one another with startling suddenness, which suggested to my somewhat moidered

> sense a mountain upheaved, with huge rocks tumbling and thundering down a slope, and the echoes rumbling and receding through space. Again and again, these loud and quick explosions were repeated, seemingly with increased violence, until they rose to the highest pitch of fury, and in unbroken continuity. All the world seemed involved in one tremendous ruin!

Stanley's comrades continued to move forward, reloading and firing all the while. "My ears tingled, my pulses beat double-quick, my heart throbbed loudly, and almost painfully."[5]

These men were dipping their feet into the holocaust, learning how to be good soldiers in their first intense combat experience. After repeated exposure to that experience, the good soldiers became better warriors. Marcus Woodcock of the Ninth Kentucky (U.S.) described his comrades as they engaged in a hard exchange of fire with advancing Confederates at Stones River on January 2, 1863. "There we stood, two solid columns within pistol shot of each other using their weapons of death with all the rapidity and precision their capacities would allow." A small group of Federals targeted the Rebel color-bearers while the rest simply tried to knock down as many men along the enemy line as possible. A bullet tore away Woodcock's ramrod and bent his musket barrel. With no replacement handy, he reloaded for his comrades and shouted encouragement. Some of the Federals stood boldly in this hail of gunfire, while others lay on the ground or bent on one knee, but all were firing intently. "Some were looking calm as death," Woodcock remembered many years later, "and without noticing anything else were rapidly loading and firing." Others laughed, while some were angry.[6]

Veterans often chattered away while firing on the battle line, both to relieve stress and to encourage their friends. "Look here, Butler, mind how you shoot, that ball didn't miss my head two inches," could be heard within the ranks of McGowan's South Carolina Brigade in Lee's army. "'Just keep cool, will you; I've got better sense than to shoot anybody.' 'Well, I don't like your standing so close behind me, nohow.'" When George D. Bowen of the Twelfth New Jersey fired thirty rounds from the rear of his regimental battle line against the Confederate attack on July 3 at Gettysburg, he was careful "to set on one knee and one foot, extending my gun as far to the front as I could reach." Yet every time he fired some men "looked back and told me I would shoot them, swearing about it, I told them they need not fear. Really think there was little danger as I was as cool as I ever was in my life."[7]

Loading

Loading the rifle musket properly was a key factor affecting both the rate of fire and whether the weapon worked as it was designed. It was still a single-shot, muzzle-loading piece, and the nine-step process of recharging had to be completed correctly. Veteran soldiers learned a number of tricks to give them an edge in reloading quickly. The most common was to lay out cartridges for easy access if they were holding an entrenched position. Sometimes they tore the loose paper off the cartridges when laying them out on a ledge of the earthworks so they would not have to do so when firing. Another trick was to stick the ramrod in the ground, or lay it down if it could not be stuck, to save the trouble of putting it back into its barrel holder. This often led them to forget the ramrod when the unit was ordered away to a new position.[8]

The easiest way to reload and fire was in a standing position, but the drill manuals offered soldiers a variety of other positions as well. Hardee described methods for reloading while lying down, and quite a few soldiers actually did this on the battlefield. Ohio troops at the battle of Kernstown reloaded while lying on their backs and then rolled over to aim and fire. Veteran soldiers of the Twelfth Connecticut did the same at the battle of the Opequon more than two years later. Most men of the relatively inexperienced Forty-fourth Massachusetts were insulted when told to reload while lying down during the battle of White Hall, North Carolina. They apparently thought it undignified for a volunteer soldier to do anything other than stand and take his chances. Sometimes men, such as those in the Fifty-third Pennsylvania at Seven Pines, fired while crouching down, assuming a halfway position between standing and lying.[9]

It was more difficult to reload while moving forward, but Civil War soldiers were quite able to accomplish this feat. The Sixth New Hampshire did so while attacking the unfinished railroad grade at Second Bull Run, and defending Confederates were very impressed when the Twelfth Connecticut did the same during an attack at the battle of Georgia Landing in Louisiana. "Your firing didn't hurt us," said a captured Rebel, "but your coming in and yelling scared us." It also "kept up the spirits of our own fellows," thought Capt. John W. DeForest, whose neck was scorched by the muzzle blast of an eager comrade who became careless of the men in the regiment's front rank.[10]

More experienced soldiers could load and fire on the move without endangering their comrades. A captain in the Third Alabama proudly described the regiment's performance during Stonewall Jackson's flank attack at Chancellorsville. "The firing of my command was executed in excellent order, the front line firing and loading as they marched on, while the rear came to the front, fired and loaded as the march continued." The veterans of the Iron Brigade loaded their guns while moving from a marching column into a battle line across the field of Gettysburg on July 1 without interrupting a step. Many of them did this on the double-quick while trailing their weapons—that is, they allowed the butt to trail along the ground as they moved. Confederate soldiers who attacked the Union left wing near the wheat field the next day reloaded in exactly the same way, although at a much slower pace. Loading on the move occurred among western troops as well—for example, the men of the Tenth Illinois who advanced against a Confederate position on the last day of the confrontation at Bentonville, March 21, 1865.[11]

Whether they reloaded while standing, lying, or moving forward, Civil War soldiers could render their gun useless if they failed to recharge it correctly. Paddy Griffith has estimated that perhaps 9 percent of muskets were improperly loaded in battle. While it is impossible to verify that, one gets the impression from various sources that it likely is a low estimate. The causes are not difficult to determine. A writer in the *Army and Navy Journal* noted that the nine-step loading process itself was a major reason. Hardee tried to warn recruits about this problem in his tactical manual. The men should "observe, when uncocking, whether smoke escapes from the tube, which is a certain indication that the piece has been discharged." If not, Hardee warned them not to reload but to see if the previous charge was still in the tube. If he forgot and rammed home another round, the soldier ought to be able to tell whether the previous round had been fired by judging how far the ramrod entered the barrel. There was no excuse, Hardee thought, for pushing yet a third round into a musket that already was useless.[12]

Yet most Civil War soldiers were unable to coolly detach themselves from their surroundings and observe their weapons as Hardee suggested. The inexperienced men of the Eighty-first Illinois nervously formed line to meet an expected Rebel raid on October 10, 1862, near Humboldt, Tennessee. One of them "was so excited, that he put two cartridges in his gun." Two months later, the same regiment again nervously assembled for battle as several members put their cartridges wrong end down in the tube. The 118th Pennsylvania fought its first battle at Shepherdstown, Maryland, on September 20, 1862. When several men came to Capt. Francis A. Donaldson, complaining that their muskets were giving them trouble, he borrowed a nipple pick and discovered they had put the bullet end of the cartridge down first. One man had rammed so many rounds that the barrel was literally filled.[13]

The most famous and best-documented case of misloading occurred at Gettysburg. The Federal army gathered up 27,574 abandoned muskets from a battlefield where some 170,000 Union and Confederate soldiers had struggled for three days. About 24,000 of the abandoned muskets were loaded, half of them had two loads each, and 6,000 had anywhere from three to ten loads per tube. In short, almost half of the recovered muskets had two rounds each and over 20 percent had more than two rounds rammed home. Most of the men who fought at Gettysburg were veterans of many battles, although one could suspect that a green soldier probably was responsible for one musket that had a total of twenty-three rounds jammed in the tube, some of which had not even been broken apart to let the powder pour in.[14]

Whether incorrect loading disabled the weapon was a point of some discussion. Soon after the war, an ordnance officer argued that tests conducted some thirty years before in Europe indicated that a smoothbore musket, at least, could be safely fired with as many as four cartridges in the tube, as long as they were placed powder end first and packed tightly together. He believed that the rifle musket could also be fired with a similar number of rounds without hurting the soldier. Few men would have confidently agreed. Although a gun enthusiast experienced in many battles, Leander Stillwell happened to put two charges in his gun during a skirmish late in the war and fired without realizing it. There was "a deafening explosion, and a kick that sent me a-sprawling on my back!" Firing with four charges in the tube undoubtedly would have injured Stillwell.[15]

Putting multiple rounds in the tube was the most common type of mistake Civil War soldiers made while reloading, but now and then they also forgot to withdraw their ramrods. An officer of the Thirty-sixth Alabama "heard an odd thud and a twang near his side" one night while on the picket line at Lookout Mountain during the Chattanooga campaign. He waved his hand in the darkness and discovered a ramrod sticking in the tree next to him, fired by a nervous Union picket into the night.[16]

There were other potential causes of misfires and improper loadings as well. The chief ordnance officer of Cheatham's Division in the Army of Tennessee wondered why so many useless muskets were picked up from the battlefield of Chickamauga, suggesting that the ammunition did not fit the bore properly or that the bullets had too little grease. Another ordnance officer in the same army suggested that the Confederate practice of using Minié ammunition in smoothbore guns was to blame. Buck and ball cartridges suited the smoothbore much better, and he believed the Minié ammunition caused unnecessary choking in smoothbores.[17]

We cannot blame inept soldiers entirely for the thousands of muskets rendered useless in the heat of battle. David Holt of the Sixteenth Mississippi, one of his unit's best gunmen, recalled that his weapon failed to detonate as his unit countercharged into the captured Mule Shoe Salient at Spotsylvania on May 12, 1864. Despite the noise and confusion, Holt realized something was wrong. "I dropped on one knee, got the picker out of my cap box, picked the tube, put on a fresh cap, and fired at the mass of yanks in the trench," he later wrote. Fortunately for Holt, the problem was only a faulty cap.[18]

Aiming

After properly reloading their muskets, Civil War soldiers tried to discharge the weapon in a way that would give them a chance to hit a man on the opposing side. We have already seen that target practice was at best haphazard in both armies, and in the often wooded environment of the battlefield, even veteran soldiers had difficulty seeing a human target. It is fair to say that the majority of firing during the Civil War took place in conditions that offered soldiers little more than an opportunity to aim generally at a poorly seen enemy.

Officers constantly tried to encourage their men to pick out targets and make their fire as effective as possible. They issued orders cautioning against "heedless aimless firing." After Shiloh, Sherman sarcastically issued an order to his division that "all discharges of muskets at the moon or tops of trees are not only wasted, but they deceive the generals, who have a right to judge of the execution by the fire of their men." Just before the battle of Prairie Grove, Confederate commander Thomas C. Hindman issued an order to his men, urging them not to fire simply because everyone else opened and not to reload hastily "for the sake of firing rapidly. Always wait until you are certainly within range of your gun, then single out your man, take deliberate aim as low down as the Knee & fire." William S. Rosecrans issued similar instructions to his Army of the Cumberland before the battle of Stones River.[19]

There certainly were battlefields where it was possible to put these instructions into effect. At Reams Station, Daniel Chisholm of the 116th Pennsylvania "*could see all,*" as he put it with astonishment, and deliberately "picked out the very largest man I could see & took sight." Other battlefields, like Fredericksburg and Franklin, also offered open panoramas to the men in the ranks. But the Wilderness, Chickamauga, Stones River, and Shiloh, with their thickets, woods, and brush, were more typical of Civil War battlefields. Union and Confederate soldiers could expect that, as a rule of thumb, more than 50 percent of any potential battleground would be obscured with vegetation so that they would have difficulty seeing more than a few tens of yards in any direction. Moreover, the accumulated smoke from rifle discharges could obscure even an open battlefield, and undulations in the ground could easily hide large troop formations.[20]

Even so, according to R. E. McBride of the 190th Pennsylvania, the men were always told to take aim, even if they simply pointed in the general direction of the enemy. They sometimes were ordered to keep firing even though there was no sign of opposing troops.[21]

Another, related problem that affected fire effectiveness lay in the parabolic trajectory of the rifle musket. Because of its high arc, the Minié ball created a huge safety zone for the enemy during much of its flight through the air. This problem was most pronounced when firing at distant targets, but this sort of long-distance firing was the key difference between the rifle musket and the old smoothbore weapon. It was incredibly difficult for the average soldier to compensate for the unusual trajectory and make his shots count at ranges longer than about 100 yards. This greatly decreased the effect of the rifle musket precisely in the area where advocates thought it might have a revolutionary impact on warfare.

A rifle musket sighted for 300 yards could be deadly at short range, but after about 75 yards the bullet sailed above the height of an average man. The next danger zone lay at the far end of the trajectory, the last 110 yards (about 240 to 350 yards from the shooter). In this last danger zone, the target could be hit at any height along his body, depending on where in the zone he happened to be when the bullet made contact. For the rest, fully 115 yards of the bullet's flight, it sailed through the air without a possibility of damaging the enemy. In short, only 185 out of 300 yards of the bullet's journey constituted a danger zone to the enemy.[22]

Soldiers often saw vivid evidence of the parabolic trajectory. Leander Stillwell heard "an incessant humming sound away up above our heads, like the flight of a swarm of bees." He came to realize it was caused by streams of bullets, 20 to 100 feet above his head. He later noticed hundreds of bullet marks on tree trunks at that elevation but chalked it up to the inexperience of the Confederate shooters. Officers tried to compensate for this problem by exhorting their men to "fire at the lower extremities; avoid overshooting." Robert Burdette called aiming too high an "unsoldierly sin," but it was a "common fault, even of the old soldier." He recalled his sergeants chanting, "Fire low, boys; fire low! Rake 'em! Shin 'em!" as they encouraged their men on the firing line. Many soldiers apparently did not understand that this was essentially a technical problem. They believed, as did Stillwell, that it resulted from human error, and thus the widespread efforts to exhort the men to take better care while aiming rather than to teach them how to use their adjustable sights. As J. F. J. Caldwell of the First South Carolina put it, "It is a universally known fact, that men fire above their own levels." Everyone accepted it; few tried to find out why.[23]

Several members of the Third Louisiana were exceptions to this rule. They picked up abandoned Federal muskets from the battlefield of Pea Ridge and took them away when the Rebel army retreated. Upon examination, they found that the muskets were sighted for a range of 200 yards even though most of the firing took place in thick woods at about 75 yards. This convinced the Louisiana men to set their adjustable sights at 70 yards, or to remove them entirely, and alter the range of their fire by sight and instinct. In essence, they abandoned all hope of hitting anything at ranges longer than those of smoothbore firings.[24]

Of course, there was an advantage in the parabolic trajectory, and it was demonstrated at Fort Donelson when the Seventy-sixth Ohio opened fire from a reserve position behind the First Nebraska. It was on slightly higher ground than the regiment in front, but even so the parabolic trajectory allowed the rounds to sail across their comrades and land more or less on the enemy.[25]

Firing Systems

Methods of organizing fire were varied and had long reaches into European military history. The drill manuals recognized firing by company, by wing, by battalion, by file, by rank, and by individual. Silas Casey, who wrote the standard Union drill manual of the Civil War, believed that "the kind of fire will be determined by the character of the ground, and the state of the action." But his assertion that file firing was the "most used in war" is not supported by the evidence. Hardee instructed the fire by file to proceed from the right of the regimental line, with each file to the left taking it up as soon as members of the former file had lowered the muzzle of their weapons. It was an automatic process, continuing until the officer shouted an order to cease.[26]

There is ample proof that firing by file occurred on many Civil War battlefields. The Thirty-fifth Indiana started with a volley at 40 paces while defending itself against John C. Breckenridge's attack at Stones River on January 2, 1863, but the men quickly reloaded and then began to fire by file. They did so for forty-five minutes before the pressure became too great, and then retreated across Stones River. The Nineteenth Ohio started firing by file at a range of 100 yards and stopped the advancing Confederates on the first day of that battle.[27]

Yet there was opposition to file firing among some commanders. G. T. Beauregard thought "it excites the men and renders their subsequent control difficult." He preferred firing by wing or company to enable officers to maintain tighter control over their units. Division commander William B. Bate in the Army of Tennessee also wanted to avoid file firing, apparently for the same reason. He preferred firing by wing or company, and, at the same time, by rank if it was practicable.[28]

Hardee specified that firing by rank should be done by the front rank first, then the second. If the battle line about-faced to the rear, Hardee still wanted the former front rank (now the rear) to fire first, in order to maintain the habitual order. Hardee also provided instructions for the rear-rank men to incline the upper parts of their bodies forward a bit when firing, so that the muzzles of their weapons would "reach as much beyond the front rank as possible."[29]

Col. Robert McAllister's Eleventh New Jersey fired by rank on the second day at Gettysburg, staggering its fire into waves along the entire regimental front rather than releasing it all at once in a single volley. A Georgia man in Alfred H. Colquitt's brigade reported that his unit opened with rank fire, but by the rear rank first, and then converted to individual fire to repel Union attacks at Cold Harbor. Many people noted the smooth rank firing by Company A, Seventeenth North Carolina at Bentonville. Capt. William Biggs ordered his rear rank to fire first, then the front rank, even though the rest of the regiment was firing by file. Observers believed Biggs had the right idea.[30]

Firing by rank posed a serious problem, for the men could endanger each other if they did not fire with discipline. Emerson Opdycke noted this with the Forty-first Ohio on the second day at Shiloh, as he saw the front rank moving farther forward than the second. He stood in front and ordered them to cease fire and close ranks again. This took place several times. Opdycke concluded that it was natural for the men to "get so earnest, and excited, they forget to think of anything, but loading and shooting their guns off, whether *at* anything or not; and a few words from an officer who is entitled to confidence, will set them thinking again."[31]

Firing straight ahead, termed direct fire, was the easiest direction to aim and it undoubtedly was the most common. To fire obliquely to right or left was more complicated. Hardee issued detailed instructions about how it was to be done. The men in the front rank should keep their feet in the proper place to face straight ahead but "throw back" either their right or left shoulder, depending on which way the oblique fire was to be delivered. The men in the rear rank were to shift either foot forward 8 inches toward the heel of the man in his front, and incline his upper body while bending the other knee a bit to fire.[32]

There is no evidence that Civil War soldiers actually did this while firing obliquely. In fact, there is relatively scant evidence that oblique firing was done, but when it occurred, observers usually thought it was effective. Col. John A. Davies described what happened to the Forty-sixth Illinois when the Confederates obtained a "raking crossfire upon my right flank" on the first day at Shiloh. The fire killed or wounded "over one-half of my right companies, badly cutting up my other companies, and 8 of my line officers, 2 color bearers, and the major wounded."[33]

There is no evidence that Civil War units practiced the famous fire by platoons, which had been honed to such a fine level of effectiveness by the Dutch and the English in the eighteenth century, although the basic concept of staggering fire by subsections of the regiment was known. Pressured by heavy enemy attacks at the battle of Allatoona, some Federal officers ordered their men to fire in relays, half a unit at a time. This allowed the rifle muskets to cool a bit and spread out the firing to keep a more continuous roll of flame spreading toward the attacker.[34]

Firing by volley was probably the most common firing system used by Civil War soldiers. They usually used the term more often than any other when referring to how they let loose on the enemy, although some officers referred to it as firing by battalion, the term used in the drill manuals. When the Twelfth Connecticut entered its first battle at Georgia Landing, it was told to open fire by file. But the men were too excited to do this. Instead, "they leveled those five hundred rifles together and sent a grand crashing volley into the hostile line of smoke which confronted them."[35]

The primary purpose of adopting a firing system was to more effectively deliver destruction on the enemy, and many officers thought it was necessary for them to maintain some degree of control over that process to maximize the effect. Brig. Gen. William B. Hazen, an officer in the U.S. Army before the war, thought volley fire was the most effective. "It further gives a careful colonel complete control of his fire," he wrote. Some historians, such as Paddy Griffith, have agreed with this view. But Griffith has also pointed out that even when Civil War units started with volleys, they inevitably gravitated toward individual fire, and he assumes that firing at will was the least controllable and least effective fire system. "Time and again we read of regiments blazing away uncontrollably, once started, and continuing until all ammunition was gone or all enthusiasm spent. Firing was such a positive act, and gave the men such a physical release for their emotions, that instincts easily took over from training and from the exhortations of officers." Griffith expressed the sentiments of that class of Civil War officers who tended to be more professionally oriented, more concerned with issues of discipline and efficiency in general.[36]

Yet there was a large segment of officers and enlisted men alike who thought that firing at will was both effective and preferable. Marcus Woodcock described his regiment's first encounter at Stones River as a volley followed "by a terrible 'firing at will' for a few minutes that seemed to almost entirely silence the rebel musketry."[37]

Over and above issues of fire effectiveness, many officers simply could not brook having their men decide for themselves when to open or close fire. Col. John Logan had a difficult time exercising this aspect of his command over the Thirty-second Illinois at Shiloh, for the regiment started firing without his orders because the rest of the brigade opened fire. Only Company B held out to the last, but finally blazed away when an aide on the division staff rode by and told the captain to join in. Logan investigated on the spot to find out why all this happened, and then told everyone to stop firing until he issued proper orders.[38]

It is generally true, however, that officers exerted far less control over their unit's firing than one would expect in the Civil War. Often it was limited to getting it started and getting it stopped, although as Logan's case indicates it could be a real job to accomplish even that much. In many cases, the role of the officer was limited to encouraging the men to do their utmost once the firing started. Lt. Col. E. F. W. Ellis and Maj. William R. Goddard repeatedly shouted to their men in the Fifteenth Illinois: "Stand firm"; "Do your duty, boys"; "Stand your ground"; "Take good aim" before both were shot at the battle of Shiloh.[39]

Rapid Firing

Given the time involved in reloading a rifle musket, many of the better soldiers adopted strategies to increase their rate of fire. A common method was to collect abandoned guns and preload them in anticipation of need. David Holt had an arsenal of seven guns at his disposal during the static fight in the Mule Shoe Salient at Spotsylvania. His comrades did the same. "We stacked them up against the breastwork with the butts in the trench, and when the Yankees came, we picked them up one by one and fired and set them down again. Many times we could not put the gun to our shoulder by reason of the closeness of the enemy, so we shot from the hip." Holt described it well when he wrote years later, "I became a rapid repeater. I stood to my guns and shot carefully."[40]

Federal troops of Alexander Hays's division did the same when opposing Pickett's Charge at Gettysburg. Many had four guns at their disposal, and Lt. John L. Brady of the First Delaware noted that officers and men alike "handled rifles with equal precision and seemed to vie with each other in scoring 'crack shots' upon the advancing column of the enemy."[41]

If spare guns were not available, another strategy to increase the rate of fire was for the less adept soldiers to reload for the crack shots and pass the weapons forward. Confederate soldiers of George B. Anderson's North Carolina Brigade did this while defending the Sunken Road at Antietam. Lt. John Quincy Adams Campbell of the Fifth Iowa tore off the loose paper end of many cartridges for his men so they would not have to do that step of the reloading process at the battle of Iuka. A man in the 118th Pennsylvania, who had not previously been a good soldier or a good shot, suddenly determined to prove himself on the second day at Gettysburg. He blazed away with intensity, shouting, "Give them hell boys." All the while his comrades, impressed with his sudden conversion, passed loaded muskets to him.[42]

The effect on the rifle musket of sustained, intense firing was marked. Many men of the Fourteenth Illinois claimed that their tubes became so hot that the powder "seemed to melt when poured into them." The 123rd New York fired so much at Peachtree Creek that the men had difficulty holding their weapons because the gun barrels were too hot. Confederate troops defending Cheatham Hill at Kennesaw Mountain fired so rapidly that the powder sometimes detonated in a flash before the bullet was rammed home. Several members of the Ninth Tennessee at Cheatham Hill told Capt. James I. Hall that shavings of lead pared off the bullet by rifling grooves were melting and running down the tube when they held their guns upright to reload. Hall advised them to hold the tube down after firing to drain the material before reloading. He watched curiously as "small round pellets of melted lead poured out of their gun barrels on the ground."[43]

Expenditure of Ammunition

The rate of firing, the length of close contact, and the morale and determination of the soldiers affected the expenditure of ammunition on any given battlefield, and that expenditure itself influenced the problem of supplying units with more ammunition. This subject has been all but ignored by historians of the Civil War. Paddy Griffith has guessed that "prolonged firefights must have been run at very low rates of fire—perhaps only one round per man per five or even ten minutes." Josiah Gorgas, chief of ordnance in the Confederate army, examined returns of ordnance officers before and after major battles and concluded that each soldier fired on average nineteen to twenty-six rounds in each engagement. He concluded that commanders need not worry as long as they had about fifty rounds of ammunition available per man before a battle. Thomas Ward Osborne, artillery chief of the Army of the Tennessee, thought that 200 rounds of infantry ammunition for each man was enough supply for Sherman's March to the Sea, for that amount would be "equivalent to four or five days steady fighting."[44]

Actually, there is quite a lot of information in the official reports and in personal accounts about the expenditure of ammunition, and it shows that all these estimates by contemporaries and modern historians alike fall midway on a spectrum that ranged widely to either extreme. Civil War soldiers were quite capable of firing off thousands of rounds of ammunition at a furious rate over a long period of time. Of course, the rate of expenditure tended to be much higher on the skirmish line, especially in situations where the opposing armies were static and close together for long periods of time. The Ninety-sixth Pennsylvania, for example, fired off 90,000 rounds in only twenty-four hours while on the skirmish line at Cold Harbor.[45]

But the expenditure of ammunition by the battle line in a variety of different engagements reaches closer to the mainstream of action in the Civil War than does firing by skirmishers. Cheatham's Division fired thirty-five rounds per man at Perryville, while Hardee's Corps fired forty rounds at Stones River and

Table 4.1. Expenditure of ammunition per man

Unit	Battle	Expenditure of Ammunition (rounds per man)
100th Pennsylvania, Welsh's brigade, Willcox's division, Ninth Corps, Army of the Potomac	South Mountain, September 14, 1862	11–15[a]
Cheatham's Division, Polk's Right Wing, Army of Tennessee	Perryville, October 8, 1862	35[b]
Hardee's Corps, Army of Tennessee	Stones River, December 31, 1862	40[b]
51st and 52nd Tennessee, Wright's Brigade, Cheatham's Division, Polk's Right Wing, Army of Tennessee	Chickamauga, September 19, 1863	20[c]
Polk's Right Wing, Army of Tennessee	Chickamauga, September 19–20, 1863	26[b]
Jackson's Brigade, Cheatham's Division, Polk's Right Wing, Army of Tennessee	Chickamauga, September 19–20, 1863	44.7[d]
Maney's Brigade, Cheatham's Division, Polk's Right Wing, Army of Tennessee	Chickamauga, September 19–20, 1863	30.9[d]
Wright's Brigade, Cheatham's Division, Polk's Right Wing, Army of Tennessee	Chickamauga, September 19–20, 1863	9.7[d]
Strahl's Brigade, Cheatham's Division, Polk's Right Wing, Army of Tennessee	Chickamauga, September 19–20, 1863	6.3[d]
Cheatham's Division, Polk's Right Wing, Army of Tennessee	Chickamauga, September 19–20, 1863	22.9 (average)[d]
Deas's Brigade, Hindman's Division, Longstreet's Left Wing, Army of Tennessee	Chickamauga, September 19–20, 1863	48.9[e]
16th Tennessee, Wright's Brigade, Cheatham's Division, Polk's Right Wing, Army of Tennessee	Chickamauga, September 19–20, 1863	7.75[f]
38th Tennessee, Wright's Brigade, Cheatham's Division, Polk's Right Wing, Army of Tennessee	Chickamauga, September 19–20, 1863	10[f]
8th Tennessee, Wright's Brigade, Cheatham's Division, Polk's Right Wing, Army of Tennessee	Chickamauga, September 19–20, 1863	9.25[f]

(*continues*)

Table 4.1. (*continued*)

Unit	Battle	Expenditure of Ammunition (rounds per man)
28th Tennessee, Wright's Brigade, Cheatham's Division, Polk's Right Wing, Army of Tennessee	Chickamauga, September 19–20, 1863	12[f]
59th Illinois, Grose's brigade, Cruft's division, 4th Corps, Army of the Cumberland	Lookout Mountain, November 24, 1863	50–80[g]
61st Georgia, Evans's Brigade, Gordon's Division, Ewell's Corps, Army of Northern Virginia	Spotsylvania, May 12, 1864	160[h]
Average of all battles		33

[a] Frederick Pettit to parents, brothers, and sisters, September 20, 1862, in Annette Tapert, ed., *The Brothers' War: Civil War Letters to Their Loved Ones from the Blue and Gray* (New York: Vintage, 1988), 88.
[b] Larry J. Daniel, *Soldiering in the Army of Tennessee: A Portrait of Life in a Confederate Army* (Chapel Hill: University of North Carolina Press, 1991), 48.
[c] John G. Hall to Leon Trousdale, October 4, 1863, ibid., 129. A. J. Paine, who served as the chief ordnance officer of Wright's Brigade, calculated the expenditure of ammunition by the consolidated 51st and 52nd Tennessee as ten rounds per man over the two-day course of the battle of Chickamauga. I have accepted Hall's statement because he served as regimental commander.
[d] John A. Cheatham to Hippolyte Oladowski, October 20, 1863, in *The War of the Rebellion: A Compilation of the Official Records of the Union and Confederate Armies*, 70 vols. (Washington, DC: Government Printing Office, 1880–1901), vol. 30, pt. 2, 82.
[e] Frederick B. Dallas to Oladowski, October 24, 1863, ibid., 332–333.
[f] A. J. Paine to not stated, October 18, 1863, ibid., 121.
[g] Entry of November 24, 1863, in Arnold Gates, ed., *The Rough Side of War: The Civil War Journal of Chesley A. Mosman, 1st Lieutenant, Company D, 59th Illinois Volunteer Infantry Regiment* (Garden City, NY: Basin, 1987), 126.
[h] G. W. Nichols, *A Soldier's Story of His Regiment* (Kennesaw, GA: Continental, 1961), 157.

Polk's Corps fired twenty-six rounds at Chickamauga. Several Confederate units provided detailed reports of the ammunition used at the latter battle, and it ranged from six rounds per man to nearly fifty. But the Fifty-ninth Illinois fired up to eighty rounds per man in a running low-intensity fight at Lookout Mountain on November 24, 1863.[46]

Rate of Fire

There is also a surprising amount of information in the official reports and the personal accounts detailing specific rates of fire; that is, how many rounds each

Table 4.2. Rate of fire

Unit	Battle	Rate of Fire
W. J. McMurray, 20th Tennessee, Statham's Brigade, Breckinridge's Corps, Army of the Mississippi	Shiloh, April 6, 1862	1 round per 2 minutes, for 1 hour[1]
27th Tennessee, Wood's Brigade, Hardee's Corps, Army of the Mississippi	Shiloh, April 6, 1862	1 round per 1.25 minutes, for 50 minutes[2]
9th Indiana, Hazen's brigade, Nelson's division, Army of the Ohio	Shiloh, April 7, 1862	1 round per 3.4 minutes, for 2 hours, and 1 round per 3 minutes, for 1 hour[3]
52nd New York, Brooke's brigade, Richardson's division, 2nd Corps, Army of the Potomac	Antietam, September 17, 1862	1 round per 30 seconds, for half an hour[4]
Maury's Division, Army of the West	Hatchie Bridge, October 5, 1862	1 round per 2 or 2.4 minutes, for 2 hours[5]
3rd Kentucky (U.S.), Hascall's brigade, Wood's division, Crittenden's Left Wing, Army of the Cumberland	Stones River, December 31, 1862	1 round per 4.6 minutes, for 3 hours[6]
83rd Indiana, Smith's brigade, Stuart's division, 15th Corps, Army of the Mississippi	Arkansas Post, January 11, 1863	1 round per 3 minutes, for 3 hours[7]
100 men of 1st U.S. Sharpshooters (Sharp's breechloaders) and 200 men of 3rd Maine (rifle muskets), Ward's brigade, Birney's division, 3rd Corps, Army of the Potomac	Gettysburg, July 2, 1863	1 round per 36 seconds, for 20 minutes[8]
Pvt. Peter H. Kipp, 1st U.S. Sharpshooters (Sharp's breechloader), Ward's brigade, Birney's division, 3rd Corps, Army of the Potomac	Gettysburg, July 2, 1863	1 round per 12.6 seconds, for 20 minutes[8]
31st Indiana, Cruft's brigade, Palmer's division, 21st Corps, Army of the Cumberland	Chickamauga, September 19, 1863	1 round per 3.2 minutes, for 2 hours[9]
9th Kentucky (U.S.), Beatty's brigade, Van Cleve's division, 21st Corps, Army of the Cumberland	Chickamauga, September 20, 1863	1 round per 30 seconds, for 30 minutes[10]
Edwards's brigade, Neill's division, 6th Corps, Army of the Potomac	Spotsylvania, May 12, 1864	1 round per 2.5 minutes, for 21 hours[11]
2nd Rhode Island, Edwards's brigade, Neill's division, 6th Corps, Army of the Potomac	Spotsylvania, May 12, 1864	1 round per 1.8 minutes, for 3 hours[12]

(*continues*)

Table 4.2. *(continued)*

Unit	Battle	Rate of Fire
Upton's brigade, Russell's division, 6th Corps, Army of the Potomac	Spotsylvania, May 12, 1864	1 round per 1.9 minutes, for 11 hours[13]
7th Indiana, Robinson's brigade, Cutler's division, 5th Corps, Army of the Potomac	Spotsylvania, May 12, 1864	1 round per 2.1 minutes, for 5 hours[14]
Pvt. James Daniel, 47th Alabama, Law's brigade, Field's division, Anderson's corps, Army of Northern Virginia	Cold Harbor, June 3, 1864	1 round per 1 minute, for 1 hour[15]
7th Minnesota, Woods's brigade, Mower's division, 16th Corps (Right Wing)	Tupelo, July 14, 1864	1 round per 1–1.2 minutes, for 1 hour[16]
Rice C. Bull, 123rd New York, Knipe's brigade, Williams's division, 20th Corps, Army of the Cumberland	Peachtree Creek, July 20, 1864	1 round per 3.4 minutes, for 4 hours[17]
W. W. Gist, 26th Ohio, Lane's brigade, Wagner's division, 4th Corps	Franklin, November 30, 1864	1 round per 1.8 minutes, for 6 hours[18]
Average of all battles		1 round per 2.1 minutes

[1] Andrew Haughton, *Training, Tactics and Leadership in the Confederate Army of Tennessee: Seeds of Failure* (Portland, OR: Frank Cass, 2000), 67.
[2] Company Commanders of 27th Tennessee to S. A. M. Wood, April 9, 1862, in *The War of the Rebellion: A Compilation of the Official Records of the Union and Confederate Armies*, 70 vols. (Washington, DC: Government Printing Office, 1880–1901), vol. 10, pt. 1, 606. This work is hereafter cited as *OR*.
[3] Gideon C. Moody to R. L. Kimberly, April 9, 1862, *OR*, vol. 10, pt. 1, 342.
[4] Paul Frank to Charles P. Hatch, September 20, 1862, *OR*, vol. 19, pt. 1, 301.
[5] James C. Bates to Ma, October 12, 1862, in Richard Lowe, ed., *A Texas Cavalry Officer's Civil War: The Diary and Letters of James C. Bates* (Baton Rouge: Louisiana State University Press, 1999), 189.
[6] Daniel R. Collier to Edmund R. Kerstetter, January 5, 1863, *OR*, vol. 20, pt. 1, 489.
[7] Record of Events, Company C, 83rd Indiana, January 11, 1863, in *Supplement to the Official Records of the Union and Confederate Armies,* 100 vols. (Wilmington, NC: Broadfoot, 1993–2000), pt. 2, vol. 18, 213.
[8] C. A. Stevens, *Berdan's United States Sharpshooters in the Army of the Potomac, 1861–1865* (Dayton, OH: Morningside Bookshop, 1972), 305.
[9] John T. Smith to W. H. Fairbanks, September 28, 1863, *OR*, vol. 30, pt. 1, 741.
[10] George H. Cram to O. O. Miller, September 26, 1863, *OR*, vol. 30, pt. 1, 814.
[11] Mason Whiting Tyler, *Recollections of the Civil War, with Many Original Diary Entries and Letters Written from the Seat of War, and with Annotated References* (New York: G. P. Putnam's Sons, 1912), 191.
[12] Elisha Hunt Rhodes to not stated, n.d., in Robert Hunt Rhodes, ed., *All for the Union: The Civil War Diary and Letters of Elisha Hunt Rhodes* (New York: Orion, 1985), 147.
[13] G. Norton Galloway, "Hand-to-Hand Fighting at Spotsylvania," in *Battles and Leaders of the Civil War,* vol. 4, ed. Robert Underwood Johnson and Clarence Clough Buel (New York: Thomas Yoseloff, 1956), 174, 174n.
[14] Merit C. Welsh to not stated, August 7, 1864, *OR*, vol. 36, pt. 1, 617.

Table 4.2. (*continued*)

[15] J. Gary Laine and Morris M. Penny, *Law's Alabama Brigade in the War between the Union and the Confederacy* (Shippensburg, PA: White Mane, 1996), 274.
[16] William R. Marshall to Henry Hoover, July 22, 1864, *OR*, vol. 39, pt. 1, 272.
[17] K. Jack Bauer, ed., *Soldiering: The Civil War Diary of Rice C. Bull, 123rd New York Volunteer Infantry* (San Rafael, CA: Presidio, 1978), 150.
[18] W. W. Gist, "The Battle of Franklin," *Tennessee Historical Magazine* 6, 3 (1920): 232.

man fired over a given period of time. This data allows us to realize how fast the rifle musket could be fired under battlefield conditions, as opposed to generic estimates of how fast a soldier on the parade ground could reload and fire under ideal conditions. The fastest rate among the sample of sixteen involving the rifle musket was one round every thirty seconds, and the slowest was one round every 4.6 minutes. The average amounted to one round every two minutes, and the duration ranged from thirty minutes to six hours. Of course, soldiers using breechloaders achieved a higher rate than those using the rifle musket, with one man firing a round every twelve seconds.

One of the most intense musketry actions of the war took place at the Mule Shoe Salient at Spotsylvania, where thousands of Union and Confederate soldiers were locked in very close range for many hours, separated only by an earthen parapet. Some units fired constantly for twenty-one hours. Fouled muskets had to be sent to the rear for cleaning while men in the front line borrowed usable guns from other units until they returned. Some men did not bother to clean their weapons but rammed the new charge down as far as it would go and considered it adequate for such a short range. Many muskets burst due to fouling because of this. The better shots in many regiments took on the responsibility of firing much more than their fair share as comrades passed loaded guns to them.[47]

The effect of this concentrated musketry was appalling to those who viewed the battlefield the next day. Head logs on the earthworks were "cut and torn until they resembled hickory brooms," in the words of G. Norton Galloway of Upton's brigade. An oak tree 22 inches in diameter was literally cut down by the stream of bullets, while "horses and men [were] chopped into hash . . . and appearing more like piles of jelly than the distinguishable forms of human life." A Federal officer counted about 150 bodies in a space that measured only 12 by 15 feet. The Confederates had evacuated the salient during the night, leaving the polluted battlefield in Federal possession. They also left thousands of muskets behind, many still sticking through firing slits beneath head logs, with hundreds of cartridge packages lining the tops of the parapets for easy access. Federal losses on May 12 amounted to 9,000 men, while the Confederates suffered about 8,000 casualties (3,000 of whom were captured).[48]

Supply of Ammunition

There was no prescribed method of bringing more ammunition to a regiment that ran out of its initial supply of forty rounds during a heated engagement, and Civil War commanders had to improvise with commonsense solutions to the problem. In the early part of the war, as Grant testified in his memoirs, they often had no idea what to do. Grant rode among the scattered remnants of McClernand's division after it had been pushed back by an unexpected Confederate attack at Fort Donelson on February 15, 1862. The men were out of ammunition, yet extra rounds were in cartridge boxes scattered across the battlefield. Grant later wrote that many regimental, brigade, and even division commanders had not yet "been educated up to the point of seeing that their men were constantly supplied with ammunition," and he had to tell the troops to scrounge around for extra rounds anywhere they could find them.[49]

At Shiloh, less than two months later, John Logan asked Col. Isaac C. Pugh what arrangements had been made to keep the men supplied with ammunition. Pugh had just taken charge of the brigade when the permanent commander had been wounded. "His reply was, none that he knew of." Pugh thought that forty rounds was enough and told Logan not to worry about it. But Logan cautioned his men not to fire unless they could clearly see a Rebel to shoot at, and before long cries of "We are out of cartridges" began to sound along the line. Later in the day, Logan had to retire because his Thirty-second Illinois ran out of rounds.[50]

Commanders on both sides of the war improvised methods of keeping up the supply. One of the most common was to rotate a fresh regiment into line to replace one whose ammunition had run out. The First Nebraska and Fifty-eighth Ohio of Col. John M. Thayer's brigade were nearly out of rounds when relieved by the Seventy-sixth Ohio at Shiloh. The two regiments took twenty minutes to refill cartridge boxes from a wagon that division commander Lew Wallace had sent up to the line for that purpose, and then returned to the action. The Third Kentucky (U.S.) had to march half a mile to the rear to find a place where it could refill its cartridge boxes and clean its muskets. Commanders often detailed available staff officers to see that more ammunition was shifted close behind their firing lines, even if those officers normally were not experienced in dealing with ordnance supplies. Horatio P. Van Cleve used his inspector general to perform this task at Stones River.[51]

Another common way was Grant's method at Fort Donelson: to encourage the men to scrounge across the field and take rounds from abandoned cartridge boxes. This, of course, was haphazard and not conducive to discipline, but in the absence of an organized supply effort it often occurred. Better if an officer detailed by a brigade or regimental commander could happen upon an ammunition wagon nearby and commandeer it if even if the wagon was not normally assigned to that particular unit. With no source of fresh ammunition available, the Twenty-ninth Tennessee retired 80 yards to a safer spot as Col. Horace Rice told the men to lie down and redistribute "what ammunition was left equally among themselves."[52]

Unit commanders often ordered the issue of extra rounds to their men when they anticipated going into a heavy fight, but that was not always a wise precaution. Lt. Francis W. Dawson, an ordnance officer in Longstreet's Corps, reported that these extra rounds could not be carried in the cartridge box and were therefore "put loose in the pockets or haversacks of the men. They are but seldom really required and are generally ordered after a few days to be returned to the ordnance officer, where three-fourths of them are found to be entirely spoiled." Dawson suggested that such prebattle issues be ordered only by the corps commander to minimize waste.[53]

The question of ammunition supply was crucial, for a regiment could go through hundreds of rounds in a heavy battle that involved extended close contact. The Seventy-fourth Indiana entered the fight at Chickamauga with sixty rounds. After firing them all, another regiment came to its relief as the Indianans went to the rear for sixty more rounds. The next day, the regiment was given sixty additional cartridges per man and exhausted all of them in the heavy fighting on the afternoon of September 20. It received more rounds at 4 P.M. but was not heavily engaged after that. Lt. Col. Henry V. N. Boynton's Thirty-fifth Ohio exhausted its supply during two hours of defending Snodgrass Hill but fortunately received a fresh supply before the Confederates attacked again. This, too, ran out, and the men had to find more cartridges from the battlefield. Then the Second Minnesota gave the Ohioans "several rounds," which allowed them to have three cartridges apiece. The regiment fixed bayonets and waited for the worst, but the fighting drew to a close. The Thirty-fifth fired one last volley as it withdrew from Snodgrass Hill, raking the battle line of a Confederate regiment that occupied the hill at a range as close as 50 yards. Boynton estimated his fire caught the opposing line at a 30-degree angle. It sent the Confederates retreating in disorder.[54]

Of course, the more organized the supply effort, the less likely that regiments like the Thirty-fifth Ohio would come so close to having empty guns when faced with the enemy. Cheatham's Division of the Army of Tennessee had an ordnance officer for each brigade who was responsible for keeping the ammunition train as close as possible behind the brigade line. Lt. A. J. Paine, ordnance officer of Brig. Gen. Marcus J. Wright's Tennessee brigade, reported that his subordinates "were employed chiefly in watching the movements of their respective regiments, so as to know all the time their exact locality, that they might be able to supply them with ammunition at any moment." It was easy to lose sight of a brigade in wooded terrain and difficult to move wagons through the brush to keep pace with advances. Cheatham's ordnance officer suggested the ammunition wagons fly a distinctive flag for each brigade train to help everyone locate them quickly.[55]

Union and Confederate troops in the eastern campaigns practiced the same methods of ammunition supply as did their comrades in the West. Sometimes the question of obtaining more rounds could be used as an excuse to get out of the fighting altogether. Col. Paul Frank was nearly out of ammunition at the Wilderness and wanted to take his brigade to the rear. Col. Robert McAllister, who commanded a supporting brigade behind Frank, tried to convince him to stay. McAllister offered to send a sergeant to guide Frank to a supply of cartridges brought up to the line by mules, but he refused. Frank pulled back and McAllister had to take his place in line. Using mules was a smart alternative to wagons in a heavily wooded environment. The Sixth Corps did this at Spotsylvania. When the ammunition got up to the firing line, enterprising soldiers distributed it by passing cartridge packets along the firing line by hand, as the Confederates did while defending the Mule Shoe Salient.[56]

Efforts to supply ammunition resulted in several high-level directives. John Bell Hood prepared for his new corps command before the start of the Atlanta campaign by forbidding units from taking themselves out of the firing line when they ran out of ammunition. He encouraged the men to obtain more rounds from the killed and wounded. George G. Meade let the Army of the Potomac know that the "question of ammunition is an important one" and encouraged his men to take abandoned rounds from the battlefield. Robert E. Lee urged more than one of his subordinate commanders to be proactive in getting ammunition to their units in a timely manner. "It will not do to wait until the ammunition is entirely exhausted before steps are taken to resupply the men."[57]

Despite the often difficult problem of moving ammunition to the firing line when it was needed, there are few documented cases of a unit having to retire from a position under fire because it had run out of rounds. Col. D. H. Hamilton, who commanded McGowan's South Carolina brigade at Chancellorsville, was nearly out of ammunition and asked for more. He received word "that it could not be just then furnished," so he asked the assistant adjutant general on the brigade staff "whether it was not possible to obtain it, for that I did not know what I should do, as the enemy were advancing on the right, and that I could not meet them with empty guns." Fifteen minutes later, before the rounds arrived, the brigade had to retire due to enemy pressure.[58]

Range of Fire

The rifle musket represented a dramatic change only in the range at which it was possible to deliver fire on the enemy. The smoothbore's effective range of 100 yards paled in comparison to the range of the rifle musket, normally given as about 500 yards. In no other area was there evidence that the rifle musket could outperform the smoothbore musket, for the loading time was similar. Contrary to the assertion of some modern-day historians, there is no evidence that the rifle musket was more accurate than the smoothbore musket at short range.[59]

Therefore, the old argument in favor of the rifle musket as a revolutionary factor in changing the face of warfare rests on the question of effective range. Unfortunately, historians who adhere to the old argument have never tried to verify that the rifle musket was used in long-range firing. Not until Paddy Griffith raised this issue in 1989 did it become apparent that most fighting in the Civil War occurred at ranges much closer to the effective range of the smoothbore than to that of the rifle musket.

Griffith catalogued a total of 113 references to range of fire in the primary literature. Only 17 of them indicated "ranges longer than 250 yards, and none beyond 500." Other historians have taken his lead and conducted searches of the primary sources themselves. Mark Grimsley found 89 references in the *Official Records,* all of them to eastern battles, and concluded that the average range of fire was 116 yards. He believes this was "an improvement over the 80–100 yards characteristic of smoothbore warfare, but at best an incremental improvement." Brent Nosworthy also falls into line with Griffith's findings. While he uncovered reports of rifle fire delivered at ranges of 220, 275, even 600 yards, Nosworthy noted that they were comparatively rare and that the "critical moments of engagements" much more often occurred at 80- to 120-yard ranges. Nosworthy has noted that this was "more than a 50% improvement" over the normal Napoleonic average of 90 yards, but that it fell very far short of the expectations of rifle advocates before the Civil War. "Although the theoretical range of the rifle musket was several times that a smoothbore, much of the fighting nevertheless occurred at ranges equal to or only slightly more than that found during previous wars."[60]

My own survey of the primary sources yielded thirty-nine references covering twenty-four battles, with ranges that varied between 15 yards and 300 yards. The average is even lower than that of Griffith and Grimsley, but quite close to that of Nosworthy, amounting to 94 yards.

Why did most Civil War soldiers fail to fulfill the potential of their rifle muskets? The answers are relatively clear. The parabolic trajectory and the difficulty of seeing clearly at ranges of 500 yards or more were the prime reasons. The only way to deal with both problems was by rigorous training in estimating distances and in how to use the adjustable sights to compensate for the curved nature of the bullet's flight. This training was never given to line infantry. As we will see in following chapters, this sort of rigorous training was offered to the sharpshooter battalions organized in Lee's army in the winter of 1864. For the rest, the vast majority of Civil War soldiers, target training was infrequent, usually consisting of little more than setting up a target and seeing if the men could hit it. There were no attempts to teach the average soldier how to estimate distance and adjust his sights accordingly.

The cluttered nature of the battlefield also impeded the full utilization of the rifle musket. Tangled thickets, dense woods, and undulating hills often limited the line of sight, making it impossible to detect targets more than 100 yards away. Most Civil War commanders thought it senseless to fire unless there was a clear target visible.

Yet another cause can be ascribed to a natural tendency for soldiers to think that close-range firing was more effective than long-range firing. Officers tried to keep Civil War combat from occurring at long range. Hood issued a general order to his corps before the Atlanta campaign, declaring, "Firing on the enemy

Table 4.3. Range of fire

Unit	Battle	Range of Fire
18th Louisiana, Pond's Brigade, Ruggles's Division, Bragg's Corps, Army of the Mississippi	Shiloh, April 6, 1862	Received initial fire at 100–150 yards[1]
26th Alabama, Gladden's Brigade, Withers's Division, Bragg's Corps, Army of the Mississippi	Shiloh, April 6, 1862	Received initial fire at 300 yards[2]
14th Iowa, Tuttle's brigade, W. H. L. Wallace's division, Army of the Tennessee	Shiloh, April 6, 1862	Colonel ordered men to wait until Confederates advanced to a point 30 paces away before regiment opened fire[3]
24th Ohio, Ammen's brigade, Nelson's division, Army of the Ohio	Shiloh, April 7, 1862	Advanced and opened fire at 50–75 yards from Confederates[4]
Wilcox's Brigade, Longstreet's Center, Army of Northern Virginia	Williamsburg, May 5, 1862	Firing occurred at less than 30 yards[5]
Gregg's Brigade, A. P. Hill's Division, Jackson's Command, Army of Northern Virginia	Antietam, September 17, 1862	Confederates started firing when Federals were 30–40 yards away[6]
108th New York, Morris's brigade, French's division, 2nd Corps, Army of the Potomac	Antietam, September 17, 1862	Firing at a range of 110–165 yards[7]
Anderson's Brigade, D. H. Hill's Division, Jackson's Command, Army of Northern Virginia	Antietam, September 17, 1862	Opened fire when Federals were 50 yards away[8]
Lee's Division	Chickasaw Bluffs, December 29, 1862	Lee told men to wait until Federals were 100 yards away before firing[9]
3rd Tennessee, Gregg's Brigade	Chickasaw Bluffs, December 29, 1862	Confederates opened fire "before full musket range," and Federals continued to advance; stopped 150 yards from Confederate position[10]
88th Illinois, Sill's brigade, Sheridan's division, McCook's Right Wing, Army of the Cumberland	Stones River, December 31, 1862	Colonel ordered regiment to wait until Confederates were 75 yards away before he allowed it to open fire[11]
21st Wisconsin, Starkweather's brigade, Rousseau's division, Thomas's center, Army of the Cumberland	Stones River, December 31, 1862	Regiment was told to wait until Confederates were within "a good hitting distance," and colonel ordered it to open fire at about 170 yards[12]

(*continues*)

Table 4.3. (*continued*)

Unit	Battle	Range of Fire
Liddell's Brigade, Cleburne's Division, Hardee's Corps, Army of Tennessee	Stones River, December 31, 1862	Opened fire on a Federal brigade at 65–70 yards[13]
22nd Indiana, Post's brigade, Davis's division, McCook's Right Wing, Army of the Cumberland	Stones River, December 31, 1862	Regiment waited until Confederates were 30 yards away before it opened fire[14]
35th Indiana, Price's brigade, Van Cleve's division, Crittenden's Left Wing, Army of the Cumberland	Stones River, December 31, 1862	Regimental commander waited until Confederates were 30–40 paces away before ordering regiment to open fire[15]
6th Indiana, Baldwin's brigade, Johnson's division, McCook's Right Wing, Army of the Cumberland	Stones River, December 31, 1862	Regimental commander waited until Confederates were 100 yards away before he allowed his eager men to open fire[16]
14th Connecticut, Smyth's brigade, Hays's division, 2nd Corps, Army of the Potomac	Gettysburg, July 3, 1863	Opened fire when Confederates were 200 yards away[17]
Kemper's Brigade, Pickett's Division, Longstreet's Corps, Army of Northern Virginia	Gettysburg, July 3, 1863	Opened fire at a range of 100 yards[17]
Garnett's Brigade, Pickett's Division, Longstreet's Corps, Army of Northern Virginia	Gettysburg, July 3, 1863	Opened fire at a range of 75 yards[17]
Armistead's Brigade, Pickett's Division, Longstreet's Corps, Army of Northern Virginia	Gettysburg, July 3, 1863	Fired at 72nd Pennsylvania at a range of 80 yards[17]
59th Ohio, Dick's brigade, Van Cleve's division, 21st Corps, Army of the Cumberland	Chickamauga, September 19, 1863	Regimental commander waited until Confederates were 50 yards away before he ordered his men to open fire[18]
9th Kentucky (U.S.), Beatty's brigade, Van Cleve's division, 21st Corps, Army of the Cumberland	Chickamauga, September 20, 1863	Regiment opened fire at 200-yard range[19]
Manigault's Brigade, Hindman's Division, Longstreet's Left Wing, Army of Tennessee	Chickamauga, September 20, 1863	Received initial fire at 90–100 yards in tree cover with little underbrush[20]

Table 4.3. *(continued)*

Unit	Battle	Range of Fire
28th Alabama, Manigault's Brigade, Hindman's Division, Longstreet's Left Wing, Army of Tennessee	Chickamauga, September 20, 1863	Regiment held for 10 minutes in a firefight with Federals at a distance of 20 yards before retiring[21]
74th Indiana, Croxton's brigade, Brannan's division, 14th Corps, Army of the Cumberland	Chickamauga, September 20, 1863	Regimental commander told men to wait until Confederates were 60–70 yards away before opening fire[22]
Wilder's brigade, Reynolds's division, 14th Corps, Army of the Cumberland	Chickamauga, September 20, 1863	Brigade, using Spencer repeaters, opened fire when Confederates were 50 yards away and repelled attack; during subsequent attack, Wilder told men to let Confederates get closer than 50 yards before opening fire and repelled them again[23]
Bratton's Brigade, Field's Division, Anderson's Corps, Army of Northern Virginia	Spotsylvania, May 12, 1864	Opened fire when Federals were 50 yards away[24]
51st Pennsylvania, Hartranft's brigade, Willcox's division, 9th Corps, Army of the Potomac	Spotsylvania, May 12, 1864	Confederates opened fire when regiment was 150 yards away[25]
Company C, 76th Ohio, Woods's brigade, Osterhaus's division, 15th Corps, Army of the Tennessee	Dallas, May 28, 1864	Company commander waited until Confederates were 50 yards away before ordering his men to fire[26]
Crawford's division, 5th Corps, Army of the Potomac	Cold Harbor, May 30, 1864	Opened fire when Confederates were 100 yards away[27]
33rd Missouri, Woods's brigade, Mower's division, 16th Corps (Right Wing)	Tupelo, July 14, 1864	Regiment opened enfilade fire on Confederates 200–300 yards away; continued firing until Confederates had retreated to a point 500–700 yards away[28]
2nd Tennessee Cavalry (C.S.), 4th Brigade, 2nd Division, Forrest's Cavalry	Tupelo, July 14, 1864	Regiment received initial fire at range of 150 yards[29]
7th Minnesota, Woods's brigade, Mower's division, 16th Corps (Right Wing)	Tupelo, July 14, 1864	Regiment told to wait until Confederates were 50 yards away before opening fire[30]

(continues)

Table 4.3. (*continued*)

Unit	Battle	Range of Fire
Manigault's Brigade, Brown's Division, Cheatham's Corps, Army of Tennessee	Atlanta, July 22, 1864	Received initial fire at 200–250 yards from Federal line[20]
Manigault's Brigade, Brown's Division, Cheatham's Corps, Army of Tennessee	Ezra Church, July 28, 1864	Received fire as close as 15 yards from Federal line before retiring[20]
Manigault's Brigade, Brown's Division, Cheatham's Corps, Army of Tennessee	Jonesboro, August 31, 1864	Received fire as close as 80 yards from Federal line before retiring[20]
46th Ohio, Walcutt's brigade, Woods's division, 15th Corps, Army of the Tennessee	Griswoldville, November 22, 1864	Delayed opening fire until Confederates were 30 yards away to let Federal skirmishers retire[31]
Pvt. Andrew Moon, 104th Ohio, Reilly's brigade, Reilly's division, 23rd Corps	Franklin, November 30, 1864	Viewed battlefield, saw bodies on ground for 400 yards before Federal works[32]
Maj. Gen. Benjamin F. Cheatham, corps commander, Army of Tennessee	Franklin, November 30, 1864	Told newspaper correspondent after the war that most Confederates were shot within 50 yards of Federal works[33]
Brig. Gen. George W. Gordon, Vaughan's Brigade, Brown's Division, Cheatham's Corps, Army of Tennessee	Franklin, November 30, 1864	Federals opened fire when his unit was 100 paces from works[33]
W. L. Truman, Guibor's Missouri Battery (C.S.), Storr's Battalion, Stewart's Corps, Army of Tennessee	Franklin, November 30, 1864	Testified that Confederate infantry was 30 steps from works when Federals opened fire[34]
Orr's Brigade, Ruger's Division, 23rd Corps	Wyse's Fork, March 10, 1865	Federals withheld fire until Confederates were 100 yards away[35]
42nd Georgia, Henderson's Brigade, Stevenson's Division, Hill's Corps, Army of Tennessee	Bentonville, March 19, 1865	Regiment waited until Federals were 40–50 paces away before opening fire[36]
Average of all battles		94.4 yards

[1] Arthur W. Bergeron Jr., ed., *The Civil War Reminiscences of Major Silas T. Grisamore, C.S.A.* (Baton Rouge: Louisiana State University Press, 1993), 35.
[2] Hugh C. Bailey, ed., "An Alabamian at Shiloh: The Diary of Liberty Independence Nixon," *Alabama Review* 11 (1958): 152.
[3] William T. Shaw to Samuel J. Kirkwood, October 26, 1862, ibid., 153.

Table 4.3. (*continued*)

[4] Frederick C. Jones to Jacob Ammen, April 8, 1862, in *The War of the Rebellion: A Compilation of the Official Records of the Union and Confederate Armies*, 70 vols. (Washington, DC: Government Printing Office, 1880–1901), vol. 10, pt. 1, 340. This work is hereafter cited as *OR*.
[5] Cadmus M. Wilcox to G. Moxley Sorrel, May 25, 1862, *OR*, vol. 11, pt. 1, 592.
[6] J. F. J. Caldwell, *The History of a Brigade of South Carolinians, Known First as "Gregg's," and Subsequently as "McGowan's Brigade"* (Philadelphia: King & Baird, 1866), 46.
[7] Oliver H. Palmer to Dwight Morris, September 19, 1862, *OR*, vol. 19, pt. 1, 334.
[8] R. T. Bennett to J. W. Ratchford, December 6, 1862, *Atlanta Journal*, April 13, 1047.
[9] Winchester Hall, *The Story of the 26th Louisiana Infantry, in the Service of the Confederate States* (N.p., n.d.), 48.
[10] Robert Ferrell, ed. *Holding the Line: The Third Tennessee Infantry, 1861–1864* (Kent, OH: Kent State University Press, 1994), 83.
[11] Francis T. Sherman to sister, January 12, 1863, in C. Knight Aldrich, ed., *Quest for a Star: The Civil War Letters and Diaries of Colonel Francis T. Sherman of the 88th Illinois* (Knoxville: University of Tennessee Press, 1999), 22.
[12] David Gould and James B. Kennedy, eds., *Memoirs of a Dutch Mudsill: The "War Memories" of John Henry Otto, Captain, Company D, 21st Regiment Wisconsin Volunteer Infantry* (Kent, OH: Kent State University Press, 2004), 83.
[13] Nathaniel C. Hughes, ed., *Liddell's Record: St. John Richardson Liddell, Brigadier General, C.S.A.* (Dayton, OH: Morningside Bookshop, 1985), 112.
[14] Michael Gooding to S. M. Jones, January 9, 1863, *OR*, vol. 20, pt. 1, 278.
[15] Bernard F. Mullen to Samuel W. Price, January 5, 1863, ibid., 611.
[16] Hagerman Tripp to P. P. Baldwin, January 4, 1863, ibid., 339.
[17] Earl J. Hess, *Pickett's Charge—The Last Attack at Gettysburg* (Chapel Hill: University of North Carolina Press, 2001), 210, 229, 228, 262.
[18] Granville A. Frambes to Charles F. King, September 26, 1863, *OR*, vol. 20, pt. 1, 832.
[19] Kenneth W. Noe, ed., *A Southern Boy in Blue: The Memoirs of Marcus Woodcock, 9th Kentucky Infantry (U.S.A.)* (Knoxville: University of Tennessee Press, 1996), 210.
[20] R. Lockwood Tower, ed., *A Carolinian Goes to War: The Civil War Narrative of Arthur Middleton Manigault, Brigadier General, C.S.A.* (Columbia: University of South Carolina Press, 1988), 98, 227, 234, 246.
[21] Joshua K. Callaway to Dulcinea B. Callaway, September 24, 1863, in Judith Lee Hallock, ed., *The Civil War Letters of Joshua K. Callaway* (Athens: University of Georgia Press, 1997), 137.
[22] Myron Baker to Charles V. Ray, September 25, 1863, *OR*, vol. 30, pt. 1, 420.
[23] Daniel H. Hill, "Chickamauga—The Great Battle of the West," in Robert Underwood Johnson and Clarence Clough Buel, eds., *Battles and Leaders of the Civil War* (New York: Thomas Yoseloff, 1956), 3:659n.
[24] John Bratton to not stated, January 1, 1865, *OR*, vol. 36, pt. 1, 1066.
[25] Joseph K. Bolton to [Charles H. McCreery], September 20, 1864, ibid., 964.
[26] Stewart Bennett and Barbara Tillery, eds., *The Struggle for the Life of the Republic: A Civil War Narrative by Brevet Major Charles Dana Miller, 76th Ohio Volunteer Infantry* (Kent, OH: Kent State University Press, 2004), 164.
[27] Samuel W. Crawford to [Gouverneur K. Warren], May 30, 1864, *OR*, vol. 36, pt. 3, 351.
[28] William H. Heath to Henry Hoover, July 21, 1864, ibid., vol. 39, pt. 1, 274.
[29] Richard R. Hancock, *Hancock's Diary; Or, A History of the Second Tennessee Confederate Cavalry, with Sketches of First and Seventh Battalions* (Nashville: Brandon, 1887), 424.
[30] William R. Marshall to Henry Hoover, July 22, 1864, *OR*, vol. 39, pt. 2, 273.
[31] Issac N. Alexander to O. J. Fast, November 25, 1864, *OR*, vol. 44, 108.
[32] Andrew Moon to Sade, December 4, 1864, in Annette Tapert, ed., *The Brothers' War: Civil War Letters to Their Loved Ones from the Blue and Gray* (New York: Vintage, 1988), 227.
[33] W. W. Gist, "The Battle of Franklin," *Tennessee Historical Magazine* 6, 3 (1920): 233, 239.
[34] W. L. Truman to editor, May 17, 1908, *Confederate Veteran* Papers, Special Collections Library, Duke University, Durham, NC.
[35] John M. Orr to Henry A. Hale, March 30, 1865, *OR*, vol. 47, pt. 1, 943.
[36] L. P. Thomas, "Their Last Battle," *Atlanta Journal*, April 13, 1901.

at long range should never be permitted, since its lack of effectiveness often gives encouragement instead of causing demoralization, as a well-directed fire at short range is certain to do." The idea was repeated by at least some division commanders in the Army of Tennessee. On the other side of the war, John Henry Otto of the Twenty-first Wisconsin bluntly reported: "We always opened fire at eighty, or at least sixty yards and fired usually twice before they began to charge. But we always took care to have a pill for them when they came to[o] close to be comfortable." Otto's assertion that 60 yards was not an "uncomfortable" distance says a great deal about the ordinary soldier's willingness to fight at ranges that seem incredibly close to a modern reader.[61]

If we need proof that long-range firing was not only wildly inaccurate but of little concern to the average soldier, consider John William DeForest's story about receiving shots from a Rebel at a range of 880 yards. DeForest was riding along the Atchafalaya River in Louisiana when a Confederate picket decided to take a potshot at him from the other side of the stream. "I did not believe he could hit me, so let my horse keep on at a walk." DeForest could hear the bullet fly about 30 feet above his head. "I pointed upward to let him know where the ball had gone, and waved my hat to signify that I entertained no hard feelings."[62]

There are numerous instances on record of officers ordering their men to wait until the enemy came quite close before initiating fire—exactly how close depended, of course, on the individual. Commanders either specified an exact distance or referred vaguely to "short musket range" or to "a good hitting distance." Grant tried to restrain some troops from opening too soon on a group of Confederates who continued to stand defiantly although the rest of their comrades had retired from the field of Shiloh on April 7, 1862. He deliberately placed himself in front of the Federals "to prevent premature or long-range firing" and got out of the way only when they had advanced "to within musket-range." Lt. Col. Benjamin H. Bristow of the Twenty-fifth Kentucky saw enemy troops 400 to 500 yards away at Shiloh and thought the distance was "too great for effective firing." Lt. Col. Frederick C. Jones of the Twenty-fourth Ohio thought 50 to 75 yards an appropriate range at which to initiate fire, while Col. William T. Shaw of the Fourteenth Iowa preferred 30 paces. At Stones River, Francis T. Sherman could see the Rebels approach from a long distance but decided to wait until they were only 75 yards away. This staggered the attackers, giving Sherman's Eighty-eighth Illinois time to reload. The second volley stopped the Rebels, who lay prone and began to fire. Sherman shouted for his men "to rise and give them h—l," and they fired so heavily that the Confederates retreated.[63]

Sometimes officers noted that their restraint was not appreciated by the rank and file. Lt. Col. Hagerman Tripp of the Sixth Indiana ordered his men to wait until the Rebels were 100 yards away before opening fire. It was possible, due to the terrain, to have opened sooner, and indeed the company officers found it difficult to restrain the men from firing before the order was given. But this is a rare instance in the primary sources; much more often one finds regiments firing at much shorter ranges and the men quite happy to obey without a murmur. Benjamin Bristow reported that his Kentuckians "readily obeyed" the order to hold their fire at targets 400 to 500 yards away.[64]

Most common soldiers seem to have understood that they could kill more of the enemy at short rather than long range. The battle of Franklin was an instructive proof of this. It took place on one of the most open battlefields of the war, with 18,000 Confederates advancing across an undulating terrain mostly clear of woods. Pvt. Andrew Moon of the 104th Ohio noted that fallen Confederates littered the ground for 400 yards before the Union earthworks, indicating that at least some Federal units opened fire at a range that was commensurate with that of the rifle musket. But Benjamin Franklin Cheatham, who visited the field fifteen years after the war, told a newspaper correspondent that most of his men had been shot within 50 yards of the Union line. No matter how much fire might have been delivered at long range, most execution was done well within the traditional range of the smoothbore musket.[65]

Fire Effectiveness

B. P. Hughes wrote a groundbreaking study of fire effectiveness in selected battles from 1630 to 1850 for which there existed enough data to estimate how many rounds were fired by one side and how many casualties were suffered on the opposing side. His task was made difficult by the paucity of detailed data, and the situation is similar for the Civil War. We have reliable data on general casualty figures but precious little on how much a particular unit suffered during a single phase of a large engagement, and there is no consistently detailed information on how many rounds any unit fired during a large battle, much less for any particular phase of it.[66]

Maj. Gen. William S. Rosecrans tried to enhance the significance of his bloody victory at Stones River by estimating the fire effectiveness of the Army of the Cumberland. He reported detailed information on the number of rounds fired by his men and estimated Confederate casualties. "We hit 165 to their 100," he boasted, but Rosecrans used a high estimate of Rebel losses, 14,560. Modern historians more accurately place it at about 10,000. If we assume that about 20 percent of the casualties were caused by Union artillery fire, a generic average for the Civil War, then the 20,000 rounds that Rosecrans's artillery fired during the battle hit about 2,000 Confederates. His infantrymen fired 2 million rounds during the battle, hitting about 8,000 Confederates. In other words, it took 10 artillery rounds to effect a Rebel loss, and 250 musket rounds to take out one Confederate soldier.[67]

The Army of the Potomac sent 27,000 troops against the stone wall at the foot of Marye's Heights at Fredericksburg, on December 13, 1862, but they failed to carry the position even though it was held by only about 3,500 Confederates. The Federals lost 3,500 men, compared to 800 Rebel casualties. In terms of initial strength in this fight, the ratio was 7.5 to 1 in favor of the Unionists, but the loss ratio was 4.5 Federals compared to only 1 Confederate casualty. This must say something about fire effectiveness, although it is only fair to point out that the terrain over which the Yankees advanced was a nearly perfect killing ground for the defenders. Some Confederates fired more than 100 rounds, using muskets loaded by their comrades and passed forward to the front rank. Their "shoulders were kicked blue by the muskets and were sore for many days," reported a North Carolina man.[68]

Another concentrated killing ground lay in front of Confederate Fort Gregg, one of many works along the Petersburg line late in the war. In the final Union attacks, which occurred on April 2, 1865, it stood before the last line of defense protecting the western approaches to the city. The Confederates hastily put 150 men of the Twelfth and Sixteenth Mississippi into the fort, and 250 men in nearby Fort Whitworth. About 100 straggling infantrymen from other units were milling about inside the fort when the Mississippians entered, and most of them were too demoralized to offer much help. They were allowed to leave but forced to give up their guns, so the Mississippians had at least two rifles each.[69]

Four thousand Federals of the Twenty-fourth Corps attacked Fort Gregg that afternoon, advancing in three waves. Each wave managed to make it to the deep, water-filled ditch of the earthwork, but it took some time for the Yankees to pry their way over the parapet and around the flank of the little fort. Once enough of them entered, the result was a relatively swift victory, but one bought at great loss of life. The Confederates who survived were marched in front of the fort to be counted, and there they saw that the "slaughter was appalling," worse than at Fredericksburg, Chancellorsville, or the Wilderness in the minds of some who had seen all of those actions. "The dead were lying two hundred and three hundred yards in front of the fort, and increased in numbers as the fort was neared, until immediately at the fort it was simply fearful," wrote Capt. Archibald K. Jones of the Twelfth Mississippi. Jones saw the dead literally piled up, and Union engineer officer John Gross Barnard, who also viewed the field right after the engagement, wrote to his wife, "I never saw so many dead in so small a space lying just where they had fallen." Federal losses amounted to a little more than 700 men, while Confederate losses probably were about 250 at Fort Gregg and Fort Whitworth combined.[70]

The Civil War historian is left mostly with impressionistic evidence to gauge fire effectiveness. We do not know how many shots the defenders of Fort Gregg fired, but we can conclude that the hardened veterans of Harris's Mississippi brigade gave a good account of themselves in this desperate action to buy time for the doomed city of Petersburg. There were many other case studies of units pouring a concentrated fire of musketry, usually at short range, into an enemy formation, and most observers assumed it was an effective way to win a battle. The Sixth Indiana "fired 20 rounds so rapidly as to make a steady storm of musketry" on the second day at Shiloh, forcing advancing Rebel troops to fall back. In contrast, the Fifteenth Illinois fired ten to fifteen rounds in a concentrated fashion at Confederate troops who had captured part of a battery it was supporting on the first day at Shiloh, and failed to drive the enemy away.[71]

The manner in which infantrymen fired their weapons was a significant factor in affecting the outcome of an engagement, but it was not the only or even the most important. Moreover, the historian should be wary of relying only on impressionistic evidence to gauge the effectiveness of the rifle musket.

A good case study of this point can be found in the actions of the Seventy-sixth Ohio at Arkansas Post, January 11, 1862. The regiment advanced toward the modest line of trench that constituted the outer works of the river fort, moving 600 yards before it began to receive Confederate artillery fire at a range of 250 yards from the earthworks. Then, at a distance of 100 yards, the Seventy-sixth began to receive small-arms fire. The Ohioans continued another 25 yards before they began to falter and then come to a halt. Col. Charles R. Woods tried to get the men to continue, but they could not bring themselves to do so. He allowed them to stay put and concentrate their fire on the Rebels. Woods was satisfied with their performance, noting in his report that they silenced two artillery pieces and killed many horses of the Rebel battery. "Quite a number of men were seen to drop as if killed or wounded, but to what extent the enemy suffered from our fire I cannot tell."[72]

Other participants were not so shy in their claims for the regiment. The Seventy-sixth had received good Springfield rifle muskets only a month before, replacing its "old French muskets," and had been bolstered by 200 recruits to a total of 900 men. Lt. Charles Dana Miller, the regimental adjutant, reported that the crack shots crawled ahead to get closer range. They were the ones who shot the artillery horses. The rest of the regiment swept the parapet so "that the Rebels did not dare to show their heads, and they just put their arms over and fired at random." The Confederate fort surrendered after three hours of futile resistance. Miller saw the Rebel dead lying in the trench and four or five dead horses and the same number of gunners lying about the battery position.[73]

Miller was certainly impressed by his regiment's performance and by the sight of what he thought were many Confederate casualties littering the earthwork. But we need to exercise caution when evaluating one-sided evidence such as this, for there is always a counterview. Col. James Deshler commanded the Confederate brigade that opposed the Seventy-sixth Ohio. It consisted mostly of dismounted Texas cavalry regiments. "A large portion of my men were armed with double-barreled shot-guns, rifles of miscellaneous caliber, & c., there being only 315 Enfield rifles in the four regiments," which included a total of 1,500 men. Some of the Enfields were so old as to be unfit for service. Deshler's men had never seen combat. The artillery was Capt. William Hart's Arkansas Battery, which had some battle experience. Among the entire Confederate force defending Arkansas Post, only about half were armed with good rifle muskets; the rest used "old flint locks, shot guns and everything else," in the words of an interested Union soldier who saw the weapons after the surrender.[74]

Deshler reported that the Seventy-sixth Ohio "kept up a very heavy and unremitting fire with long-range rifles upon us" (without noting that the range at which it fired was about the same as for the smoothbore musket), but he asserted that his men "kept up a slow and deliberate but effective fire from our sharpshooters along the line and with marked effect." In fact, the Seventy-sixth Ohio lost sixty-eight men in the battle, lending credence to Deshler's report. Most likely, those few men in the Texas brigade who were armed with good Enfields were responsible for this Union loss. Deshler can easily be discounted when he reports that the Federals repeatedly attacked, and "his ranks seemed actually to wither under our fire." He also probably underreported his losses at twenty men for the entire engagement.[75]

Despite the posturing in the personal accounts, one can safely conclude that the fire of the Seventy-sixth Ohio was not as effective as its members wished to believe. Regimental commander Woods wisely refrained from claiming too much for his regiment's fire in his report. Even if it were demonstrably true that the Seventy-sixth devastated Deshler's brigade with its musketry, it could not be considered a major triumph for the rifle musket, given the poor weapons the Texans were forced to use. Moreover, even though Deshler strained credulity in some parts of his report, it is clear that Confederate losses were not so devastating as the Federals assumed. It is possible to argue that the Seventy-sixth Ohio dominated the musket battle on its sector of the Arkansas Post battlefield, but that was mostly due to the mismatch of weaponry between it and its opponents. The few good rifle muskets among the Texans exacted a surprisingly high casualty list in the regiment. While the Ohio unit had 900 good Springfields in action, Deshler had less than one-third that number of good Enfields to use in the battle, yet the Ohio unit was damaged more than a little at a range of 100 yards, even though it lay on the ground for protection.

At least one conclusion can be drawn from this case study: that we should be wary of assuming too much about the actual battlefield performance of a weapon simply because it had the potential to perform well. Combat is one of the most complex of human activities; a million factors can affect its course and outcome, and it is dangerous to assume that the level of military technology is the chief factor. Weapons are tools, and their level of effectiveness depends on how they are used. The human factor, in all its various manifestations, determines that aspect of battle.

5

THE ART OF SKIRMISHING

While the rifle musket had only a limited impact on the operations of the battle line, it had an important effect on more specialized tasks associated with Civil War combat. Anything that involved individual skill, such as skirmishing and sniping, was well suited to the enhanced capabilities of the new rifle. Skirmishing was a relatively new feature of combat, having come into its own only during the wars of the French Revolution some seventy years before the firing on Fort Sumter. Sniping, as we define it in the modern world, did not become prevalent until the Civil War itself. Neither activity became the centerpiece of battle, but both were important adjuncts to the operations of the battle line.

Civil War soldiers were fully aware of the importance of the skirmisher to modern combat. Maj. Gen. Lew Wallace referred to skirmishers as those "who are to the main body what antennae are to insects." Civil War contemporaries, however, often used the term *pickets* interchangeably with that of *skirmishers,* although technically there was a clear difference between them. Pickets constituted a line of sentinels for the main body, usually while lying in camp between battles. They were to give an alarm at the approach of danger and maintain the security of the main force by screening everyone passing nearby. Skirmishers were a battle force sent out from the main line when within range of the enemy during an engagement. They were usually more numerous than a line of pickets and geared for immediate action in support of the battle line. Of course, both pickets and skirmishers were, to a degree, interchangeable in their functions. That is why, especially in the West, soldiers often used the two terms interchangeably.[1]

There was a similarity between skirmishing and sniping as well, even though the two functions were quite different in most respects. Likewise, Civil War contemporaries often failed to differentiate between the two terms, all too frequently using the words *skirmishing* and *sharpshooting* to cover a wide range of activities. Skirmishing was performed by units, dispersed in open order but in a line, preceding and supporting the operations of the battle line. Sometimes skirmishers, if locked in a stationary position, could simulate the functions of snipers in that they harassed the enemy by firing from cover. But they still did this as a unit and as part of their job to protect the battle line. Snipers operated individually or in small groups, often using specialized, long-range target rifles. They had freedom to go anywhere and practice their stalking craft as they pleased. Modern sniping was a creation of the rifle; without the new technology, there was little possibility of an individual effectively harming an enemy beyond about 100 yards. The best sniper was not only adept at using the new weapon but often a loner with the skills of a hunter and the nerves to take life impersonally at a distance.[2]

This chapter is a discussion of the theory and art of skirmishing and the creation of specialized skirmish units in the Union and Confederate armies. A detailed examination of sniping in the Civil War follows in the next chapter.

Antecedents

The art of skirmishing developed slowly for many decades before the onset of the French Revolution in 1789. The imperial French army experimented with detaching men from the battle line to cover the front of a given formation, sometimes sending out entire companies of infantrymen. The functions of a skirmisher often duplicated that of a light infantryman, although they were two different types of soldiers. The Austrian army specialized in the development and use of light infantry, troops who were trained to range across the countryside, often armed with a lighter musket, to interdict enemy transportation and communications and to gather intelligence. Trained in loose-order tactics, similar to those skirmishers employed, light infantrymen could in fact make good skirmishers when their operations were tied to the needs of a battle line during an engagement. But light infantrymen normally were not wasted in exposed positions in front of the main force; they contributed their best when allowed to roam independently while the battle line detached its own units to do the job of skirmishing.[3]

Because the tactics employed by light infantry and skirmishers were so similar, observers often used the two terms interchangeably. In fact, after the Seven Years' War, many French writers began to call for all line infantry to receive training in loose-order tactics so they could be employed in a variety of roles as needed. George Washington's army during the American Revolution organized impressive units of light infantry armed with rifles, and it is possible that European observers learned from this experience to push for a more important role to be played by skirmishers in their own national armies. The French used the term *tirailleurs* to denote soldiers who fought with more freedom of movement on the skirmish line than was typical of the strictly ordered actions of the battle line. During the early period of the French Revolutionary Wars, the French fielded a mass of poorly trained skirmishers in front of their battle lines. On some occasions, this mass operated with little discipline or order, swarming in waves against enemy positions. Their spirited firing and unpredictable movements often intimidated opponents, but eventually the French came to add discipline and order to the chaos. Before the end of the Napoleonic Wars, the French and many of their enemies had developed compact, well-disciplined bodies of skirmishers who adhered to systematized methods of covering the larger formations, disrupting enemy positions, finding weak spots in the enemy position, and sending back a flow of information on terrain and troop placement to the main line. Civil War soldiers would have been very familiar with the operations of the Napoleonic skirmish line.[4]

American observers continued to believe in the old European concept of light infantry into the early nineteenth century. William Duane, author of the first true military treatise in the United States, listed ten duties normally assigned to light infantrymen as he clarified the differences between them and skirmishers. He wanted skirmishing to be performed by companies armed with rifles, sending out half their men while retaining the rest as a reserve. Duane believed that skirmishers should be positioned 120 paces in front of the battle line with the reserve 60 paces behind.[5]

The concept of light infantry had completely disappeared by the time of the Civil War, but skirmishing remained an important aspect of military operations. In fact, Paddy Griffith has asserted that "each successive American infantry manual from 1812 to 1867 placed a greater emphasis than its predecessor on skirmishing." Methods for deploying skirmishers had evolved into a drill that was more intricate than that of the main battle line. Skirmishers operated in dispersed, open-order formations, demanding more individual self-assurance of the soldiers and officers alike. Lacking the assurance of shoulder-to-shoulder formations, skirmishers had to learn how to keep a formation while separated from their neighbors by several yards. They also had to maneuver around the natural and man-made impediments that dotted every battlefield, such as rocks, trees, and buildings, without losing contact with their comrades. Skirmishers were an exposed advanced guard, operating within striking distance of enemy skirmishers and vulnerable to attack by the opposing battle line as well. Yet they played a vitally important role in protecting the battle line by giving it advance warning of enemy movements and were expected to resist any pressure by the opposing side short of a full-scale attack.[6]

Brig. Gen. Silas Casey wrote the standard drill manual adopted by the Federal army in 1862, duplicating much that was found in Hardee's drill manual, originally published in 1855. He noted that open-order tactics were needed to move skirmishers "in any direction with the greatest promptitude." Casey did not expect all troops to conduct their skirmish drill "with the same precision as in closed ranks, nor is it desirable, as such exactness would materially interfere with their prompt execution." Casey differed from Duane in that he suggested placing the skirmish line much farther away from the battle line, and he wanted commanders to deploy larger reserves for the skirmishers. In fact, Casey suggested that the skirmish line ought to be supported by two, not one, reserve forces, and it could rely on the battle line itself as a reserve if circumstances forced the skirmish line commander to place his men very close to the main force.[7]

Casey allowed skirmishers to carry their weapons "in the manner most convenient to them." He urged them to practice loading and firing while "kneeling, lying down, and sitting, and much liberty should be allowed in these exercises, in order that they may be executed in the manner found to be most convenient." He cautioned skirmishers that, while loading on the move, they should always keep the barrel vertical while ramming and pause a second while priming the gun. Casey also urged skirmishers to take aim at a target while firing, trying to properly estimate distance beforehand.[8]

A number of drill manual authors reinforced what Casey explained. Col. James Monroe, a former captain in the regulars and commander of the Twenty-second New York State Militia, published a company manual that included skirmish drill. Monroe emphasized the need for the skirmisher to take full advantage of cover on the battlefield and wisely advised him to reload before moving away from cover. He suggested that tapping the butt of the musket on the ground "for an instant, before ramming" would hasten the process of reloading. Monroe stressed that a good skirmisher needed lots of practice at firing long range.[9]

Civil War soldiers possessed a great deal of instructional material to guide them on how to skirmish, but the level at which any unit practiced the specialized drill varied according to the commander. The Sixty-seventh Ohio not only practiced skirmish drill but did so across rugged terrain in the mountains of northwestern Virginia. In contrast, a modern historian has concluded that the Army of the Ohio did not begin training in skirmish drill until the fall of 1862.[10]

Whether they drilled early or not, most Civil War soldiers came to fully appreciate the demands placed on good skirmishers as well as the need to drill them rigorously. Many soldiers also came to realize that skirmishing demanded particular characteristics of those who wished to excel in the art. William H. Chamberlin of the Eighty-first Ohio noted that a skirmisher needed to be good "in the exercise of skill, endurance, ingenuity, and audacity." On the skirmish line, "a man literally offered himself to death. Almost every skirmisher was a target." Yet there were opportunities as well as dangers on the skirmish line. Capt. John William DeForest of the Twelfth Connecticut pointed out that skirmishers usually could find and take advantage of cover, and that they had the opportunity of shooting at their own discretion. "Now to fire at a person who is firing at you is somehow wonderfully consolatory and sustaining; more than that, it is exciting and produces in you the so-called joy of battle."[11]

Many observers praised the emotional skills of a good skirmisher, noting that he possessed "nerve and reckless daring," "peculiar courage," and "sagacity." Wilbur Fisk believed that the Vermont brigade commanded by Brig. Gen. Lewis A. Grant was all too often called upon for skirmish duty, but he took pride in its accomplishments. "With a good General to lead, we can string out the whole brigade of us, in a line five feet apart, or ten if they want; and advance straight ahead without pulling apart here, or crowding together there, keeping a straight line and going straight ahead, until we come upon the enemy's works, and find out all we wish to know."[12]

Staff officer Francis A. Walker thought Brig. Gen. Francis C. Barlow's First Division, Second Corps had also made a special effort to perform skirmish duty more effectively than any other unit. Walker disparaged the skirmish skill of most units. "It is a melancholy fact that three men out of four who entered the service of the United States left it, if alive, without ever having seen a really good piece of work of this character. Indeed, most regiments in the service had as little idea of skirmishing as an elephant." Barlow's division, however, had taken to it well. For Barlow's men,

> the very life of military service was in a widely extended formation, flexible yet firm, where the soldiers were thrown largely on their individual resources, but remained in a high degree under the control of the resolute, sagacious, keen-eyed officers, who urged them forward, or drew them back, as the exigency of the case required; where every advantage was taken of the nature of the ground, of fences, trees, stones, and prostrate logs; where manhood rose to its maximum and mechanism sank to its minimum, and where almost anything seemed possible to vigilance, audacity, and cool self-possession.[13]

Whether an entire division could become expert at skirmishing is doubtful, but there is no question that smaller units and individuals were well suited to this specialized task. The density and frontage of a skirmish line could vary widely. Casey cited the example of one company covering the front of a parent battalion and another example of an entire battalion deploying on a wider front. Casey, like Hardee before him, assumed that any and all companies on the battle line could do skirmish duty now that the rifle musket was standard issue. Early in the war, the tendency was for commanders to follow the manual and send out two companies to cover the regimental front. Historians have noted a change as the war progressed; by 1864, commanders tended toward sending out more units to the skirmish line. This was due to reduced numbers of troops but also, some historians argue, to a greater faith in loose-order tactics as a way to lessen casualties.[14]

An even more important reason for the increased numbers of troops sent out to skirmish probably lies in a practical admiration for the skirmish line, a growing awareness that it had the potential to be more useful on the battlefield than anyone could have anticipated before the war. By the end of the conflict,

Civil War commanders North and South were exploiting those advantages to the full. In fact, the American Civil War represents the apogee of the skirmisher in world military history. Never before and never after would the skirmish line be used so vigorously and consistently as in the conflict that raged in America from 1861 to 1865.

Federal Skirmish Units

The rifle corps of the Revolutionary War set a precedent for the creation of specialized skirmish units in the Civil War, although George Washington had created those Continental army units by detailing men from preexisting regiments well after the start of the war. Civil War armies were unique in that they began organizing specialized skirmish units from the start of the conflict, inspired in part by prewar enthusiasm for the rifle and a desire to create units that could maximize the potential of the new weapon. Ironically, even though all of these units were armed with different kinds of rifle muskets, most of them were never used in a dedicated skirmish role. They simply were put into the mainstream of army units and more often held a part of the battle line than skirmished.

The number of units that adopted words like "Sharpshooters" or "Rifles" as their nickname indicates how widely men wanted to be part of an elite unit of skirmishers. The members of the Sixty-fourth Illinois hoped for breech-loading rifles and called themselves the Yates Sharpshooters, after Gov. Richard Yates. The Sixty-sixth Illinois called itself the Western Sharpshooters and received what someone called "the Demmick American deer and target rifle" upon its organization. Neither regiment was officially known by its nickname, and neither was used as a specialized skirmish unit.[15]

Initially, the Federal government hesitated to accept specially organized units of marksmen. A company of "hunters and practical marksmen" had been recruited from the mountainous section of western Massachusetts in the spring of 1862. They were "anxious to be accepted . . . as sharpshooters and be armed with target rifles." But officials in Washington refused the recruits, telling Gov. John Andrew that the men would have to join an ordinary volunteer regiment and accept whatever weapons were issued to them. By the early fall of 1862, the government was more open to accepting specialized skirmish units. The secretary of war Edwin M. Stanton told Gov. William Sprague of Rhode Island that sharpshooters "are regarded as a very important arm of the service, especially

Table 5.1. Federal independent skirmish units

Designation	Dates in Service	Subunits
1st Battalion, Maine Sharpshooters	October 1864–April 1865	Companies A, B, C, D, E, F
1st Company Sharpshooters, Massachusetts Volunteers	September 1861–September 1864	None
2nd Company Sharpshooters, Massachusetts Volunteers	October 1861–October 1864	None
Hall's Independent Battalion, Michigan Sharpshooters	November 1864–February 1865	Companies A, B
1st Michigan Sharpshooters	November 1862–June 1865	10 companies
1st Battalion, New York Sharpshooters	September 1862–June 1865	6th, 7th, 8th, and 9th Companies
5th Independent Company, Ohio Sharpshooters	December 1862–June 1865	None
6th Independent Company, Ohio Sharpshooters	December 1862–June 1865	None
7th Independent Company, Ohio Sharpshooters	January 1863–June 1865	None
8th Independent Company, Ohio Sharpshooters	March 1863–June 1865	None
1st United States Sharpshooters	August 1861–November 1864	10 companies
2nd United States Sharpshooters	October 1861–January 1865	Companies A, B, C, D, E, F, G, H

Source: Record of Events, in *Supplement to the Official Records of the Union and Confederate Armies,* 100 vols. (Wilmington, NC: Broadfoot, 1993–2000), pt. 2, vol. 25, 166–169; vol. 27, 449–452, 452–457; vol. 30, 420–421, 421–437; vol. 42, 492–498; vol. 50, 529–531, 532–533, 533–535, 536–538; vol. 79, 303–332, 332–363.

for the protection of batteries." General in Chief Henry W. Halleck thought they were best organized into independent companies "and employed as circumstances shall require." Stanton issued a general order specifying that recruits entering such companies had to demonstrate their skill at target practice. He set precise standards and mandated that a certificate proving the result had "to be written on the target used at the test."[16]

There were only twelve units in the Union army officially designated as sharpshooters (five regiments, three battalions, and four companies). A clerk in the First Michigan Sharpshooters sadly noted in late 1864 that his regiment, "though organized and raised as sharpshooters, . . . is armed with Springfield rifles and has been used as infantry of the line ever since it joined the Army." Col. Charles V. DeLand tried unsuccessfully to obtain Sharps and Henry rifles for the First Michigan. He recruited his regiment according to Stanton's regulations and had many crack shots in the ranks. Pvt. William J. Ross refused to accept the Springfield, protesting that he was not a common infantryman, and was sentenced to fourteen days in the guardhouse for his insubordination.[17]

Only two of the twelve units officially designated as sharpshooters managed to fulfill their aspirations, and both were raised by the Federal government. In short, they were Federal volunteers rather than state volunteers, and that probably had a great deal to do with the special attention paid them. The man who organized the two regiments, Hiram Berdan, also played a large role in keeping his regiments uppermost in the mind of government officials and the public. Berdan wanted his initial regiment, which was designated the First United States Sharpshooters, to be recruited from across the Northern states and armed with "the most reliable rifle made," in the words of one of his officers. Proficiency at target shooting was a requirement for admission into the regiment. Aspiring recruits had to place ten consecutive rounds into a target 200 yards away so that the average distance from the center did not exceed 5 inches. What was termed the "string measurement" of the ten shots could not be more than 50 inches. One of the best performances in this was that of Charles H. Townsend of Wisconsin, who placed five rounds with a string measurement of only 3.5 inches. The average of all recruits was close to 20 inches per ten rounds, less than half the leeway that was allowed.[18]

Berdan, a mechanical engineer from New York State, led the first regiment as colonel. He assembled his new command in September 1861 with men recruited from five states. Berdan adopted a uniform of "dark green coat and cap with black plume, light blue trowsers (afterward exchanged for green ones) and leather leggins." He wanted Sharps rifles for his men and had to fight resistance on the part of the Washington bureaucracy to get them. Before the Sharps were ready, both regiments were temporarily armed with Colt's revolving rifles. Berdan himself was a crack shot, demonstrating his skill for the regiment in the Camp of Instruction near Washington during the fall and winter of 1861–1862. Two companies were armed with James target rifles, equipped with a four-power telescopic sight. It had a range of 500 yards. Once, when Lincoln and McClellan visited the regiment, 100 men in a trench fired one round each at two man-sized targets 600 yards away, scoring hits every time. Berdan actually hit the eye of a man-sized figure (labeled as Jefferson Davis) at that distance, greatly impressing the spectators. Target competitions among the men drew interest both for show and for the $5 prize that was offered. The best long-range shooting was done by men using the James rifles. These weapons were superb for stationary fighting, as in fortified positions, but too cumbersome for skirmishing across rugged terrain.[19]

Berdan was as careful in training his men to perform skirmish drill as he was in honing their marksmanship. He followed the drill manuals exactly in using bugle signals to maneuver them. Berdan put the Sharpshooters through physical conditioning by forcing them to cross a "high board fence" that "required a good jump and a secure hold on top."[20]

The two regiments took the field in the spring of 1862, the First United States Sharpshooters assigned to Brig. Gen. Fitz John Porter's Fifth Corps, Army of the Potomac. It gained notoriety for the sniping and skirmishing done by its members at Yorktown during May, but after that the regiment served both as line infantry and as skirmishers like most other infantry units. Both regiments gradually faded into obscurity as their roles changed from specialized skirmish outfits to that fulfilled by any other regiment, despite the accumulation of shooting talent concentrated in each unit. Still, the spirit of competition remained high among the men. After the battle of Fredericksburg, a "Grand Shooting Match" was held, with a silver medal awarded to the best shot in each regiment and a gold medal given to the better of the two. Competitors did five "off-hand shots at 100 yards, open sight," with Sharps rifles, and a member of the first regiment won the grand prize. Berdan's Sharpshooters were mustered out in stages by companies beginning in August 1864. By February, just weeks before the war ended, both regiments ceased to exist. Two Massachusetts sharpshooting companies originally formed in 1861 for Berdan were attached to Massachusetts infantry regiments instead, armed mostly with telescopic rifles.[21]

Berdan's two regiments were in a class of their own because of their high level of skirmish training and long-range target practice. They were impressive units in the makeup of their personnel, all of whom were proven marksmen of a high order. But the United States Sharpshooters were never consistently used as dedicated skirmishers because skirmishing could be performed by any number of regiments armed with rifle muskets. Even though some units knew the skirmish drill better than others, commanders found it difficult to dedicate entire regiments to that task alone.

By 1864, the Confederate Army of Northern Virginia successfully organized smaller, battalion-sized units of dedicated skirmishers. The only evidence that the Federals did this is contained in general orders issued by divisions of black regiments serving in the Eighteenth and Twenty-fifth Corps of the Army of the James. Beginning in June 1864 and extending through February 1865, detachments of dedicated skirmishers were drawn from the Fifth, Sixth, Eighth, Twenty-second, 107th, 116th, and 127th United States Colored Troops. These detachments consisted of 100 men and three officers, but there is precious little evidence to document their activities.[22]

Confederate Sharpshooters

Many Southern officials understood the importance of selecting the right kind of man armed with the right kind of weapon. Pres. Jefferson Davis advised Lt. Gen. Leonidas Polk to accept recruits who came to the army with "their hunting-rifles; they will make your best skirmishers if properly organized and commanded." The Confederate government encouraged the creation of many specialized units for skirmishing. On April 21, 1862, at the same time that it instituted conscription, the Confederate Congress passed a resolution authorizing the army to create battalions of sharpshooters. The resolution specified that no more than one battalion of three to six companies would be allowed for each brigade, but it was vague on other matters. Congress indicated that the battalions could be created either by detailing men from preexisting regiments (as was done in the Revolution) "or otherwise" (presumably by recruiting men from civilian life). While officers had to be appointed by the president and approved by the Senate (suggesting an intention to create regiments and battalions that were independent of preexisting units), the act also specified that the battalions "shall constitute parts of the brigades to which they belong." In short, they were administratively attached to their parent units. Congress further specified that the battalions were to be "armed with long-range muskets or rifles."[23]

The Confederate War Department authorized commanders to act on the resolution by May 3, 1862. The ambiguities in the congressional resolution gave rise to two different kinds of sharpshooter units, and both were organized simultaneously in 1862. Independent units, organized from scratch like Berdan's two regiments in the North but under the authority of state governments, began to crop up throughout the year. In addition, several field army commanders encouraged the creation of sharpshooter battalions by detailing men from preexisting brigades. These battalions did not have independent status but remained attached to their parent units.[24]

The Confederates organized twenty-three independent sharpshooter units; three were regiments but the rest were battalions. Almost every state in the Confederacy (plus Missouri, which never formally seceded) contributed this type of sharpshooter unit. Most of them were organized in 1862, but four were put together the following year. After that, the urge to create new, independent sharpshooter units virtually disappeared. The battalions consisted of from two to six companies each. Some of the members were drawn from preexisting infantry regiments or artillery batteries and in one case from a pioneer company, but most were recruited or conscripted directly from the civilian population. In the case of Maj. John E. Austin's Fourteenth Louisiana Battalion and the First Georgia Battalion Sharpshooters, there was some effort to handpick good marksmen and give them special training, but that seems to have been comparatively rare. The sharpshooters were generally armed with the same kind of rifle muskets as infantry units. Some of the units were broken up well before the end of the war, probably because all were small to begin with and attrition, combined with the difficulty of finding replacements, doomed their futures.[25]

The quality of these independent sharpshooter units seems to have been at least moderately high. Because they were so small there are relatively few personal accounts by surviving members, and only now and then do these units appear in the records. Company A of the Fifteenth Mississippi Battalion had initially been organized as pioneers in the Army of the Mississippi at Corinth during May 1862. In fact, the company continued to act as pioneers for two months while converting to sharpshooters, practicing skirmish drill between bouts of digging wells and working on roads. "The men comprehend the skirmish drill very readily," asserted a member, going on to state that the company "is well drilled and well armed and well disciplined and ready to do very efficient service as sharpshooters."[26]

There is no evidence to suggest that any independent Confederate sharpshooter unit approached the level of training or marksmanship seen in Berdan's two regiments. Instead, they all shared the same fate as the other independent sharpshooter units in the Union army, serving in the dual role of line infantry and occasional skirmisher. The units simply merged into the mainstream order of battle, assigned to brigades and fighting like any other state-sponsored unit with the same type of weapons. A member of the Ninth Mississippi Battalion put it well when he wrote that his comrades had "fought as skirmishers and in

Table 5.2. Confederate independent skirmish units

Designation	Dates in Service	Subunits
17th Battalion, Alabama Sharpshooters	June 1862–November 1863	Companies A, B[1]
23rd Battalion, Alabama Sharpshooters	December 1863–February 1865	Companies E, F, G[1]
12th Battalion, Arkansas Sharpshooters	January–June 1863	Companies A, B, C, D[1]
Stirman's 1st Arkansas Sharpshooters	June 1862–July 1863	10 companies[2]
1st Battalion, Georgia Sharpshooters	July 1862–August 1864	Companies A, B, C, D[1]
2nd Battalion, Georgia Sharpshooters	June 1862–August 1864	Companies A, B, C, D, E[1]
3rd Battalion, Georgia Sharpshooters	June 1863–January 1865	Companies A, B, C, D, E, F[1]
4th Battalion, Georgia Sharpshooters	May 1863–October 1864	Companies A, B, C[1]
1st Georgia Sharpshooters	Summer 1862–April 1865	Companies A, B, C, D[3]
14th Battalion, Louisiana Sharpshooters	July 1862–April 1865	Companies A, B[1]
15th Battalion, Louisiana Sharpshooters	July 1864–May 1865	5 companies[2]
1st Battalion, Mississippi Sharpshooters	October 1861–February 1865	Companies A, B, C, D[1]
9th Battalion, Mississippi Sharpshooters	June 1862–August 1864	Companies A, B, C[1]
15th Battalion, Mississippi Sharpshooters	June 1862–December 1863	Companies A, B[1]
9th Battalion, Missouri Sharpshooters	October–November 1862 (incomplete data)	Companies A, B, C[1]
Searcy's Battalion, Missouri Sharpshooters	Data not available	Data not available[1]
1st Battalion, North Carolina Sharpshooters	April 1862–April 1865	2 companies[4]
1st Battalion, South Carolina Sharpshooters	August 1862–October 1863	Companies A, B, C[1]
2nd Battalion, South Carolina Sharpshooters	July–October 1862	Companies A, B, C[1]

(*continues*)

Table 5.2. (*continued*)

Designation	Dates in Service	Subunits
Palmetto Sharpshooters, South Carolina	April 1862–April 1865	10 companies[4]
24th Battalion, Tennessee Sharpshooters	September 1861–April 1864	Companies A, B, C[1]
1st Battalion, Texas Sharpshooters	May 1862–October 1863	Companies A, B, C, D, E[1]
30th Battalion, Virginia Sharpshooters	September 1862–March 1865	Companies A, B, C, D, E, F[5]

[1] Record of Events, in *Supplement to the Official Records of the Union and Confederate Armies,* 100 vols. (Wilmington, NC: Broadfoot, 1993–2000), pt. 2, vol. 1, 490–493, 553–558; vol. 2, 483–485; vol. 6, 69–77, 112–116, 158–163, 193–200; vol. 24, 327–333; vol. 32, 647–659; vol. 33, 162–175, 385–391; vol. 38, 594, 658–659; vol. 64, 396–399, 460–461; vol. 67, 21–28; vol. 68, 527–536; vol. 67, 21–28; vol. 68, 527–536.
[2] Stewart Sifakis, *Compendium of the Confederate Armies,* 10 vols. (New York: Facts on File, 1991), 4: 125–126; "Col. Erasmus I. Stirman," *Confederate Veteran* 22 (1914): 226; Ras Stirman to sister, September 24, 1862, in Pat Carr, ed., *In Fine Spirits: The Civil War Letters of Ras Stirman* (Fayetteville, AR: Washington County Historical Society, 1986), 48; Joseph J. Crute Jr., *Units of the Confederate States Army* (Midlothian, VA: Derwent, 1987), 41–42.
[3] Russell K. Brown, *"Our Connection with Savannah": History of the First Battalion Georgia Sharpshooters, 1861–1865* (Macon, GA: Mercer University Press, 2004), 1–5.
[4] Sifakis, *Compendium of the Confederate Armies, Louisiana,* 7: 98, 5: 78–79, 10: 110–111.
[5] Record of Events, in *Supplement to the Official Records,* pt. 2, vol. 72, 153–155; Crute, *Units of the Confederate States Army,* 377.

line of battle. It has acted as provost guard and as scouts at various times and places[,] at other times as flankers, all of which has been done cheerfully and to the best of their ability."[27]

The other type of unit to emerge from the congressional resolution of April 1862 were attached sharpshooter battalions, which were created wholly by detailing men from preexisting regiments and administratively linked to their parent brigades. Officers commanding these units retained commissions in their original regiments, and thus government officials in Richmond and the state capitals were not involved in commissioning them for this duty. These attached units were organized more easily, for there was a captive manpower pool to draw from, and officers could handpick the best marksmen from the ranks. Despite these advantages, the process of organizing attached units was slow and sporadic during 1862–1863. Only in Lee's Army of Northern Virginia, during the early months of 1864, were such units apparently created in a uniform fashion for an entire field army.

In the West, Gen. Braxton Bragg ordered in April 1863 that one regiment of every brigade and two companies of every regiment in the Army of Tennessee should thoroughly practice the skirmish drill. But this fell very far short of creating dedicated skirmish units. We know that some brigades in the Army of Tennessee did create such units. Maj. Gen. Patrick R. Cleburne organized sharpshooters in his division by early 1863, training them in judging distance and dealing with different kinds of terrain. Dedicated skirmish detachments among several units of Bragg's army were mentioned in official reports regarding the battle of Chickamauga in September 1863. Gen. Joseph E. Johnston urged the creation of more attached sharpshooter units in the Army of Tennessee during the winter of 1864. These battalions were exempt from normal camp duties and drill but received specialized training in skirmishing. Even so, not all brigades or divisions in Johnston's army created these units, and the combat record of those that did exist is shrouded in mystery. There apparently was no real commitment to making the sharpshooter battalions a reality in the Confederacy's main field army in the West.[28]

The main Confederate army in the East, however, began to experiment with attached skirmish units as early as April 1862, just a few weeks before Congress authorized their creation. After the Seven Days' campaign, Lee urged the creation of more battalions for a few brigades, but it is unclear why he was content to leave the implementation of this recommendation up to his subordinate officers. Among the most aggressive in forming skirmish battalions was Maj. Gen. Robert Rodes, who may have been the first to create a division-level organization of sharpshooters, in the early months of 1863. Three other division commanders in Lee's army tried to follow his example. Rodes's scheme was to combine the brigade-level battalions for unified action when necessary.[29]

Maj. Eugene Blackford of the Fifth Alabama was delighted to take command of the battalion attached to Rodes's original brigade in January 1863. He admitted the job was dangerous work, but it also presented a marvelous opportunity for distinction. Blackford had a low opinion of the performance of line units on the skirmish line, claiming that "as a general rule Skirmishers have had a poor chance, being brushed off the field early in the action." By intensively training his handpicked men with bugle calls, he hoped to show what a dedicated corps of skirmishers could do.[30]

Brig. Gen. William T. Wofford encountered stubborn resistance among the members of Cobb's Georgia Legion when he tried to organize a battalion in

April 1863. The Georgians refused to volunteer, but Wofford insisted on detailing soldiers to the battalion anyway. "It is mean to take men from the company of their choice against their wishes and put them in a company whose officers they do not like," commented Samuel A. Burney. "The feeling against it in the Brigade is bitter, and if Wofford organizes the battalion he will do it in the face of universal opposition." Wofford was "more stubborn than human," according to Burney, and "there is no escape."[31]

The sharpshooter battalions were detachments of their parent organizations and thus had no administrative history of their own. The battalions often were created and then disbanded for unknown reasons, only to be re-created again, often when a commander suddenly took an interest in them. Even in Rodes's Alabama brigade, a general order went out after the battle of Chancellorsville to create a battalion, as if one had never existed before. The order specified that one-twelfth of the men were to be detailed to the battalion, amounting to twenty-five men from the Third, Fifth, and Sixth Alabama and fifteen men from the Twelfth and Twenty-sixth Alabama. Some of Lee's officers tried to create a corps-level sharpshooter unit after Chancellorsville, although there is no evidence that they succeeded. Organizational efforts picked up before and after Gettysburg, with more commanders creating sharpshooter battalions. Brig. Gen. Junius Daniel shifted his brigade from North Carolina to join Lee's army in June and received orders to form a battalion, detailing one man out of six for duty. Some of the battalions organized in early 1863 were disbanded only a few months later.[32]

The winter of 1864 marked a watershed in the history of Lee's sharpshooter battalions. For reasons that are unclear, an order went out to create one for each brigade. Cadmus M. Wilcox later claimed credit for this development. One of the chief rifle advocates before the war, Wilcox apparently petitioned Lee to move forward on this innovation. Wilcox also arranged to have a portion of his 1859 book reprinted in Richmond so that his detailed instructions for estimating distance could be available for training. Another contender for credit is Rodes, who had experimented with full-scale mobilization of sharpshooters in his division a year earlier. Lee never spoke or wrote on this subject, and ironically his order implementing the plan has never been found.[33]

By late April 1864, Lee's directive seems to have produced a battalion for each brigade in the Army of Northern Virginia. The status of these battalions was a bit anomalous. The members were excused from routine camp duty and general drilling. They slept in separate groups and held separate roll calls, yet they received their rations and equipment from the regular brigade channels. At best, they could be called semi-independent, although they held no separate status in the order of battle. The number of men as well as the number of companies in each battalion varied. Some brigade commanders detailed one man for every ten on duty, others averaged four men per company. Some battalions had three companies, some four, and others five or six. Brig. Gen. Edward L. Thomas's battalion consisted of 160 men, divided into four companies. In Brig. Gen. Alfred M. Scales's North Carolina brigade, also in Wilcox's Division, the battalion consisted of one commander, eight commissioned officers, ten noncommissioned officers, 160 privates, four scouts, and two buglers. It also was divided into four companies. The commander reported to Scales, and the sharpshooters ate and slept in squads of four men each.[34]

Historians have assumed that each of the thirty-four brigades in Lee's army created a sharpshooter battalion, even though there is no evidence to prove it. At full strength, that could mean anywhere from 4,000 to 7,000 dedicated skirmishers for an army of 65,000 men. Tactical manuals assumed that one or two companies per regiment would do the skirmishing when needed; based on that proportion, Lee would need between 6,500 and 13,000 skirmishers.[35]

The sharpshooters preferred the short version of the English Enfield rifle because it was "lighter and handier" than the version with the longer barrel. In fact, to supply the battalions, short Enfields were sometimes taken from men in the parent unit and given to the sharpshooters. Most battalions eventually came into the possession of a few long-range target rifles, although these were very scarce in the Confederacy. The most common type was the English Whitworth, imported through the blockade. The battalions of both Brig. Gen. John B. Gordon's Georgia brigade and Brig. Gen. William Mahone's Virginia brigade had only two Whitworths each. They were equipped with globe sights and were used "on special occasions" by the best shots in each battalion—to pick off officers and artillerymen. In short, a few men in each battalion served as modern snipers.[36]

Officers put a lot of effort into the selection process while organizing these battalions, for the purpose of the sharpshooters would be nullified if the ranks were filled with inept marksmen or men constitutionally unfit for the dangerous work of skirmishing. In some units, such as the Forty-second Mississippi of Col. John M. Stone's brigade, the entire command was assembled and the purpose explained. Volunteers were called for, and 90 men stepped forward out of the 240 who were present. Capt. Robert F. Ward later admitted that "the desire to avoid the various duties from which sharpshooters were to be exempt influenced many of the ninety volunteers," but they were soon weeded out and only 24 were selected. In Thomas's brigade, officers selected men only if more than the allotted number volunteered for sharpshooter duty. Brig. Gen. Nathaniel H. Harris put prospective candidates through daily target practice to weed out the poor shots. Jerome B. Yates of the Sixteenth Mississippi had always believed he was the best shot in his company, but this regimen proved that he was mistaken. "We all do very badly shooting," he admitted. The daily practice in Mahone's battalion forced some men to go back to their original units every day. Wilcox's Division, to which Harris and Mahone belonged, seems to have used shooting far more intensively as a qualification for recruiting the battalions than any other division.[37]

Selection of officers was even more important than selecting the men, for the effectiveness of specialized units such as these rested heavily on the commanders. Much of the selection process pivoted on personal observation by superiors, their assessment of the grit and determination of the subordinates. Col. William A. Feeney of the Forty-second Mississippi selected Robert F. Ward for a responsible position in Stone's sharpshooter battalion because he had known Ward ever since the two served as lieutenant and sergeant in the Ninth Mississippi. "He had doubtless observed then my fondness for skirmish drill," Ward later commented of his superior. Maj. James M. Crow of the Ninth Alabama and Lt. Elias Davis of the Tenth Alabama had developed a mutual respect for each other during the battle of Antietam, and months later Crow selected Davis for an officer's slot in Brig. Gen. Abner Perrin's sharpshooter battalion.[38]

The battalions engaged in an intensive program of training before the opening of the spring campaign in 1864. They practiced the skirmish drill using bugle signals and engaged in bayonet exercises. The battalions attached to Wilcox's Division used the brochure printed by their commander as a drill manual. Stone's sharpshooters drilled six hours each day. In Thomas's battalion, the men created intervals of 8 paces and counted off from right to left to create groups of four. Combining four groups into a section, their commander could maneuver the unit by group, section, or battalion, adding another layer of flexibility to the drill. Adhering to the skirmish training to be found in tactical manuals of the day, the sharpshooters "were instructed in drill to take advantage of every tree, stump, or inequality in the surface, and were not required to preserve perfect alignment when moving to the front." William Dunlop praised their training further by asserting that the sharpshooters were "instructed in all the rules of rapid and extraordinary formation" and "were equipped for the conflict, whether in field or forest, or in the streets and lanes of a town or city."[39]

For many of Lee's sharpshooter battalions, the first drill involved practice at estimating distance. These veterans already had some familiarity with open-order tactics and marksmanship, but few had been exposed to scientific methods of ascertaining distances under field conditions. In McGowan's battalion, a man or an object about the size of a man was placed at varying distances, and each member of the unit was required to guess how far away it stood. "A few . . . were naturally and hopelessly deficient in their powers of estimating distance, and hence were exchanged for others." Thomas's battalion spent a month intensively training to gauge distances. The men were repeatedly told to describe what part of a man's dress or appearance they could discern at different distances. After a soldier gave his estimate, a tape measure was brought out to determine how close he came to the actual distance. In Stone's battalion, the target to be examined was placed from 50 to 800 yards away. Officers took into account that at the longer distances weather conditions could alter the men's perception.[40]

The sharpshooters engaged in intense target practice, with Wilcox's battalions firing three rounds at 100 yards, five at 300 yards, ten at 600 yards, and four at 900 yards. Wilcox allowed them to use a rest for the gun barrel and forbade firing at 600 and 900 yards in windy weather. The ranges varied from 50 to 1,000 yards among all the battalions in Lee's army. Sometimes "offhand" shooting was encouraged. Thomas's battalion fired at a plank 6 feet long and 1 foot wide at 100 yards' distance. Instructors painted a "three inch ring in the center." Bigger planks with larger rings were used at longer distances, so that at 900 yards the board was 16 feet wide, 6 feet tall, and with a ring 3 feet in diameter. The ring had a series of subrings 3 inches wide at the smallest. At 900 yards, the men who tended the target could not hear the name of the shooter, so each member of the battalion was assigned a number and fired four shots before the tenders evaluated the result and recorded his success. McGowan's battalion more or less used the same process, except that its man-sized target had a more complex bull's-eye in the center. It was 5 inches in diameter but with two outer circles 14 and 24 inches wide. This was painted on a pine plank 1 inch thick and 2 feet by 6 feet at 100 yards, and up to 6 feet by 6 feet at 900 yards. The rather large bull's-eye was correspondingly enlarged as well. The men used a tripod weighted down with a sandbag as a barrel rest. Thomas awarded the best shooter in these training exercises a rare English target rifle as a prize.[41]

"Sharpshooters, like fiddlers, are born and not made," said A. P. Hill. Many men in his numerous battalions agreed. "Skirmishing is ticklish business at times," wrote Marion Hill Fitzpatrick of the Forty-fifth Georgia, "but I like to do it when it has to be done." Fitzpatrick volunteered to be a part of Thomas's sharpshooter battalion. On their own, many battalions adopted insignia to denote their membership in elite units. In Kirkland's North Carolina battalion, it was a gold cross sewn onto the men's sleeve. Mimicking a common practice for regimental flags, the sharpshooter also sewed the names of battles in which he had participated.[42]

Soon after the start of the Overland campaign, battalion commanders began to select their best men to serve as scouts. Their job was to gather information by sneaking about between the opposing forces, sometimes even entering Federal lines. In Wilcox's Division, two men were detailed for this hazardous duty from every sharpshooter company. The scouts usually were armed with Whitworth rifles so they could do double duty as snipers. Scouts were the true daredevils of the sharpshooter battalions, risking their lives more than their comrades but providing valuable intelligence for higher-level commanders.[43]

Lee's sharpshooter battalions became elite, specialized units within the Army of Northern Virginia. No other units in the Civil War, except Berdan's regiments, were trained so intensively in estimating distance, in marksmanship, and in open-order skirmish drill. At best, only 34 battalions and 2 regiments out of some 3,000 Union and Confederate units received modern training to use their rifle muskets and other improved weapons. And most of those units were not regular members of the battle line, which was the main formation in Civil War engagements. Moreover, there is scant evidence that these specialized units actually performed better than their opponents in a consistent way during field operations.

This is not to suggest that their training was useless or that their battlefield performance was dismal. The performance of Lee's battalions in 1864–1865 will be addressed in the following chapter, but for now it is useful to note that there is no evidence that these units were able to make much of a difference on the battlefield, except in a handful of special circumstances. No matter how good the skirmisher, skirmishing always remained an adjunct to the operations of the battle line. The units on the battle line, not those in the skirmish force, consistently determined which side won and which lost Civil War engagements.

Soldier of the Thirty-third New York, proudly displaying his bayoneted musket. Prints and Photographs Division, Library of Congress, LC-USZ62-99872.

Regimental battle line of 110th Pennsylvania, Third Corps, at Falmouth, Virginia, April 24, 1863. Andrew J. Russell photograph, Prints and Photographs Division, Library of Congress, LC-DIG-ppmsca-07268.

Line of Company E, Fourth USCT, at Fort Lincoln, Washington, DC. Prints and Photographs Division, Library of Congress, LC-DIG-cwpb-04294.

Stacked muskets at Petersburg, after the fall of the city in April 1865. Prints and Photographs Division, Library of Congress, LC-DIG-cwpb-02647.

"Austrian Musket in Action." Unlike other Union soldiers from the West who were naturally comfortable using firearms, J. W. Gaskill of the 104th Ohio never warmed up to his musket. Frontispiece of J. W. Gaskill, *Footprints through Dixie*.

Empty ammunition boxes on the battlefield at Petersburg, soon after the fall of the city in April 1865. Prints and Photographs Division, Library of Congress, LC-DIG-cwpb-02635.

Snipers of the Eighteenth Corps at Petersburg. Alfred R. Waud sketch, Prints and Photographs Division, Library of Congress, LC-USZ62-7053.

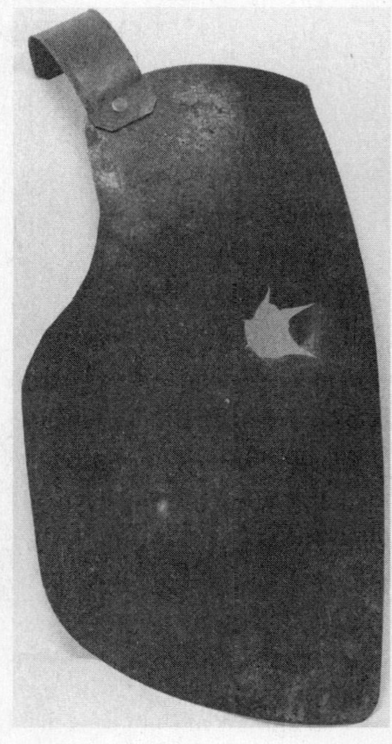

Breastplate of Col. William P. Rogers, Second Texas. Rogers was shot and killed during the attack on Battery Robinette at the battle of Corinth on October 4, 1862, his body armor proving ineffective in safeguarding his life. Wisconsin Veterans Museum, Madison.

6

SKIRMISHING IN BATTLE

During the first half of the Civil War, commanders consistently sent out two companies of skirmishers to cover the front of each regiment, deploying them in loose-order fashion as prescribed by drillmasters. There was relatively little variation on the standard use of skirmishers because commanders were learning their craft and battles were set-piece affairs lasting one or two days and separated from each other by months of inaction.

Only when the armies came to rest within striking distance of each other for some period of time did tactical conditions encourage a more intensive use of skirmishers. At Fort Donelson, Yorktown, Corinth, and the sieges of Vicksburg and Port Hudson, skirmishers had more opportunity to ply their trade at close range, because the opposing sides were locked in static positions for days or weeks at a time. This tactical situation sometimes allowed them to utilize the long-range firing capability of the rifle musket.

During the last year of the war, the skirmisher came into his own in the campaigns of 1864–1865, in part due to increased experience but mainly because those campaigns saw the most intense continuous contact of the war. During the Overland and Atlanta campaigns, skirmish lines became stronger and commanders tended to rely more heavily on them. Skirmishers not only sought to locate the enemy and provide information on his position, but they also tried to push opposing skirmishers away and try to break or move opposing battle lines. The real reason for this is not the use of the rifle musket but the policy of continuous contact that Grant introduced in the spring of 1864, which resulted in extended campaigning that kept the armies in contact with each other for months at a time. This extended contact encouraged aggressive brigade and division commanders to push the capacities of their skirmish lines farther than ever before, proving the worth of loose-order tactics for specialized tasks.

Moreover, by 1864, most Union and Confederate soldiers were hardened veterans fully capable of pushing the envelope when it came to skirmishing. This specialized military craft demanded aggressiveness, self-confidence, agility, and sharp reflexes, all key characteristics of successful veterans. In no other campaign did skirmishing become as intense as during the Atlanta campaign. For four months, from early May until early September 1864, the Union and Confederate armies were in constant contact with each other. The skirmishing increased in intensity as the campaign progressed, until it reached a level similar to a never-ending minibattle as the troops were locked in static positions north and west of the city for more than a month. Nothing like this evolved during the Overland or Petersburg campaigns in Virginia, despite the presence of Lee's dedicated skirmish units and the longer period of contact. The best possible explanation for this is the larger presence of gun-adept soldiers in the western armies.

The French revolutionary armies had experimented with new tactics for skirmishers, creating clouds or swarms of ill-disciplined tirailleurs to cover their front and confuse the enemy. Civil War commanders, in contrast, did not innovate in form or procedure, but they maximized the well-developed forms and procedures handed down to them by previous generations that had already standardized the use of skirmish lines. As a result, the Civil War witnessed the high point of skirmishing in world military history.

The Western Campaigns, 1861–1864

Not surprisingly, regimental and brigade commanders paid a great deal of attention to the prescribed use of skirmishers found in the drill manuals during the early years of the war. They deployed two companies from each regiment and sent out more when needed, sometimes taking their entire unit forward.[1]

The skirmishers became more prominent at Fort Donelson, where the opposing sides held static positions from February 13 to 16, 1862. Observers were fascinated with the sight. Charles C. Nott described the skirmishers of the Fourteenth Iowa as "crawling up behind trees and stumps, sometimes dragging themselves along the ground, sometimes on their hands and knees. Their shots were frequent, and sounded as though a sporting party were below us. It was hard to believe that they were shooting at men." Their task was to silence Confederate batteries by picking off the gunners and to generally keep the enemy quiet. "A rebel could not show his head above the breastworks without getting a bullet after it," commented Abner Dunham of the Twelfth Iowa. Allen Morgan Geer of the Twentieth Illinois actually saw one of his targets fall after firing.[2]

The Confederates failed to send out a skirmish line and suffered as a result. Flavel C. Barber of the Third Tennessee called it "a new species of danger." Federal skirmishers, concealed in timber, "swept the crest of the hill behind our trenches, so that it became quite hazardous to pass and repass from the rifle pits to the camp. Every tent within their range was riddled with bullets, and nearly every tree and bush on the crest was well marked." The enemy skirmishers sometimes exposed themselves, "and our men would return their fire with musketry, but they were out of the range of our guns and we had to endure a provoking fire all day without being able to return it with much effect."[3]

Aggressive Union skirmishing was a prominent factor in Union victory at Fort Donelson, but commanders employed skirmishers in a fashion more typical of the early war years at Shiloh. Col. David A. Enyart of the First Kentucky (U.S.) started the day on April 7 by deploying Companies A and G 300 yards ahead of his regimental battle line. He then advanced half a mile before encountering Confederate skirmishers, and drove them a mile away. Later, when ordered to support another brigade, Enyart reinforced his skirmish line with Companies C and I. By the end of the day's fighting, he had established a picket line half a mile ahead of his regiment to serve warning in case the Confederates threatened his position.[4]

Maj. Gen. Don Carlos Buell thought it necessary to lay down the "habitual" order of battle for his Army of the Ohio right after Shiloh. It included a prescription for using the two flank companies of each regiment as skirmishers, in accordance with established principles. But Buell thought it would be wise to use only one company if the purpose was merely to locate or test the enemy position. Buell's order was out of synchronization with how his units were already conducting their skirmishing, for brigade and regimental commanders never seem to have used only one company.[5]

At Chickasaw Bluffs, the Confederates had a fine view of the Yazoo River bottomland crossed by Sherman's troops as they advanced toward their fortified position. Flavel C. Barber admired the Federal use of skirmishers in the attack of December 29, 1862. Skirmishers preceded each brigade column. They plopped onto the ground to fire, then rose and ran ahead, going to ground again to reload. "Behind this cloud of skirmishers the unbroken battalions still advance," Barber later wrote. The attack was stopped, and during the night the Confederates collected more than 1,000 abandoned Enfield and Belgian rifles from the battlefield. They used them to good effect the next day in answering the aggressive Union skirmish fire.[6]

The Army of the Cumberland, as Buell's Army of the Ohio came to be designated, continued to deploy two companies per regiment on the skirmish line at Stones River. There was some room for variation on that model, for the Sixth Indiana sent out only the first platoons of its Companies A and B. One of the few sharpshooter battalions in the Confederate Army of Tennessee, that belonging to Col. A. J. Vaughan's brigade of Maj. Gen. Benjamin F. Cheatham's Division, was placed in the battle line and not used for skirmishing at all.[7]

After a terrible battle on December 31, 1862, the armies at Stones River lay quietly within a few hundred yards of each other the next day. This gave rise to miniskirmish battles along the line, such as were evident at Fredericksburg two weeks before and seen at Gettysburg the next year. Confederate skirmishers occupying some cabins 300 yards from the Fifth Kentucky (U.S.) so annoyed the Federals that Lt. Col. William W. Berry sent two companies across open fields to attack them. With the help of a third company, they captured the cabins and burned them, leaving five dead Confederates.[8]

During the battles of the Vicksburg campaign, the Federals continued to send only two companies from each regiment forward to skirmish. But when the siege commenced, Union commanders greatly intensified their skirmishing to harass and wear down the defenders. Initially, the Federal skirmishers took shelter in the abundant cane that filled ravines in this rugged landscape, firing as much as 300 yards to reach their targets, but the range steadily decreased as the besiegers dug siege approaches to get nearer the Confederate line. During the course of the six-week siege, skirmishers had ample time to ply their trade, firing 100 to 200 rounds per man each day. Regimental commanders often sent out several companies rather than the traditional two, and at times entire regiments skirmished. Repeated practice made tolerable marksmen of many Federals. Charles Dana Miller of the Seventy-sixth Ohio claimed that his men "became almost perfect in their range and skimmed the Rebel parapets so they were unable to man their guns or put their heads above the works. They often put their hats out on sticks to see our boys fire at them." The siege spoiled many Federals when it came to fire discipline. Maj. Gen. Edward O. C. Ord, commander of the Thirteenth Corps, wanted to carry at least 240 rounds per man with him when the corps was shipped to Louisiana after the fall of Vicksburg, because "my men have been accustomed in the sieges to fire away much ammunition."[9]

The Confederates who were on the receiving end of this nearly continuous firing suffered from the very start of the siege. The Federal skirmishers began before dawn, lessened their fire about noon, but intensified it by the evening. After a few days, they even began to fire at night. By early June they were within 60 to 100 yards of the Rebels. "Sharpshooting very bitter," reported Brig. Gen. Francis A. Shoup. The Confederates were subjected to "a perfect rain of Minie balls, which prevented any one from showing the least portion of his body," wrote Brig. Gen. Stephen D. Lee. On another part of the line, Col. Ashbel Smith of the Second Texas described how "an incessant stream of Minies swept just above the upper slope of our parapet, increasing in strength day after day, as rifle-pit after rifle-pit was constructed. This stream was kept up from dawn till dark, whether any one was seen on our works or not. Let, however, a head appear, and the whir of a hundred Minies instantly hissed around it. This constant firing rendered the position in our rear unsafe; several men were thus wounded." Smith's men desperately wanted to return fire, but he received numerous orders to save his ammunition. "We used our ammunition with a strict regard to economy," Brig. Gen. John C. Moore reported. "This enabled the enemy to approach more rapidly and with greater impunity than they otherwise could have done."[10]

Similar skirmishing took place at the siege of Port Hudson, although not as persistently or effectively as at Vicksburg. During the Federal attack of May 27, 1863, the Twelfth Connecticut was ordered to silence Confederate batteries. It sent four companies out to form a skirmish line while the other six remained as a reserve. The four companies deployed in loose order for half a mile and took cover behind stumps and tree trunks, shooting at gunners, at tents behind the artillery, "or at whatever else seemed worth hitting."[11]

Table 6.1. Expenditure of ammunition on the skirmish line

Unit	Battle	Expenditure of Ammunition (rounds per day)
Allen Morgan Geer, 20th Illinois, Wallace's brigade, McClernand's division	Fort Donelson, February 14, 1862	10[1]
48th Indiana, Sanborn's brigade, Crocker's division, 17th Corps, Army of the Tennessee	Vicksburg, May 24, 1863	15,000–20,000[2]
John Quincy Adams Campbell, 5th Iowa, Matthies's brigade, Smith's division, 17th Corps, Army of the Tennessee	Vicksburg, June 5, 1863	100[3]
Edmund Newsome, 81st Illinois, Stevenson's brigade, Logan's division, 17th Corps, Army of the Tennessee	Vicksburg, June 7, 1863	70[4]
Eugene Blackford, Sharpshooter Battalion, O'Neal's Brigade, Rodes's Division, 2nd Corps, Army of Northern Virginia	Gettysburg, July 3, 1863	84[5]
96th Pennsylvania, Upton's brigade, Russell's division, 6th Corps, Army of the Potomac	Cold Harbor, June 2, 1864	90,000 rounds in 24 hours[6]
Hazen's brigade, Wood's division, 4th Corps, Army of the Cumberland	Atlanta, July 22–August 17, 1864	Up to 5,000[7]
4th Corps, Army of the Cumberland	Atlanta, July 31, 1864	150,000 rounds in 24 hours[8]
96th Illinois, Whitaker's brigade, Stanley's division, 4th Corps, Army of the Cumberland	Atlanta, early August 1864	3,000–5,000 rounds fired by one-third of regiment every 24 hours[9]
10th Illinois, Morgan's brigade, Davis's division, 14th Corps, Army of the Cumberland	Atlanta, August 9, 1864	Over 2,000 rounds fired by one company in one day[10]
17th New York, Morgan's brigade, Davis's division, 14th Corps, Army of the Cumberland	Atlanta, August 14, 1864	1,500 rounds fired by one company in 24 hours[11]
Sharpshooter Battalion, McGowan's Brigade, Wilcox's Division, 3rd Corps, Army of Northern Virginia	Reams Station, August 25, 1864	160 rounds per man for 5 hours (1 round every 1.8 minutes)[12]

Table 6.1. (*continued*)

Unit	Battle	Expenditure of Ammunition (rounds per day)
Manigault's Brigade, Anderson's Division, Lee's Corps, Army of Tennessee	Atlanta, August 1864	Over 6,000 rounds fired by 175 men every 24 hours (35 rounds per man, or 1 round every 41 minutes)[13]
Johnson's Division, Department of North Carolina and Southern Virginia	Petersburg, December 1864	18 rounds per man every 24 hours[14]
81st Illinois, Geddes's brigade, Carr's division, 16th Corps	Mobile, March 27, 1865	13,000[15]
Army of Northern Virginia	Petersburg, March 1865	37,000[16]

[1] Entry of February 14, 1862, in Mary Ann Andersen, ed., *The Civil War Diary of Allen Morgan Geer, Twentieth Regiment, Illinois Volunteers* (New York: Cosmos, 1977), 18.
[2] Samuel E. Snure to Sir, June 21, 1863, in William A. Russ Jr., "The Vicksburg Campaign as Viewed by an Indiana Soldier," *Journal of Mississippi History* 19, 4 (1957): 268–269.
[3] Entry of June 5, 1863, in Mark Grimsley and Todd D. Miller, eds., *The Union Must Stand: The Civil War Diary of John Quincy Adams Campbell, Fifth Iowa Volunteer Infantry* (Knoxville: University of Tennessee Press, 2000), 105.
[4] Edmund Newsome, *Experience in the War of the Great Rebellion* (Carbondale, IL: E. Newsome, 1879), 32.
[5] Eugene Blackford memoir, in Noah Andre Trudeau, ed., "Taking Aim at Cemetery Hill," *America's Civil War* 14, 1 (2001): 51.
[6] Isaac O. Best, *History of the 121st New York State Infantry* (Chicago: W. B. Conkey, 1921), 157.
[7] William B. Hazen, *A Narrative of Military Service* (Boston: Ticknor, 1885), 419–420.
[8] Entry of July 31, 1864, in Arnold Gates, ed., *The Rough Side of War: The Civil War Journal of Chesley A. Mosman, 1st Lieutenant, Company D, 59th Illinois Volunteer Infantry Regiment* (Garden City, NY: Basin, 1987), 252.
[9] Charles A. Partridge, ed., *History of the Ninety-sixth Regiment Illinois Volunteer Infantry* (Chicago: Brown, Pettibone, 1887), 387.
[10] Entry of August 9, 1864, in Janet Correll Ellison, ed., *On to Atlanta: The Civil War Diaries of John Hill Ferguson, Illinois Tenth Regiment of Volunteers* (Lincoln: University of Nebraska Press, 2001), 73.
[11] William B. Westervelt diary, August 14, 1864, in George S. Maharay, ed., *Lights and Shadows of Army Life: From Bull Run to Bentonville* (Shippensburg, PA: Burd Street, 1998), 204.
[12] W. S. Dunlop, *Lee's Sharpshooters; Or, The Forefront of Battle* (Dayton, OH: Morningside Bookshop, 1982), 193.
[13] R. Lockwood Tower, ed., *A Carolinian Goes to War: The Civil War Narrative of Arthur Middleton Manigault, Brigadier General, C.S.A.* (Columbia: University of South Carolina Press, 1988), 239.
[14] Bushrod R. Johnson to Major Duncan, December 2, 1864, in *The War of the Rebellion: A Compilation of the Official Records of the Union and Confederate Armies*, 70 vols. (Washington, DC: Government Printing Office, 1880–1901), vol. 42, pt. 1, 918.
[15] Newsome, *Experience in the War,* 132.
[16] J. F. J. Caldwell, *The History of a Brigade of South Carolinians, Known First as "Gregg's," and Subsequently as "McGowan's Brigade"* (Philadelphia: King & Baird, 1866), 206.

No matter how many battles a field army endured, there always was a need to perfect skirmishing capabilities among its units. Gen. Braxton Bragg ordered that one regiment in every brigade and two companies in every regiment were to undergo intensive skirmish drill in the Army of Tennessee during the spring of 1863. The level of effectiveness in this army, the Confederacy's premier force in the West, always lagged behind expectations, and that might account for the extraordinary order to beef up its knowledge of loose-order drill so long after its organization. Bragg's army had only eight sharpshooter battalions employed in the battle of Chickamauga the following September, and they mostly skirmished during the engagement rather than acted as line infantry. The Fourteenth Louisiana Battalion of Sharpshooters was reinforced by a company detailed from each of four regiments at one point in the battle, indicating one problem with relying on dedicated skirmish units—they often did not have enough manpower to do the job. Federal units continued to send out two companies from each regiment at Chickamauga, adding more when needed.[12]

An incident in the experience of Manigault's South Carolina brigade at Chickamauga demonstrated that Bragg's army still needed to learn how to use skirmishers properly. The brigade began the day on September 20 by placing its skirmish line 150 yards in front of the battle line. It then advanced 300 yards before encountering Federal skirmishers. The battle line could not wait to let the Confederate skirmishers do their job. It advanced enthusiastically but in disorder, "ran over our skirmishers," and pushed the Federal skirmishers half a mile, unnecessarily exhausting the men in performing a task that their own skirmishers probably could have accomplished.[13]

The Federals made very good use of their skirmish line in the capture of Fort McAllister on December 13, 1864. Skirmishers from several regiments took shelter within 200 yards of the Rebel fort, hidden mostly by tall grass, and found easy targets because most of the fort's twenty-two guns were mounted en barbette, exposing the gunners to fire. The Federals silenced the Confederate artillery, allowing the attacking units to form "in full view of and not more than 500 yards from the parapet" before their successful assault.[14]

The Eastern Campaigns, 1861–1863

In the East as well, Union and Confederate armies skirmished by the book as a rule but intensified their efforts whenever locked near the enemy for a few days. At Yorktown, where the opposing forces were stuck for a month, Federal skirmishers had orders "to fire when they see any body," and "they pop away all day." But at the battle of Williamsburg, on May 5, 1862, both sides consistently deployed two companies per regiment onto the skirmish line. By the time of the battle of Chancellorsville in May 1863, some commanders were deploying an entire regiment to do the skirmishing for their brigade. The distance between the skirmish line and the main line varied widely according to circumstances, sometimes only 50 yards and at other times 400 yards. When the Confederates took position outside Suffolk in the late spring of 1863, they established a picket line manned by six men detailed from each company. These troops carried 100 rounds each and sheltered themselves in rifle pits 25 yards apart, three men occupying each pit, and fired all day.[15]

The three-day battle of Gettysburg gave rise to intense skirmish battles between the lines that drew in many troops and produced significant casualties. One of the most severe occurred on the Bliss Farm, where a large, well-constructed barn served as an ideal blockhouse for Confederate skirmishers about halfway between the opposing armies. The Union and Confederate battle lines were some 1,200 yards apart. The Federals repeatedly sent troops out to seize the farm and suppress harassing skirmish fire, extending the minibattle over the course of thirty-two hours on July 2–3, 1863. A total of 2,163 Federal and 2,313 Confederate soldiers took part in the struggle. The result was that the Federals finally seized the farm long enough to burn the buildings and put an end to the harassment, but they suffered 17 percent losses, compared to the Confederate loss of 20 percent of the men involved.[16]

Maj. Eugene Blackford, who commanded the sharpshooter battalion attached to O'Neal's Brigade, took directions from his division commander to annoy Federal troops on Cemetery Hill at Gettysburg. Maj. Gen. Robert Rodes wanted Blackford to establish his skirmish line as close as possible to the Federals and "annoy them within all my power" on July 3. After studying the ground, Blackford posted his sharpshooters in the houses at the southern edge of town about 500 yards from the Union position. His men fired so rapidly that Blackford had to improvise an ammunition supply, sending men to pick up abandoned cartridge boxes and distributing them along his line by using baker's carts found in the town. "The men soon complained of having their arms & shoulders very much bruised by the continual kicking of the muskets but still there could be no rest for them." Blackford himself fired eighty-four rounds that day, using at least two guns to let one cool. "My shoulder pad became so sore that I was obliged to rest."[17]

Table 6.2. Range of skirmishing

Unit	Battle	Range
5th Kentucky, Baldwin's brigade, Johnson's division, McCook's Right Wing, Army of the Cumberland	Stones River, January 1, 1863	Received Confederate skirmish fire at 300 yards[1]
Shoup's Brigade, Smith's Division, Army of Vicksburg	Vicksburg, June 9, 1863	Received Federal skirmish fire at 60–70 yards[2]
William Henry Harrison Clayton, 19th Iowa, Orme's brigade, Herron's division, 17th Corps, Army of the Tennessee	Vicksburg, June 17, 1863	Set sights at 300 yards and 800 yards and came close to hitting targets[3]
2nd Missouri Infantry (C.S.), Cockrell's Brigade, Bowen's Division, Army of Vicksburg	Vicksburg siege	Received Federal skirmish fire at 200–300 yards early in the siege[4]
76th Ohio, Woods's brigade, Steele's division, 15th Corps, Army of the Tennessee	Vicksburg siege	400 yards[5]
Baldwin's Brigade, Smith's Division, Army of Vicksburg	Vicksburg siege	Received Federal skirmish fire at 100 yards[6]
Sharpshooter Battalion, O'Neal's Brigade, Rodes's Division, 2nd Corps, Army of Northern Virginia	Gettysburg, July 3, 1863	500 yards[7]
David Holt, 16th Mississippi, Harris's Brigade, Mahone's Division, 3rd Corps, Army of Northern Virginia	Cold Harbor, June 1864	75 yards[8]
Manly's North Carolina Battery, Cabell's Battalion, 1st Corps, Army of Northern Virginia	Cold Harbor, June 1864	Received Federal skirmish fire at 125 yards[9]
10th Illinois, Morgan's brigade, Davis's division, 14th Corps, Army of the Cumberland	Atlanta, August 9, 1864	75 yards[10]
Sharpshooter Battalion, McGowan's Brigade, Wilcox's Division, 3rd Corps, Army of Northern Virginia	Reams Station, August 25, 1864	400 yards[11]
Manigault's Brigade, Anderson's Division, Lee's Corps, Army of Tennessee	Atlanta, late August 1864	80 yards[12]
French's Division, Stewart's Corps, Army of Tennessee	Atlanta, late August 1864	200 yards[13]

Table 6.2. (*continued*)

Unit	Battle	Range
Harrison's brigade, Ward's division, 20th Corps, Army of the Cumberland	Atlanta, late August 1864	50 yards[14]
Sharpshooter Battalion, McGowan's Brigade, Wilcox's Division, 3rd Corps, Army of Northern Virginia	Sutherland Station, April 2, 1865	400 yards[15]
Hazen's division, 15th Corps, Army of the Tennessee	Fort McAllister, December 13, 1865	200 yards[16]

[1] William W. Berry to William Mangan, January 8, 1863, in *The War of the Rebellion: A Compilation of the Official Records of the Union and Confederate Armies*, 70 vols. (Washington, DC: Government Printing Office, 1880–1901), vol. 20, pt. 1, 342. This work is hereafter cited as *OR*.
[2] Francis A. Shoup to J. G. Devereux, July 8, 1863, *OR*, vol. 24, pt. 2, 408.
[3] William Henry Harrison Clayton to father and mother, June 18, 1863, in Donald C. Elder III, ed., *A Damned Iowa Greyhound: The Civil War Letters of William Henry Harrison Clayton* (Iowa City: University of Iowa Press, 1998), 74.
[4] Ephraim McD. Anderson, *Memoirs: Historical and Personal; Including the Campaigns of the First Missouri Confederate Brigade* (Dayton, OH: Morningside Bookshop, 1972), 328.
[5] Stewart Bennett and Barbara Tillery, eds., *The Struggle for the Life of the Republic: A Civil War Narrative by Brevet Major Charles Dana Miller, 76th Ohio Volunteer Infantry* (Kent, OH: Kent State University Press, 2004), 96, 98.
[6] William E. Baldwin to Devereux, July 10, 1863, *OR*, vol. 24, pt. 2, 403.
[7] Eugene Blackford memoir, in Noah Andre Trudeau, ed., "Taking Aim at Cemetery Hill," *America's Civil War* 14, 1 (2001): 50–51.
[8] Thomas D. Cockrell and Michael B. Ballard, eds., *A Mississippi Rebel in the Army of Northern Virginia: The Civil War Memoirs of Private David Holt* (Baton Rouge: Louisiana State University Press, 1995), 278.
[9] Unidentified member of Manly's Battery to editor, June 6, 1864, *Daily Confederate*, Raleigh, NC, June 10, 1864.
[10] Entry of August 9, 1864, in Janet Correll Ellison, ed., *On to Atlanta: The Civil War Diaries of John Hill Ferguson, Illinois Tenth Regiment of Volunteers* (Lincoln: University of Nebraska Press, 2001), 73.
[11] W. S. Dunlop, *Lee's Sharpshooters; Or, The Forefront of Battle* (Dayton, OH: Morningside Bookshop, 1982), 191.
[12] R. Lockwood Tower, ed., *A Carolinian Goes to War: The Civil War Narrative of Arthur Middleton Manigault, Brigadier General, C.S.A.* (Columbia: University of South Carolina Press, 1988), 253.
[13] Samuel G. French to William D. Gale, December 6, 1864, *OR*, vol. 38, pt. 3, 905.
[14] Benjamin Harrison to John Speed, September 14, 1864, ibid., vol. 38, pt. 2, 349.
[15] Susan Williams Benson, ed., *Berry Benson's Civil War Book: Memoirs of a Confederate Scout and Sharpshooter* (Athens: University of Georgia Press, 1992), 189.
[16] Orlando M. Poe to Richard Delafield, October 8, 1865, *OR*, vol. 44, 61.

After Gettysburg, commanders reverted to their normal practice of deploying two companies at Bristoe Station. But at the Federal attack on Rappahannock Station, November 7, 1863, they experimented with using a much heavier skirmish force to lead the assaulting columns against a fortified Confederate bridgehead. Brig. Gen. David A. Russell ordered the entire Sixth Maine to cover the advance of his Sixth Corps division. Another regiment, the Fifth Wisconsin, acted as the skirmish reserve. The Fifth Corps, assigned to support Russell's advance and demonstrate to take Confederate attention away from it, deployed 932 skirmishers detailed from three divisions and under the command of Brig. Gen. Kenner Garrard to screen its front. An entire division of the corps, led by Brig. Gen. Joseph J. Bartlett, acted as the skirmish reserve. Bartlett's command "kept within supporting distance of General Garrard, and was governed in its operations by the action of the skirmishers," surely a unique occurrence on a Civil War battlefield. Bartlett formed his division into a line with each regiment in double column, intervals between each column closed to make a more compact formation. But when he came within range of Rebel artillery, Bartlett sent forward the right and left regiments of each brigade to form a battle line 200 yards ahead of his division so they could more readily move forward to support the skirmishers if needed. The division continued forward another half mile in this formation until Garrard's skirmishers stopped close enough to the Rebel earthworks to pick off troops. The advanced regiments of Bartlett's division took cover close behind Garrard's skirmishers and 600 yards from the Confederates, while the rest of the division remained 400 yards to their rear as a general reserve. When Russell struck the defenses, Garrard's men also attacked and entered the fortifications. The Federals captured the position in fine style, taking more than 1,500 Confederates prisoner.[18]

What happened at Rappahannock Station was an innovative way to use a heavy skirmish force to precede a well-coordinated attack. It foreshadowed what was to come in the concentrated campaigns of 1864 in both East and West.

Atlanta

More than any other campaign of the Civil War, Sherman's drive toward Atlanta from May to September 1864 offered skirmishers their best opportunity to shine. It involved heavy, aggressive skirmishing that helped to shape the contour of operations. Line commanders relied more heavily on their skirmishers to find, weaken, and drive away enemy skirmishers than in any other campaign. Near the end of the long offensive, Federal skirmishers were seriously degrading the stamina and fighting effectiveness of troops on the Confederate battle line. As a result, in the last battle of the campaign, at Jonesboro on August 31 and September 1, many Rebel brigades were too weakened to fight effectively. It is impossible to argue with Maj. William H. Chamberlin of the Eighty-first Ohio: "In the western army, the skirmish line was undoubtedly seen at its best during the Atlanta campaign." A Union officer from a Kentucky regiment also mused that the Atlanta campaign "might properly be termed the skirmishers's war."[19]

From the beginning of the campaign, at Rocky Face Ridge, commanders tended to send out an entire regiment to skirmish for its parent brigade. Col. Adolphus Bushbeck, faced with the task of advancing up the steep, tall slope of the ridge, also sent detachments from each regiment to reinforce the skirmishers before he hit the Rebel line. But Rocky Face Ridge saw relatively light fighting. At the second confrontation near Resaca on May 13–15, Federal skirmishers severely harassed the defending Confederates, who were poorly sheltered behind ill-constructed earthworks. Lt. Rene T. Beauregard's South Carolina battery was targeted by at least twenty-five Federals, who hid so well behind logs and trees that he could not see any of them. "Balls come flying around constantly," reported an observer, "and scarcely a man of his battery but has bullet holes through his clothes."[20]

The skirmish war intensified as Sherman drove deeper into northwest Georgia. By late May, the armies were locked in static positions for many days along the Dallas, New Hope Church, Pickett's Mill line. "We think the Federal Sharpshooters are 'terribly' inconvenient," wrote John S. Jackman of the Ninth Kentucky (C.S.). The "stream of bullets coming over the hill" prevented his comrades from leaving their trench. "We could not even stretch our blankets as dog tents" complained another Ninth Kentuckian, named Johnny Green, "but had to just squat in the trenches or behind trees with our blankets wrapped around us & take the rain & catch what sleep we could in this squatting posture." Jackman and his comrades gathered up smoothbore muskets abandoned by some Tennessee troops and loaded them as fully as possible with buck and ball cartridges, "trying to see who could shoot the largest loads out of them." Of course, they could not reach the Federal skirmishers with this ordnance, but it provided some amusement tinged with a bit of danger.[21]

The skirmishing was so intense during the Atlanta campaign partly because of aggressive unit commanders in Sherman's army group. Col. William Grose consistently pushed his brigade skirmishers forward at every opportunity. "Grose is a daisy," wrote an admiring subordinate. "If Rebel pickets cease firing at night it seems to waken him at once and he is so darned inquisitive that he starts his [skirmish] line right out to learn what is the trouble with them." A "star brigade commander," Lt. Chesley A. Mosman of the Fifty-ninth Illinois paid him the ultimate tribute: "Old Grose shoves things."[22]

While Federal commanders seem to have used their skirmishers as fully as possible, Confederate commanders had difficulty knowing how to do the same. Corps commander William J. Hardee, author of the famous drill manual, issued a general order on May 31 reminding his subordinates how to deploy and when to use a skirmish line. The order included the most basic information on the subject, specifying that each skirmisher was to be no more than 5 paces from his neighbor, that a skirmish line should be deployed every time the parent unit assumed a position, and that the skirmish line should fight as hard as possible, retiring only in the face of an advance by the enemy's battle line. It is astonishing that such an order had to be issued after three years of fighting by the Confederacy's main field army in the West.[23]

Brig. Gen. Alpheus Baker described how his skirmish line lost a fight against Federal skirmishers in mid-June. He placed his picket line half to three-quarters of a mile in front of his main line, "too far to be either supported or withdrawn in case of sudden attack," Baker later admitted. When the Yankees advanced without warning, they overwhelmed his men. Baker lost 130 troops, mostly from the Fortieth Alabama, because those men "stood their ground and fired till the enemy swarmed all over their rifle pits." Another reason for the comparatively poor performance of Confederate skirmishers during the Atlanta campaign was concern about a potential shortage of ammunition in the Army of Tennessee. Commanders were subject to periodic orders to "enforce economy in its use—especially by skirmishers."[24]

As a result, the Atlanta campaign was characterized by repeated instances of Federals dominating their opponents on the skirmish line. Lt. Andrew Jackson Neal of the Marion Light Artillery Battery from Florida reported the punishment taken the whole day of June 19, when Federal skirmishers relentlessly punished his position. The flag was torn with thirty-one bullet holes, and seven additional bullets splintered the staff. "The trees behind us were riddled with balls on one little sapling I counted about eighty balls on the body. The face of the pieces upper part of axels [*sic*] & wheels have hundreds of marks made by

Table 6.3. Distance between skirmish line and battle line

Unit	Battle	Distance
1st Kentucky, Bruce's brigade, Nelson's division, Army of the Ohio	Shiloh, April 7, 1861	300 yards[1]
31st Illinois, McClernand's brigade	Belmont, November 7, 1861	Half a mile[2]
Sharpshooter Battalion, O'Neal's Brigade, Rodes's Division, 2nd Corps, Army of Northern Virginia	Chancellorsville, May 3, 1862	400 yards[3]
19th Mississippi, Wilcox's Brigade, Longstreet's Center, Army of Northern Virginia	Williamsburg, May 5, 1862	50 yards[4]
Manigault's Brigade, Hindman's Division, Longstreet's Left Wing, Army of Tennessee	Chickamauga, September 20, 1863	150 yards[5]
Sharpshooter Battalion, Stone's Brigade, Heth's Division, 3rd Corps, Army of Northern Virginia	Wilderness, May 5, 1864	60 yards[6]
Sharpshooter Battalion, Stone's Brigade, Heth's Division, 3rd Corps, Army of Northern Virginia	Wilderness, May 6, 1864	40–50 yards[7]
Sharpshooter Battalion, Mahone's Brigade, Anderson's Division, 3rd Corps, Army of Northern Virginia	Wilderness, May 6, 1864	150 yards[8]
Brooke's brigade, Barlow's division, 2nd Corps, Army of the Potomac	Spotsylvania, May 12, 1864	Brigade commander told skirmishers not to go farther than 30 yards ahead in attack[9]
Sharpshooter Battalion, McGowan's Brigade, Wilcox's Division, 3rd Corps, Army of Northern Virginia	Cold Harbor, June 13, 1864	200 yards[10]
Baker's Brigade, Stewart's Division, Hood's Corps, Army of Tennessee	Kennesaw Mountain, June 14–15, 1864	Half to three quarters of a mile[11]
Ward's brigade, Mower's division, 16th Corps (right wing)	Tupelo, July 14, 1864	550 yards[12]
Manigault's Brigade, Anderson's Division, Lee's Corps, Army of Tennessee	Atlanta, July–August 1864	825 yards initially, Federals drove it back to 250 yards within three weeks[13]

(*continues*)

Table 6.3. (*continued*)

Unit	Battle	Distance
Sharpshooter Battalion, McGowan's Brigade, Wilcox's Division, 3rd Corps, Army of Northern Virginia	Reams Station, August 25, 1864	600 yards[14]
Anderson's Division, Lee's Corps, Army of Tennessee	Atlanta, August 1864	500 yards[15]

[1] David A. Enyart to S. T. Corn, April 9, 1862, in *The War of the Rebellion: A Compilation of the Official Records of the Union and Confederate Armies*, 70 vols. (Washington, DC: Government Printing Office, 1880–1901), vol. 10, pt. 1, 350. This work is hereafter cited as *OR*.
[2] John A. Logan to John A. McClernand, November 11, 1861, ibid., vol. 3, 288.
[3] Eugene Blackford to Mary, May 21, 1863, Gordon-Blackford Papers, Maryland Historical Society, Baltimore.
[4] L. Q. C. Lamar to W. A. Harris, May 13, 1862, *OR*, vol. 11, pt. 1, 598.
[5] Joshua K. Callaway to Dulcinea B. Callaway, September 24, 1863, in Judith Lee Hallock, ed., *The Civil War Letters of Joshua K. Callaway* (Athens: University of Georgia Press, 1997), 136.
[6] Robert F. Ward to William S. Dunlop, December 30, 1898, in W. S. Dunlop, *Lee's Sharpshooters; Or, The Forefront of Battle* (Dayton, OH: Morningside, 1982), 369.
[7] Dunlop, *Lee's Sharpshooters*, 381.
[8] John E. Laughton Jr., "The Sharpshooters of Mahone's Brigade," *Southern Historical Society Papers* 22 (1894): 101.
[9] John R. Brooke to Assistant Adjutant General, 1st Division, November 1, 1865, *OR*, vol. 36, pt. 1, 409.
[10] J. F. J. Caldwell, *The History of a Brigade of South Carolinians, Known First as "Gregg's," and Subsequently as "McGowan's Brigade"* (Philadelphia: King & Baird, 1866), 159–160.
[11] Alpheus Baker to Joseph E. Johnston, July 4, 1874, Joseph E. Johnston Papers, Special Collections, College of William and Mary, Williamsburg, VA.
[12] Lyman M. Ward to J. B. Sample, July 24, 1864, *OR*, vol. 39, pt. 1, 277.
[13] R. Lockwood Tower, ed., *A Carolinian Goes to War: The Civil War Narrative of Arthur Middleton Manigault, Brigadier General, C.S.A.* (Columbia: University of South Carolina Press, 1988), 253.
[14] Dunlop, *Lee's Sharpshooters*, 191.
[15] Patton Anderson to J. W. Ratchford, February 9, 1865, *OR*, vol. 38, pt. 3, 770.

balls shot through the embrazures [*sic*] of our work while our canteens blankets & c just in rear of the portholes were shot to pieces." After losing four men to this fire, Neal sent the rest back to the rear with the caissons and horses and waited out the storm, evacuating his position that night.[25]

By the end of June, the Army of the Cumberland was expending 200,000 rounds of small-arms ammunition every day, most of it on the skirmish line. Maj. Gen. Oliver O. Howard was convinced that stronger, more aggressive skirmishing gave greater confidence to the men holding the battle line and turned the loose-order formations into "a veritable line of battle" as well.[26]

When mobile operations slowed to a crawl, and most Union and Confederate units became rooted to one spot for three or four weeks in late July and throughout August, the skirmishers had a wonderful opportunity. The so-called siege of Atlanta "was in the main a gigantic battle of skirmishers," wrote William H. Chamberlin. "The picket firing never ceased, day nor night. Sometimes it was lazy, scattered, and weak, and again swelling into volleys like the beginning of a battle, and now and then being followed by the roar of artillery." What had transpired from early May to late July was but preparation for the thoroughly dominating role played by Sherman's skirmishers in the maze of trenches north, west, and southwest of Atlanta during August.[27]

The Ninety-sixth Illinois in the Fourth Corps kept one-third of its number on the picket line every day, firing between 3,000 and 5,000 rounds. One company of the Seventeenth New York fired 1,500 rounds in its twenty-four-hour tour of duty on the skirmish line, a volume exceeded by the Tenth Illinois in the Fourteenth Corps. Fourth Corps skirmishers tended to fire at least 150,000 rounds each day, nearly double that number when the firing grew more intense. Brig. Gen. William B. Hazen's Fourth Corps brigade fired 5,000 rounds each day for three weeks in a steady effort to punish the Rebels opposite its position. Hazen learned after the war that the Confederate units lost three to five men of each regiment every day, so he estimated that perhaps 1 out of 500 rounds fired from his skirmish line hit a human target. Hazen thought this a cost-effective way to damage the enemy. Meanwhile, the ammunition shortage in the Army of Tennessee worsened. Rebel deserters told Federal officers that they were ordered to hold fire on the skirmish line unless they were sure of their target, greatly reducing Confederate ability to counter the increasingly aggressive Union skirmishing.[28]

For a time in August, Federal skirmishers in front of Patton Anderson's and Henry Clayton's divisions fired and pushed ahead so vigorously as to amount "to almost an engagement," in the words of corps commander Stephen D. Lee. Anderson placed his skirmish line 500 yards ahead of his battle line and heavily fortified it with head logs and loopholes. The opposing Union skirmishers were anywhere from 60 to 100 yards away, with the Union main line 200 yards to their rear.[29]

Arthur M. Manigault, who commanded a South Carolina brigade in Anderson's Division, had established his initial skirmish line 800 to 825 yards in front of his main line right after the battle of Ezra Church. During the next three weeks, Federal skirmishers drove that line back to a second and later to a third position, ending up only 200 to 250 yards from Manigault's battle line. At this point, the interval between the opposing skirmish lines was only 80 yards. The intervening space had been "densely wooded" to begin with, but by the end of three weeks the vegetation had been so chipped and denuded by constant rifle and artillery fire that one could see for hundreds of yards. Trees as thick as a man's leg were so weakened by bullets that they fell over when a strong wind came up. The head logs were "filled with bullets, much splintered, and in shreds." The Federals planted a battery only 100 yards from Manigault's skirmish line and protected the embrasures with iron mantlets so that the Confederates could not pick off the gunners.[30]

Manigault's skirmish line was 450 yards long. He kept 175 men on it at all times, firing on average more than 6,000 rounds each day, or 35 rounds per man. There was no rest while on duty, necessitating a rotation to the main line for four days between periods of duty on the skirmish line. "The mind was in a constant state of tension as well as the body," noted Manigault. Union skirmish fire was worse during the day, slacked off a bit in the evening, then continued at a moderate pace during the night, the bullets often overshooting and falling on the Confederate main line. Manigault lost on average ten men each day, sometimes up to fifty-five men. Ironically, most of the troops who were hit had carelessly exposed themselves while on the main line. The skirmishers were always on the alert to take cover. When needing water, the skirmishers loaded their canteens onto the shoulders of one man, who sprinted to the rear as his comrades concentrated their fire on the Federals, repeating the process as he returned. There was, however, a certain degree of honor on the skirmish line. If a man was wounded, and his comrades alerted the Federals to the fact, they often ceased firing to let him go to the Confederate rear for treatment.[31]

Those divisions that were careful to keep the Federal skirmishers as far from their main line as possible enjoyed more daily comfort and freedom of movement. William H. Young's brigade, in Samuel G. French's division, maintained one or two regiments on its skirmish line at all times to keep the enemy at bay. French kept the Federal skirmishers 200 yards from his main line and lost about 370 men due to skirmishing from July 18 to the end of August. As with Manigault's Brigade, the ground in front of French's skirmish line was devastated by rifle fire. All the trees withered and died from the constant nicking of bullets, and some trees 4 inches thick were chopped down by balls. After the Federals evacuated their lines, French's men sallied forward to find no-man's-land littered with spent bullets covering "the ground like hail." The balls already were "oxidized white like hailstones." One of his brigades picked up 5,000 pounds of lead and sold it to Confederate ordnance officers.[32]

The Federals developed an abiding confidence in their ability to skirmish by the end of the Atlanta campaign. Maj. Henry Hitchcock, who joined Sherman's staff after the city's fall, remarked, "Our officers say our skirmishers always beat the rebs." Their success came with a toll. Samuel W. Price, colonel of the Twenty-first Kentucky (U.S.), believed that half of all Union losses during the Atlanta campaign occurred on the skirmish line. Maj. J. C. Thompson, an officer on A. P. Stewart's division staff, thought that skirmish losses would at least total the equivalent of a major battle's casualties. Some Federal units suffered from vigorous Confederate skirmishing as well. Many members of Absalom Baird's Fourteenth Corps division were "imprisoned in their trenches, not daring to show their heads above the parapet." Baird lost up to thirty men per day due to skirmish fire. Col. Benjamin Harrison's brigade in the Twentieth Corps also was subject to effective Confederate skirmishing. His men "were compelled to keep continually under cover and suffered great constraint by being kept so continuously in the ditches, which were frequently very wet and muddy." Various units in the Fourth and Sixteenth Corps also suffered, especially in the area immediately behind the main line, due to balls overshot from the opposing skirmishers. "Men were killed in their camps at their meals," wrote division leader David S. Stanley, "and several cases happened of men struck by musket-balls in their sleep, and passing at once from sleep into eternity. So many men were daily struck in the camp and trenches that men became utterly reckless, passing about where balls were striking as though it was their normal life, and making a joke of a narrow escape, or a noisy whistling ball."[33]

Overland

The intense campaigns of 1864–1865 gave the eastern skirmishers in both armies an opportunity to prove themselves. Lee's sharpshooter battalions, many of them newly organized, especially wanted to demonstrate their abilities against Federal line units, which had less training and usually less experience at loose-order tactics. From the Wilderness to Appomattox, the Army of Northern Virginia and the Army of the Potomac were rarely out of contact with each other for eleven months.

The two-day battle at the Wilderness witnessed deadly skirmishing by both sides. Capt. Robert F. Ward, who commanded the sharpshooters detailed from the Forty-second Mississippi, later recorded what happened to the battalion of

Stone's Mississippi brigade in the Third Corps of Lee's army. At the start of the engagement on May 5, Ward's men remained in the ranks of the Forty-second Mississippi until called on for duty. Then they "stepped out quickly in double rank, in front of the right wing of the regiment, and counted twos." Ward gave the order to spread out by guiding on the right group of four while still in place, but his regimental commander overrode that directive because of the rapid approach of Federal skirmishers. He instructed Ward "to take intervals as we advanced." The sharpshooters positioned themselves 60 yards ahead of the battle line as Ward took charge of the battalion's right wing. Told to hold there until he could see the Federal flag, Ward followed his instructions to the letter. He retired to the Forty-second, where the colonel had created room on the left of his regimental line for the sharpshooters.[34]

Here Ward's twenty-four men were crowded into a space "of not more than fifteen steps." Nineteen of them were shot during the first hour of fighting. Losses among the companies of the Forty-second Mississippi were nearly as heavy, especially among the officers, and Ward was told to leave his remaining sharpshooters to fend for themselves while he took charge of the regiment's left wing. He instructed the sharpshooters to rejoin their companies for the rest of the day.[35]

Not every sharpshooter battalion in Lee's army fared so badly as did Ward's contingent, but his story demonstrates that even expertly selected, well-trained skirmishers could be mishandled or devastated by the circumstances of battle. On May 6, skirmishing at the Wilderness continued to take place at relatively short ranges—50 yards and 150 yards, to cite two examples—and casualties continued to be heavy. McGowan's South Carolina Brigade lost 20 percent of its sharpshooters during the two days of fighting. The absence of dedicated skirmish units did not hurt the Union battle effort. This was true even in the case of Smyth's brigade of the Second Corps, in which nearly all regiments were still armed with smoothbores—the Twenty-eighth Massachusetts drew skirmishing duty simply because it had rifle muskets.[36]

By the time of Spotsylvania, with the campaign lengthening out, Lee's sharpshooter battalions developed a habit of manning the skirmish line only during the day and letting details from line units hold it during the night to give them some rest. Mahone's sharpshooter battalion preferred to stay on duty during the night as well, being relieved one day a week to rest. Another feature of the history of Lee's sharpshooters began to develop at Spotsylvania: to detail the most qualified men for scouting duty. This was the most dangerous role expected of the sharpshooters. Berry Benson of McGowan's battalion began to go out on his own, banking on his privileges as a sergeant, until his exploits came to the attention of higher officers, who then specifically ordered him to go forward to find out specific bits of intelligence about the Yankees. Benson enjoyed this duty a great deal until he was captured on May 17. Held for five months, he then escaped to rejoin his unit.[37]

For the Second Corps attack on the Mule Shoe Salient at Spotsylvania on May 12, Brooke's brigade of Barlow's division deployed one regiment as skirmishers. The men were placed 1 pace apart to cover the constricted front of the brigade, which was formed as a concentrated column, and told not to get more than 30 yards ahead of the unit during the advance.[38]

In the fierce Confederate counterattacks into the captured section of the Mule Shoe Salient on May 12, McGowan's sharpshooters volunteered to go in as part of their brigade battle line. They "fought like tigers" and lost heavily, leading to a division-level order against such employment except in cases of dire emergency. The survivors of McGowan's battalion rested two days after their exertion.[39]

Ward's sharpshooters of the Forty-second Mississippi found themselves in another tight spot farther to the right of the Salient on May 12, stationed in an open pine woods that offered little shelter and with a slashing of trees 80 yards deep between them and the friendly battle line. It was very difficult to crawl through the felled trees to safety, and the sharpshooters had to take their punishment. Within one week of the opening of the Overland campaign, Ward and one other man were the only ones left out of twenty-four sharpshooters from the Forty-second Mississippi. The sharpshooter contingent from the Eleventh Mississippi, also in Stone's Brigade, lost five commanders during the month-long course of the Overland campaign.[40]

During the North Anna phase of the Overland campaign, the sharpshooter battalions of Mahone's Brigade and Sanders's Brigade, totaling 300 men, held off the skirmishers of the Union Fifth Corps for two hours as the Federals crossed the North Anna River and established a bridgehead on the south side near Jericho Mills. The Confederates dug in, using fence rails to make a breastwork for each group of three sharpshooters. The men in each group staggered their fire to keep one gun always loaded and shot down Union skirmishers as close as 30 yards away. On other parts of the line at the North Anna, skirmish firing seemed to grow more intense than at previous engagements. One Union gunner called it "exceedingly severe and murderous. We were greatly annoyed by it." Often men were hit while relieving themselves, causing soldiers on both sides to increasingly despise their opponents. A recent historian has estimated that up to 600 Federal losses out of some 2,000 suffered at the North Anna were due to Confederate skirmish fire. There was no general engagement or determined local attacks by the Federals during this phase of the campaign, thus accounting for the higher percentage of loss due to skirmishing.[41]

The Cold Harbor phase of the Overland campaign extended for many days, from about May 26 through June 12, 1864. In Lysander Cutler's division of the Fifth Corps, a brigade skirmish line consisted of 360 men and was supported by a reserve of four companies. The skirmishers were pushed out 400 yards in front of the main line, only 200 to 300 yards from the main Confederate line. After the failed Union attack on the morning of June 3, the armies remained locked in static positions as close as 40 yards from each other for more than a week, producing conditions akin to siege warfare. Skirmishers on both sides annoyed their opponents with deadly consistency. "All day long we have to huddle close to the breastworks," reported Daniel Chisholm of the 116th Pennsylvania. With both sides dug in, the battlefield had a deceptive appearance of emptiness. "The air was filled with whistling bullets yet not a man could be seen," marveled a Union engineer officer, "nothing but a dull line of fresh yellowish earth from which during the day asscended [*sic*] a succession of little white clouds of smoke and at night, bright flashes of fire."[42]

Jaded by the long campaign, and a little bored, too, David Holt of the Sixteenth Mississippi tried some fancy target shooting on the skirmish line. Positioned only 75 yards from the Federals, he fired two rounds at a flagstaff and cut it in two. Men on both sides cheered his prowess, but Holt agreed with some of his comrades that "it was an accident." Other jaded men along the Cold Harbor lines raised their hats on the tips of swords and bayonets to tempt enemy skirmishers. Robert Stiles, a Confederate artilleryman, was amused that Federal skirmishers shot ninety holes in a shelter tent behind the Rebel line.[43]

The gunners of Capt. Basil C. Manly's North Carolina battery were not happy at the treatment accorded them by the Yankees at Cold Harbor. They were under tremendous fire for four days and nights at a range of 125 yards. The Yankees "shot the wheels off" one of their Napoleon guns and nicked the spokes in the wheels of all other pieces. Even the bronze tubes were nicked by bullets. An anonymous member of the battery, writing to a hometown newspaper, concluded that Grant's skirmishers "do much more execution than all his 'charges.'" They hit "any thing from the size of a quarter dollar, up to a man. Some of our boys put a tin cup on the breastworks yesterday, and in less than a minute they put three holes through it." Manly lost one killed and nine wounded at Cold Harbor, a level of loss comparable to the battery's casualties at Gettysburg and Spotsylvania, and almost all of his Cold Harbor losses were due to skirmish fire.[44]

When Grant disengaged at Cold Harbor and moved south across the James River, he covered the maneuver with a cavalry division. McGowan's Brigade was among the Confederate units sent to follow up and find out what the Yankees were doing. McGowan's sharpshooter battalion handily lead the brigade advance, 200 yards in front, yelling as they moved through woods and driving the dismounted Union cavalry skirmishers before them. "Our sharpshooters drove them so easily and continuously that the line of battle had only to follow up and keep in readiness to assume their burden of battle," an admiring officer later wrote.[45]

Petersburg

The contending armies remained in heavily fortified positions near Petersburg from mid-June 1864 until April 2, 1865. In places, they were as close as 125 yards apart. Yet along almost the entire line, which eventually extended to more than 30 miles in length, commanders maintained a fortified skirmish or picket line. Firing waxed and waned, depending on the sector and the period of time during this long campaign. It tended to be heavier from the Appomattox River to the Jerusalem Plank Road, where the lines were close together, and almost nonexistent west of the road because the Federals deliberately planted their works more than a mile away from the enemy as they extended to outflank the Confederate right.

Perhaps the heaviest skirmish fire occurred early in the campaign at places like Hare's Hill and Pegram's Salient. "There is not a day passes when at least one member of our number is not struck with a Minnie ball," commented George D. Bowen of the Twelfth New Jersey. "Have not been in what one would call an engagement either. This picking us off one at a time gets on one's nerves." Another Second Corps member wrote home, "If you dare to show your head above the rifle pits you are a gone goose." A Ninth Corps soldier named

Clarion Miltmore warned his mother that three of his comrades were shot near him as he wrote a letter home. The Eighty-third Pennsylvania lost four to seven men each day on June 19–20.[46]

Some Confederate units kept up a steady expenditure of ammunition on the skirmish line. Bushrod R. Johnson detailed 1,100 men from his division as skirmishers, firing a total of 20,000 rounds every twenty-four hours. That amounted to 18 rounds per man. Johnson was pressured by superiors not to waste ammunition, but he argued that he had to maintain firing at this level to counter the Federal skirmishers. Deserters told him that Yankee commanders issued 100 rounds to each picket, with orders to fire them all during their shift on the skirmish line. Johnson sometimes risked censure and encouraged his men to fire all night. While not all Confederate units kept up heavy firing, rumor had it that the Army of Northern Virginia was expending 37,000 rounds each day along the Petersburg line by the end of the campaign.[47]

The Federals also maintained a heavy volume of skirmish fire at Petersburg. Capt. Joab Goodson of the Forty-fourth Alabama lost his brother Rufe to a Federal skirmisher who shot him when Rufe reached for a haversack to get something to eat. Goodson was heartbroken. "I wept as I never wept before. I bent over the dear boy, and called him back, but in vain." Rufe's body remained in the trench for several hours until darkness offered an opportunity to remove it.[48]

The Petersburg campaign also involved several Union offensives to extend the line and a number of sharply fought battles. In the failed Union attack at Pegram's Salient, after the explosion of a mine that ripped a hole in Confederate defenses, Rebel counterattacks once again sealed the breach. The sharpshooters of Mahone's Virginia brigade happened to be off duty when the unit was ordered to the scene of action, and they participated in the charge. The battalion fought hard, losing 94 of its 104 men in vicious hand-to-hand combat. Capt. Wallace Broadbent, its commander, was killed, with up to fifteen bayonet wounds in his body.[49]

At Reams Station on August 25, 1864, the sharpshooters of Lee's army had their finest day of the war. Several Confederate brigades tried to eject the Union Second Corps from its lodgment on the Weldon Railroad, Lee's principal supply line from the south. The initial Rebel attack failed. To pave the way for a second attempt, the sharpshooter battalions of McGowan's and Scales's brigades drove in the Union skirmishers along the west side of Hancock's fortified position and established themselves on a slight ridge 400 yards from the main Union line. The sharpshooters estimated the distance, adjusted their sights, and began to fire at will for five hours, with comrades bringing up extra ammunition. McGowan's skirmishers fired 160 rounds each, or 1 round every 1.8 minutes. At first, the Federals responded vigorously, but the return fire slacked off until it virtually disappeared. Some Confederate sharpshooters even began to fire while standing, and all of them zeroed in on the Union artillery horses and gun crews. The Tenth Massachusetts Battery lost nearly all its horses; some of them were "riddled by a dozen balls." This effective skirmish fire reduced the spirit and fighting capacity of the Federals, allowing the infantry attack to lodge inside the Union position. After hard fighting, Hancock's troops were driven away.[50]

During many of the engagements at Petersburg, commanders deployed an entire regiment to cover their brigade front, a decision necessitated by the reduced size of most regiments by this stage of the war. The Seventeenth Michigan, for example, had only 100 men on duty when it was sent to the skirmish line during the battle at Fort Stedman on March 25, 1865. At least one Union regiment, the 190th Pennsylvania, seemed to enjoy frequent assignments to the skirmish line. Armed with Spencer repeaters, the Pennsylvanians "were uneasy if placed in line of battle." They preferred to go forward in loose order where they could "stir up every thing in front."[51]

The sharpshooters of Lane's North Carolina brigade, led by Maj. Thomas J. Wooten, developed a trench-raiding tactic designed to scoop up Union prisoners and supplies before the enemy could react. Wooten called it "seine hauling" after a common fishing technique. He quietly advanced his men in line across most of no-man's-land, and then ran the last few yards, sweeping right and left. McGowan's and Scales's battalions also conducted trench raids to secure supplies and camp equipment from the Yankees.[52]

The Federals apparently did not conduct trench raids at Petersburg, having little need to capture Confederate supplies, but they mounted pushes now and then to take sections of the Rebel skirmish line. One such operation involved the Twentieth Indiana, the Second United States Sharpshooters, and three companies of the Ninety-ninth Pennsylvania, which advanced along Jerusalem Plank Road to drive back the Confederate skirmishers and relieve intense sharpshooting in that sector. The Federals struck at 1 A.M. on September 10 and easily accomplished their objective, employing the element of surprise. Confederate counterattacks drove back the three companies of the Ninety-ninth Pennsylvania, but the other two units held firm and dug in to permanently secure their gains. The Federals captured ninety prisoners and lost forty men in this operation.[53]

Confederate skirmishers had a final opportunity to display their talents in the final rounds of fighting at Petersburg. Union skirmishers advanced to capture a long stretch of the Confederate picket line west of the Weldon Railroad on March 25, 1865. This gave them an important advantage in assembling assault columns, but the Confederates could not hope to retake all that was lost. Instead, they concentrated on recapturing McIlwaine's Hill, a vantage point deemed worth the effort. All the sharpshooter battalions of Wilcox's Division, some 400 men, set out after dark on the night of March 26. Two battalions constituted the first line, wheeling right and left when they reached the Federal pickets atop the hill, while the second line moved forward to hold the crest. By concentrating power on a small spot and moving unexpectedly, the Confederates easily retook the hill with as few as eight men wounded, while injuring twenty-three Federals and capturing twenty-six more. After arranging for a truce to collect their dead from the March 25 action, the Confederates retired from the crest before dawn but maintained their picket line close to the hill.[54]

As the Petersburg campaign drew to a close, Berry Benson was "elated" to be assigned permanent scout duty "with a few men under me." When McGowan's Brigade advanced against the Federals from the White Oak Road Line, on the far Confederate right, Benson and two comrades led the way. The three men maintained 100-yard intervals, with the sharpshooter battalion to their rear and the brigade battle line behind the battalion. One of the last skirmish actions of McGowan's sharpshooters occurred at the battle of Sutherland Station, April 2, 1865. Here Benson recalled setting the sights of his rifle at 400 yards and enjoying "some right lively sharpshooting" with the Federals across this largely open battlefield.[55]

The Significance of Skirmishing in the Civil War

Historians have only recently begun to pay attention to the sharpshooter battalions of Lee's army during the last year of the war, leading to the conclusion that they were elite, specialist units that proved their worth in action. It is generally believed that the sharpshooters "tended to dominate the picket line" during the Overland and Petersburg campaigns. The best battalion commanders developed innovative tactics well suited to trench warfare. Siphoning off the best marksmen from parent units tended to decrease the combat effectiveness of some brigades, but the qualitative superiority on the skirmish line made up for it, according to the most recent historian of the battalions. "The organic sharpshooter battalions gave the Confederates a degree of tactical flexibility that the Federals simply could not match," Fred Ray has recently written, as he compares the battalions to the specially selected attack battalions organized by the German army late in World War I.[56]

Modern analysis of the sharpshooter battalions is influenced mostly by the fact that they were among the few Civil War units that received specialized training in estimating distance and target shooting, and by the accounts written by survivors of the battalions. Capt. John D. Young, who commanded the sharpshooters of Scales's North Carolina brigade, argued that rotating line units on and off the skirmish line often forced officers who knew little of that duty to operate in loose-order formations, leading to inefficiency. Lee's sharpshooter battalions could, "by discipline and association," become "equal in strength and effectiveness to twice or even thrice their number," Young argued.[57]

Capt. William S. Dunlop, who led McGowan's sharpshooters, called these battalions the "case hardened steel points attached to the brigades, whose duty it was to crush the outer lines and admit our columns to the inner lines and strongholds of the enemy." Dunlop even suggested that Lee allowed his skirmishing battalions to operate almost without orders from the battle line by the late summer of 1864. Curbing the men's enthusiasm as well as the fierce spirit of rivalry between battalions became the only discipline problem experienced by sharpshooter officers, according to Dunlop. Casualties were heavy, to be sure. McGowan's battalion lost half its men by late August 1864, and 125 more had to be detailed from parent regiments the following February when numbers dipped to a mere fifty of the original. Yet Capt. John E. Laughton Jr. thought the sharpshooters were "the most effective troops . . . for the numbers of men engaged."[58]

It is easy to give the veterans their just due for performing the arduous and dangerous duty of skirmishing, and also easy to admit that Lee's sharpshooter battalions were unique and impressive examples of Civil War specialist units. But there is no convincing evidence that they consistently dominated the skirmish line or regularly outshot their opponents during the Overland or Petersburg campaigns. In fact, as Robert Ward admitted in his postwar account, his sharpshooter battalion was roughly handled by the Yankees at both the Wilderness and Spotsylvania. The Confederate sharpshooters performed brilliantly at Reams Station and smartly reclaimed McIlwaine's Hill, but they lost a huge chunk of the skirmish line on March 25 and, as one example of a smaller-scale version of that loss, on September 10 along the Jerusalem Plank Road. They were not invincible. Every time they were beaten in the Overland and Petersburg campaigns, it was at the hands of nonspecialist infantrymen rotated for a tour of duty onto the skirmish line from the main battle formation, fighting without any training in estimating distance or target shooting. In fact, by far the best skirmishing ever done in the Civil War was by the Federals during the Atlanta campaign, especially during the last month of the operation in August 1864. This is well and thoroughly documented by surviving accounts on both sides, and it was performed entirely by Federal units without any specialized training for the role of skirmishers.

This is not to suggest that all the training laboriously given the Confederate sharpshooter battalions was wasted. Lee's units excelled the Federals in scouting, trench raiding, and perhaps sniping. Union regiments operating on the skirmish line seem to have performed these operations rarely, if at all, although there was a general tendency to detail the best shots to snipe now and then. The assistant inspector general of Hardee's Corps in the Army of Tennessee thought the sharpshooter battalion of Cleburne's Division was "a most efficient Corps in operating against artillery and mounted men, preceding as well as during engagement." Andrew Haughton has recently noted that the sharpshooter battalions were "the most adaptable and utilitarian units in the Confederate forces" because they could perform as skirmishers, snipers, and support for batteries as well as fight in the battle line. Ironically, all these roles were regularly and effectively performed by ordinary line units as well. Some of the roles were undoubtedly performed better by the sharpshooters, but traditional line infantry still remained the true all-purpose units of the Civil War.[59]

In short, there is every reason to be cautious when evaluating the effectiveness of the sharpshooter battalions. The Federal army won the war, and turned in the best skirmishing performance of the conflict in the Atlanta campaign, without them. A few specially detailed skirmish units appear in the Union army in the East. The Fifth, Sixth, Ninth, and Eighteenth Corps had division-level sharpshooter companies for a time in 1864. Although termed "sharpshooters," one wonders if these were actually details for sniping duty in the trenches. They were all created by corps- or division-level initiatives, and many were only temporary.[60]

Generally, the Federals did not think it necessary to create specialized skirmish units. Sherman noted in his memoirs that prewar tradition assigned available rifles to the first and tenth companies of a regiment for their use as skirmishers, while the other eight companies retained smoothbores. These two skirmish companies "were never used exclusively for that special purpose" in the Civil War, "and in fact no distinction existed between them and the other eight companies" because all of them were now armed with rifle muskets. When Silas Casey published the official U.S. Army tactical manual in 1862, he retained the old idea of designating two companies as skirmishers. The secretary of war Edwin M. Stanton, however, inserted a clause suspending that provision of this otherwise impressive manual. In the long run, it might have been the wiser course for the Federals to insist that all their regiments be able to skirmish rather than rest that important duty on a small cadre of elite troops.[61]

Paddy Griffith has noted that even though the skirmish line was increased in size and more heavily used than ever before, it was still only a skirmish line and could never replace the function of the battle line. Other historians, including those who firmly adhere to the old interpretation of the rifle musket's revolutionary impact on warfare, have agreed with that conclusion, adding that loose-order formations were not suitable for main line units in the Civil War due to the inability of officers to control large numbers of men without close-order tactics.[62]

Yet there is a strain in the historiography, a kind of mystique about skirmishing in the Civil War, that portrays it as the wave of the future. Writers who yearn to see some degree of innovation, especially to see signs of a commitment to loose-order tactical formations for entire armies, tend to grasp onto every straw they can find in Civil War accounts to argue that a few wise men were beginning to get away from the foolishness of fighting in close-order lines. Sherman unwittingly provided fuel for this historiographical tendency in his memoirs. "Very few of the battles in which I have participated were fought as described in European text-books, viz., in great masses, in perfect order, manoeuvring by corps, divisions, and brigades. We were generally in a wooded country, and, though our lines were deployed according to tactics, the men generally fought in strong skirmish-lines, taking advantage of the shape of ground, and of every cover."[63]

This oft-noted passage from the first memoir to be published by a major figure of the war is fraught with misapprehensions. If Sherman had the idea that European warfare was characterized by precise maneuvering over open, parade-ground-like terrain, he could only blame writers and historians for it. Actually, the typical European battlefield was, like its American counterpart, cluttered with obstacles, including woods, hedgerows, hills and, unlike in America, small farming villages and fortified farmsteads. All of these impediments just as readily broke up formations as did similar obstacles in the United States. Based mostly on the experience of the Atlanta campaign, Sherman generalized far too recklessly about his men fighting "in strong skirmish-lines." The effect of the good skirmishing as they drove toward Atlanta was to better prepare for success when the next pitched battle took place by the main line. The intense skirmishing of August so deteriorated some Confederate brigades that the Rebel attempt to counter Sherman's seizure of the Macon and Western Railroad at Jonesboro was severely retarded. The result was that the campaign ended on the battlefield of Jonesboro with a whimper instead of a bang by some of Hood's formerly best units.

In short, good skirmishing was never a substitute for good fighting by the battle line, and the latter still performed in close-order linear formations rather than loose order. Some type of loose-order formation did lay in the future of infantry warfare, but skirmishing would rapidly decline in significance as a result. The Civil War not only represented the high point in the history of skirmishing, it was also among the last major wars in which large-scale skirmishing took place. The rapid advance in weapons technology—converting from rifle muskets to repeater, or magazine rifles—spurred experiments in applying loose-order tactics to the battle line that gradually developed during the decades after Appomattox. In the process, army tacticians employed numerous scouts to precede the formation and find the enemy. Once found, the battle line deployed into three successive lines of comparatively loose-order formations—the first to fire on the enemy (and fulfill some of the function of skirmishers), the second to attack, and the third to act as a reserve. Doctrine dictated that the battle line should absorb the functions of the skirmish line and begin to adapt to the loose-order concept. No longer was there a distinction between the functions and deployment of the skirmish line versus the battle line.[64]

In the end, the Civil War vindicated prewar notions that the rifle musket made it possible for all line infantry to skirmish effectively. This occurred even though line infantry did not receive training to use the rifle for long-range firing or any special training for physical conditioning. The average soldier learned loose-order tactical formations, and that tended to be all the special skirmish training he received. For the rest of it, line infantrymen relied on native talent and hard experience earned on the skirmish line to achieve whatever level of proficiency they accomplished as skirmishers.

SNIPING

The Civil War became the first conflict in history in which men were specifically detailed to perform the modern role of sniper. Something like sniping had occurred in warfare since the introduction of small arms in the sixteenth century, but in a haphazard, individual fashion. As the first rifle war, it is not surprising that the Civil War would see a significant change in the history of sniping. Furthermore, Americans led the world in the development of telescopic sights for long-range rifles before 1861. Morgan James of Utica, New York, and William Malcolm of Syracuse were among the prominent promoters of rifle target shooting with telescopic sights. Horace William Shaler Cleveland praised the use of telescopic sights for picking off individual targets on the battlefield in an article published in 1862. Exaggerating a bit, Cleveland argued that "with the telescope-rifle the question is not, whether an enemy shall be hit, but what particular feature of his face, or which button of his coat shall be the target."[1]

Confederate congressmen worded the resolution authorizing the creation of sharpshooter battalions in such a vague way that it is difficult to know if they intended these units to serve as organized snipers or as dedicated skirmishers. As it turned out, the units gave rise to both types of specialized troops, for individuals among the sharpshooters were detailed on regular duty to prowl the battlefield and pick off enemy targets at long range. Even though they never systematically created an equivalent to the sharpshooter battalions, the Federals also detailed individuals and squads of men for dedicated sniping duty.[2]

It is not always easy to differentiate a sniper from a skirmisher in Civil War accounts, for their duties sometimes overlapped, and snipers often were detailed from the skirmish line. Skirmishers at times laid down concentrated fire on the enemy for long periods of time or took potshots at individuals when the lines were stabilized within range of each other. Those men on the receiving end could not know if the bullet was fired by a skirmisher performing a sniper-like deed or someone who had earned the right to be given special duty as an expert, lone marksman.

Nevertheless, it is important to differentiate between skirmishing and sniping, for the two were indeed separate military functions. The word *sniping* has its origin in nineteenth-century India, where British officers often amused themselves by hunting the snipe, "a small, fast-flying game bird" that was difficult to hit in midair. The term *snipe* therefore became a byword for a crack shot, but the term was apparently not widely used until World War I, when newspapers adopted it to describe what Civil War contemporaries more commonly referred to as sharpshooting.[3]

For our purposes, it is useful to define sniping in the Civil War as strictly as possible. The ideal sniper was not only adept at handling his weapon, he was a particular personality and psychological type. A loner who shunned the limelight and often the company of his comrades alike, the typical sniper of the twentieth century was capable of spending hours alone waiting for his target to expose himself for a few seconds. He was a hunter more than a soldier, often working with other snipers in pairs or squads for greater efficiency in drawing out and bagging their game.[4]

There definitely were men of this type in the Civil War, most of them obscure individuals, though their presence was noted by various observers. The only one who became widely famous was California Joe, a member of the First United States Sharpshooters, who plied his trade much to the enthusiasm of Northern journalists during the Yorktown phase of the Peninsula campaign.[5]

A few Civil War contemporaries understood the modus operandi of true sniping and commented on it in their postwar writings. Lew Wallace admiringly wrote of Birge's Sharpshooters, formally designated the Sixty-sixth Illinois. "In action each was perfectly independent. They never manoeuvred as a corps. When the time came they were asked, 'Canteens full?' 'Biscuits for all day?' Then their only order, 'All right; hunt your holes, boys.' Thereupon they dispersed, and, like Indians, sought cover to please themselves behind rocks and stumps, or in hollows. Sometimes they dug holes; sometimes they climbed into trees. Once in a good location, they remained there the day. At night they would crawl out and report in camp."[6]

Wallace was bending the truth here, for there is little evidence that Birge's men ever performed exactly in this way except perhaps during the Union advance on Corinth, when the Federals were within striking distance of the Rebels for several days at a time. Nevertheless, Wallace provided a concise description of sniping methods completely understandable to any Civil War soldier.

William Y. W. Ripley also paid close attention to his comrades in the First United States Sharpshooters who became good snipers. "In fact, sharp shooting is the squirrel hunting of war; it is wonderful to see how self-forgetful the marksman grows—to see with what sportsman-like eyes he seeks out the grander game, and with what coolness and accuracy he brings it down. At the moment he grows utterly indifferent to human life or human suffering, and seems intent only on cruelty and destruction; to make a good shot and hit his man, brings for the time being a feeling of intense satisfaction."[7]

1862

In the early part of the war, individual commanders sometimes took the initiative to detail their best marksmen on specialized target duty, using whatever weapon they happened to possess. This was always done when the armies were temporarily close to each other and only for limited times, often just to reduce incoming fire at a particular moment in a campaign. Col. William T. Shaw of the Fourteenth Iowa kept "a few of my best marksmen sufficiently advanced to keep the enemy from coming outside their intrenchments to annoy my men by their marksmen" at Fort Donelson. These detailed men were under the temporary direction of Capt. Warren C. Jones of Company I, indicating that Shaw must have designated a sizeable number of them. Col. Carroll C. Marsh of the Twentieth Illinois also detailed some of his troops to pick off the Confederates "as they exposed themselves above the breastworks" at Donelson.[8]

On the other side of no-man's-land, the Confederates were annoyed enough by these sharpshooters that individuals who happened to possess good rifles took it upon themselves to counter the fire. Capt. Flavel C. Barber of the Third Tennessee owned a Maynard rifle, which he loaned to his men and sometimes used personally to fire at the Yankees. Barber's lieutenant colonel told him on the evening of February 14 to select thirty of his best shots and report for duty. He assumed he was to lead a true sharpshooter detachment to pick off the Federals who were harassing a nearby Rebel battery, but his superior merely wanted him to relieve the skirmish line of the Fourteenth Mississippi. Barber was a bit frustrated with this more mundane assignment, but night put an end to the sharpshooting anyway by the time he positioned his troops on the skirmish line.[9]

The battle of Shiloh led the commander of the Seventieth Ohio to form a company of marksmen by detailing five men from each company "to pick Officers and Commanders [off] when the line of battle formed." They were to operate "in small squads near enough to make sure work" and return to their usual place in the ranks when not needed. There is no evidence that this innovative experiment lasted long or was tested in battle.[10]

In the East, the Peninsula campaign afforded opportunity for sniping to develop. The First United States Sharpshooters detailed two companies with James target rifles some 800 yards from the Confederate defenses to fire on gun crews. They contributed to the wear and tear on Confederate nerves during the month-long confrontation at Yorktown. A Confederate soldier wrote home that the Yankees "pick a fellow off if he shows so much as his head." The most famous member of the regiment, however, performed impressive feats of marksmanship with a Sharps rifle. California Joe, the nickname of Truman Head, was described as "an unassuming man, past the middle age, short in stature, light in weight, and a true gentleman in every sense of the word." He derived his nickname because he traveled from his home in California to join the army. "Joe was a dead shot," reported Pvt. Robert Knox Sneden, "and had often kept the enemy from firing a gun for half an hour at a time by shooting five, six, and seven one after another . . . through the embrasures."[11]

The stories that circulated about Joe were not hyperbole. Sneden documented an incident in which a Confederate sharpshooter lodged himself in the chimney of a house and hit several Federals 500 yards away. Joe established himself in an advantageous spot and fired three rounds, being able to see where they hit the bricks. On the fourth try, he hit the Rebel. Several days later, when the Federals advanced skirmishers to the house, they pried the body out of the chimney and discovered it was a Native American who had been shot "between the eyes[,] and the back of his skull was all blown out!"[12]

Truman Head was discharged from the army in the fall of 1862 because of old age and "failing health in which his eyes were affected." He returned to California, where he died in 1888, and the citizens of San Francisco erected a monument to his memory.[13]

Charles A. Stevens, a member of the First United States Sharpshooters, recorded the efforts of an African American who tried his hand at sniping on the Confederate side of the line at Yorktown. Berdan's men let the would-be sniper alone for a few days to encourage him to expose himself. As the man crept closer to the Union position, shielding himself by using trees and the lay of the land, the Federals stationed a sniper in a position to hit him when an opportune moment arrived, which soon happened.[14]

Stevens also wrote of a comrade he identified only as Seth, who took a James rifle and dug into the ground between the lines at Yorktown. Because it took time to properly reload the weapon, his comrades covered him with their fire, allowing Seth to silence a nearby artillery piece for two days in a row. His friends brought forward food, water, and extra ammunition at night, crawling along the ground.[15]

Sniping was possible even during a pitched battle lasting only a few hours, although not as common as during a static confrontation. Maj. Gen. Winfield S. Hancock detailed men from his division of the Union Second Corps on the morning of September 18, 1862, the day after the battle of Antietam, when Confederate sharpshooters annoyed the command. Hancock reported that these men were detailed from the battle line rather than the skirmishers, and in fact they supplemented the return fire of the skirmishers.[16]

1863

Sniping became more common in 1863 with the increased use of long-range target rifles among Confederate troops, most of which were smuggled through the Union naval blockade, and with the growing experience of veteran troops in developing new ways to conduct war. Some Confederate sharpshooters climbed trees and targeted Union officers at the battle of Champion Hill on May 16, but the siege of Vicksburg provided a greater opportunity than any yet for snipers

Table 7.1. Range of sniping

Unit	Battle	Range
1st United States Sharpshooters, Porter's division, 3rd Corps, Army of the Potomac	Yorktown, April 1862	800 yards with target rifles, against Confederate artillery[1]
Sharpshooters, Cleburne's Division, Hardee's Corps, Army of Tennessee	Liberty Gap, June 26, 1863	700–1,300 yards with Whitworth rifles, against Federal skirmishers[2]
12th Connecticut, Weitzel's brigade, Augur's division, 19th Corps	Port Hudson, June 1863	Received Confederate rifle musket fire at 800 yards[3]
Sharpshooters, Cheatham's Division, Hardee's Corps, Army of Tennessee	Atlanta campaign	Set Whitworth rifle sights at 2,200 yards and claimed to have driven off a Federal wagon train[4]
1st United States Sharpshooters, Crocker's brigade, Birney's division, 2nd Corps, Army of the Potomac	Spotsylvania, May 9, 1864	1,500 yards with Sharps rifles and improvised sighting mechanism, against a Confederate signal party in a tree[1]
1st United States Sharpshooters, Walker's brigade, Birney's division, 2nd Corps, Army of the Potomac	North Anna, May 24, 1864	300–1,000 yards with Sharps rifles, against Confederate skirmishers[1]
1st United States Sharpshooters, Walker's brigade, Birney's division, 2nd Corps, Army of the Potomac	North Anna, May 1864	400–600 yards with Sharps rifles, against Confederate artillery[1]
Franklin M. Goff, 14th Connecticut, Smyth's brigade, Gibbon's division, 2nd Corps, Army of the Potomac	Cold Harbor, June 1864	At least 800 yards with a Sharps rifle, against infantrymen[5]
James Ragin, 1st United States Sharpshooters, Walker's brigade, Birney's division, 2nd Corps, Army of the Potomac	Petersburg, June 1864	Dueled with two Confederate snipers at 300 yards and 400 yards on consecutive days, using telescopic rifle[1]
Edgeworth Bird, 15th Georgia, DuBose's Brigade, Field's Division, 1st Corps, Army of Northern Virginia	Petersburg, July 1864	Reported that Federal snipers killed cattle at a distance of 1,100 yards, then shot Confederates who tried to cut up the carcasses[6]
Unidentified member of Sharpshooter Battalion, Mayo's Brigade, Heth's Division, 3rd Corps, Army of Northern Virginia	Petersburg, July 1864	2,250 yards with a Whitworth rifle, reportedly hitting a mounted Federal officer[7]

Table 7.1. *(continued)*

Unit	Battle	Range
Blackwood Benson, Sharpshooter Battalion, McGowan's Brigade, Wilcox's Division, 3rd Corps, Army of Northern Virginia	Petersburg campaign	Randomly fired a Whitworth rifle at a Federal camp more than a mile away, reportedly hitting two men[8]
Francis Bass, Sharpshooter Battalion, Mayo's Brigade, Heth's Division, 3rd Corps, Army of Northern Virginia	Petersburg campaign	500 yards with an Enfield rifle, shooting a Federal sniper out of a tree[7]
Isaac N. Shannon, Sharpshooters, Brown's Division, Cheatham's Corps, Army of Tennessee	Franklin, November 30, 1864	Half a mile with a Whitworth rifle against a Federal battery[4]

[1] C. A. Stevens, *Berdan's United States Sharpshooters in the Army of the Potomac, 1861–1865* (Dayton, OH: Morningside, 1972), 38, 47, 435, 440, 460–461.
[2] Patrick R. Cleburne to Archer Anderson, August 3, 1863, in *The War of the Rebellion: A Compilation of the Official Records of the Union and Confederate Armies*, 70 vols. (Washington, DC: Government Printing Office, 1880–1901), vol. 23, pt. 1, 587. This work is hereafter cited as *OR*.
[3] John William DeForest, "Port Hudson," in James H. Croushore, ed., *A Volunteer's Adventures: A Union Captain's Record of the Civil War* (New Haven, CT: Yale University Press, 1946), 117.
[4] Isaac N. Shannon, "Sharpshooters with Hood's Army," *Confederate Veteran* 15 (1907): 126, 125.
[5] Theodore G. Ellis to H. J. Morse, August 9, 1864, *OR*, vol. 36, pt. 1, 458.
[6] Edgeworth Bird to Sallie Bird, July 17, 1864, in John Rozier, ed., *The Granite Farm Letters: The Civil War Correspondence of Edgeworth and Sallie Bird* (Athens: University of Georgia Press, 1988), 176.
[7] F. S. Harris, "Fine Shots in the Virginia Army," *Confederate Veteran* 4 (1896): 73–74.
[8] Berry Benson, "How General Sedgwick Was Killed," *Confederate Veteran* 26 (1918): 115.

to ply their trade. James H. Wilson, who served on Grant's staff during the siege, later wrote of the "voluntary practice of the good marksmen, many of whom selected advantageous positions behind stumps and head logs, either to the front or in the main line of works, and, after covering themselves effectively from observation and crossfire, made it their daily practice to watch the enemy and, whenever a head or even a hand showed itself above the defenses, to fire at it singly or in groups." Wilson noted that this mode of warfare "seemed to have a strange fascination for men of a sporting turn of mind." As a result, the Federals found many Rebel soldiers wounded "in the head, arms, and hands" in Vicksburg's hospitals after the siege.[17]

Union snipers often devised special protection for themselves, such as "wooden turrets" in key locations from which they could loose "a plunging and enfilading fire" on the enemy. Wilson marveled at how few moral qualms they exhibited about killing at long distance, treating it as if they were shooting wild game. Even so, when Union and Confederate marksmen conversed across the lines, they quickly made friends of sorts and "observed unwritten rules of informal truce." Both sides were careful to shout warnings before resuming fire.[18]

Like Truman Head at Yorktown, a single Union sniper came to symbolize all the rest at Vicksburg. Lt. Henry C. Foster of Company B, Twenty-third Indiana came to be called Coonskin because he favored this peculiar headgear. Foster was known as "an unerring shot." One night he dug a rifle pit and constructed a firing port, staying there for several days with a stockpile of food and water. Not satisfied with this burrow, Foster gathered crossties from the demolished Jackson and Vicksburg Railroad and constructed a tower in crib fashion, like a log cabin. It reached a sufficient height so that he could see the Confederates behind their parapet and open fire at them through cracks in the wall. Foster used his famous landmark, located about 200 yards from the enemy along the sap leading toward the Third Louisiana Redan, to harass the Confederates. By this time, Rebel artillery was all but silent in his sector, and the railroad ties were adequate protection against small-arms fire.[19]

The Confederates could also lay down effective sniper fire at Vicksburg. In the Union attack of May 19, Col. Thomas Kilby Smith's brigade halted behind slight cover only 75 yards from the Rebel line and suffered from Confederate sniping that was delivered "with devilish skill." Smith reacted by rotating companies from each regiment to keep up a steady skirmish fire and by detailing "the most accurate marksmen . . . with *carte-blanche* to select the best cover." His men fell back to a safer position after dark. To counter Union siege approaches on the position held by the Thirty-first Louisiana, Brig. Gen. William E. Baldwin obtained "a dozen hunting rifles" and put them "in the hands of experienced marksmen." He dug a trench in front of the main Confederate line but behind a line of palisades, from which his snipers could harass Union diggers and slow their approach.[20]

Federal troops at the siege of Port Hudson, like their contemporaries at Vicksburg, had ample opportunities to develop sniping tactics. Members of the Twelfth Connecticut established a line 150 yards from the enemy and made loopholes by combining two logs with portals cut through them. The men then waited "with the patience of cats," as Capt. John William DeForest put it. Three companies at a time were detailed to this position for a duty rotation that lasted one day, with rest for two more days. The men could often see dust kicked up and splinters fly due to their near misses. Only after the Confederates surrendered was DeForest able to get some insight into the effect of this fire when a Confederate officer told him that most losses were due to bullets "between the brim of the hat and the top of the head." The Rebel officer displayed a hoe handle that had been shot three times in three minutes. Among other things, DeForest's story demonstrates how sniping and regular skirmishing tended at times to blend together in battlefield experience.[21]

DeForest refused to participate in sniping, even though his fellow officers often tried their hand at it to relieve boredom. "I could never bring myself to what seemed like taking human life in pure gayety," DeForest later confessed. The Confederates returned the fire his men so methodically delivered, dropping bullets into the gully where the Twelfth Connecticut bivouacked. The regiment lost up to sixty men during the siege due to this fire. One Confederate sharpshooter even plugged a rubber blanket hanging in the center of the gully on a stump. It was about 5 feet above ground and at least 800 yards away from the sharpshooter.[22]

Contemporary with the sieges of Vicksburg and Port Hudson, Union forces conducted a small-scale but effective siege of Confederate Battery Wagner on Morris Island, near the entrance to Charleston Harbor along the South Carolina coast. Engineer Capt. Thomas B. Brooks became dissatisfied with the organization and performance of infantrymen detailed for sharpshooting duty, for they were not adequately protecting the men assigned to dig approach trenches. "The present so-called sharpshooters are inefficient," Brooks sarcastically reported to Maj. Gen. Quincy A. Gillmore, commander of Union forces operating against Battery Wagner. "First, they are not good shots; second, their arms are not in good condition; third, they are not in sufficient numbers, even if they were efficient; fourth, they are not properly officered." Brooks offered a simple but effective solution: make sure that regimental commanders carefully selected only the best marksmen by testing them in target-shooting competitions involving at least five shots. Unit leaders selected no more than one-third of the best men in these competitions, along with two good officers. "These men to be organized into a company, encamped by themselves, and provided with the best arms that can be procured." Gillmore immediately approved of the scheme and it was implemented. The men, however, used Springfield rifle muskets because specialized target weapons were not available, but Brooks later complimented the sharpshooters as having "proved themselves efficient."[23]

The Confederate defenders of Battery Wagner agreed with Brooks that the

Federal snipers did their job. They also were able to retaliate a bit when a shipment of six Whitworth rifles made it through the blockade; two of them were sent to Battery Wagner. Capt. Samuel A. Ashe fashioned two loopholes with sandbags and detailed his best shots to use the weapons. Comrades raised their hats on a ramrod to draw Union sniper fire and gave the Whitworth men an opportunity to retaliate. The gun had long range and a reputation for accuracy, but with only two weapons available they apparently had little effect on the Federals.[24]

The Whitworth rifle became the weapon of choice among Confederate snipers. More than any other target rifle, it represented the ultimate in achieving the long-range fire capabilities of rifled arms. The Whitworth also contributed to a heightened sense of professionalism among Confederate sharpshooters. With the specialized selection process (involving target competitions) and the organization of snipers into separate units, sharpshooting squads evolved into true elite units by the latter half of the Civil War. Some 250 Whitworth rifles were smuggled into the Confederacy through the blockade from England. It was an impressive weapon for target competition, sporting, or military use, able to throw a bullet 1,800 yards, a bit more than a mile. Some observers noted, however, that one could not hit individual targets at that distance, but at 1,500 yards a skilled marksman could guarantee a hit on a man-sized target.[25]

At a hefty price tag of at least $1,000, the Whitworth employed an "elevated iron sight" similar to the Springfield and Enfield for distances of less than 1,200 yards. Beyond that range, the Davidson telescopic sight was necessary. The weapon had a "recoil pad" on its butt to reduce the kick, although there is evidence that the Whitworth sharpshooters of Cleburne's Division "threw them away as useless." The shooter had to rest the end of the telescopic sight on his eye to achieve maximum visual effect, and the recoil pushed it against the eye socket, causing a considerable amount of bruising. The Davidson telescope was 10 inches long and "fitted with lenses of great power." It was hinged to the barrel at the breech end of the weapon and a sliding apparatus secured the other end to allow the shooter to adjust the elevation for greater distances. The 40-caliber cartridge was "wrapped in parchment and coated with paraffine."[26]

When the Whitworths began to trickle in through the blockade early in 1863, officers started to organize sniper squads. Charles F. Vanderford created several of them in the western Confederate armies, first for Cleburne's Division at Wartrace, Tennessee, and then "a corps comprising all of the Whitworth Rifles in Bragg's army" near Chattanooga. The third was in Johnston's command at Meridian, later led by Leonidas Polk and termed the Army of Mississippi. Polk's army later was transferred to become a corps of the Army of Tennessee during the Atlanta campaign.[27]

While there is ample documentation to prove that several divisions in the Army of Tennessee had Whitworth sniping squads, there is no evidence of an armywide organization. The number of Whitworths in the Army of Tennessee was quite small; Cleburne's squad had only five of them. But the Whitworth sharpshooters were intensely drilled in estimating distance. They first saw action at the battle of Liberty Gap during the Tullahoma campaign, on June 26, 1863. Cleburne employed his five Whitworth sharpshooters to counter heavy Union skirmish fire on his position, and they "appeared to do good service. Mounted men were struck at distances ranging from 700 to 1,300 yards."[28]

Confederate snipers at the battle of Chickamauga may have used Whitworth rifles to harass their opponents on September 20, 1863. Maj. Gen. John M. Palmer reported that they were "busy, and killed and wounded several officers." Some of Palmer's more "adventurous men" countered this fire and shot several Confederate snipers "from the trees upon which they were perched."[29]

Later in the year, during the Knoxville campaign, Longstreet's two divisions from the Army of Northern Virginia possessed twenty Whitworth rifles. A western Confederate named John W. Minnich of the Sixth Georgia Cavalry was enthralled by the gun when he met one of Longstreet's sharpshooters, who assured him that the snipers "were exempt from the usual soldier's routine . . . and that every man of them acted on his own free will when there was anything doing, without restraint, subject only to orders from the division commander or Longstreet himself." Minnich naively offered to trade his Enfield musket for the weapon. "Good Lord, boy! I wouldn't dare go back to camp without my gun. I'd be court-martialed and shot if I lost it that way," retorted Longstreet's man.[30]

1864–1865

In the early months of 1864, a shipment of Kerr rifles made it through the blockade and found its way to the Army of Tennessee. A muzzle-loading target rifle made in England, the Kerr had a range of about a mile but needed special powder and careful loading. The Confederates tried the same powder used in the rifle musket but found they had to clean the barrel after every fourth or fifth round. The Kerr had an adjustable rear sight similar to the Enfield, but it also had an adjustable globe sight on the other end. A sniper squad was created in Brig. Gen. Joseph H. Lewis's Kentucky brigade by holding target competitions to select the ten best shots. This squad operated for the rest of the war, although half its original members were killed, and most of the rest were wounded. With replacements, the total number of men who served in the squad amounted to seventeen.[31]

Lt. George Hector Burton commanded the Kentucky sniper squad. Described as "a superior marksman" himself, and a "dare-devil fighter," Burton possessed "cool judgment, quick apprehension of whatever would give advantage of position, and a dogged resolution that made him proof against sore discomfort and unshaken by disaster." Given general instructions as to the employment of his men, Burton was also told not to go closer than a quarter mile to the Yankees. His command ranged along the front of Hardee's Corps during the Atlanta campaign and sometimes was loaned to other corps if they had particular trouble with enemy artillery.[32]

More Whitworth rifles made it through the blockade, enabling the organization of another sniper squad in Cheatham's Division, Hardee's Corps. Target competitions in each regiment of the division involved shooting at 500 yards, with each man allowed three shots at a board with a bull's-eye painted on it.[33]

The Whitworth sharpshooters of Cleburne's Division, also a part of Hardee's Corps, had their own wagon and reported directly to division headquarters, usually camping near Cleburne's quarters. The squad, commanded by Lt. Abraham B. Schell of the Second Tennessee, received some of the Kerr rifles smuggled through the blockade early in 1864, enough to arm ten of its men. One of Cleburne's staff members later considered its services "as equal to a light battery." Two of the Whitworth sharpshooters were used as scouts, and the squad lost 60 percent of its numbers during the Atlanta campaign.[34]

Lt. Isaac N. Shannon of the Ninth Tennessee led the snipers of Cheatham's Division. He had great faith in the long-range capabilities of the Whitworth rifle, arguing that they "were sighted up to over two thousand yards, and I always believed they would throw their balls five miles." Shannon claimed he drove away a Federal wagon train by setting his Whitworth sights at 2,200 yards and then aiming at the tops of several "tall pine trees." "I always believed the distance to be near three and a half miles," he wrote after the war.[35]

The Atlanta campaign witnessed a great deal of activity by Confederate snipers. Burton's squad of the Kentucky brigade was active every day of the four-month-long campaign except for one, when Burton requested an opportunity for his men to rest and wash themselves. The historian of the brigade explained their mode of operation. "The general plan was to work themselves at night between the lines, reconnoiter, fix upon a rallying base, and then cover the front of the army, and keep a lookout for opportunities to kill off pickets, men who exposed themselves along the lines of Federal breast-works, and officers who came in view beyond while directing the operations of their troops." The snipers paid particular attention to Union artillerymen.[36]

The Federals also employed snipers during the Atlanta campaign, but less is known of their operations. Members of Col. Ellison Capers's Twenty-fourth South Carolina captured a Federal sniper on June 24 along with his rifle. It "had a small looking-glass attached to the butt, . . . so that he could sit behind his breast-work, perfectly protected, with his back to us, and by looking into his glass, sight along the barrel of his piece." This must have been a workable arrangement. Lt. Henry O. Dwight of the Twentieth Ohio reported in mid-August 1864, "Sharp-shooters play an important part in the operations of our army." They secured themselves in rifle pits and used tricks to fool the enemy into exposing themselves. "Nothing short of an actual attack in force will dislodge these sharp-shooters; and it is rarely that one of them is killed." Dwight was not describing the work of skirmishers, for he went on to talk of the stalking instinct displayed by the sharpshooters in question. "They take the same pride in their duty that a hunter does in the chase, and tally their victims in three separate columns—the 'certainly,' the 'probably,' and the 'possibly' killed—thinking no more of it than if it were not men they hunt so diligently."[37]

That Federal sharpshooters had an effect on their enemy is shown by a dispatch written by Leonidas Polk to army headquarters on June 1. He was supposed to forward a report of operations during the campaign thus far but could not do so because of a number of factors, one of which was "the constant occupation of the troops in the trenches (my whole line being very much exposed to the Enemy's Sharp Shooters)." In addition, the Eleventh Tennessee of Vaughan's Brigade lost six men due to Federal sniping in one hour at New Hope Church, and other brigades counted up to eleven casualties per day during the campaign as a result of enemy sharpshooting.[38]

Casualty figures represent one level of loss, but an individual victim, with a life story altered by the sniper's bullet, hits home on a more emotional level. John T. Bell of the Second Iowa was looking at a comrade, George Norris, when a Confederate sniper shot Norris in the face on August 10. The victim stood up, blood gushing from both nostrils and his mouth. "The doctors said he could not live," later reported Bell, but he lost only one eye "and the personal beauty which had formerly made him one of the noticeable men in the company has disappeared forever." Twenty-two-year-old Norris "had been a model soldier, always winning the premiums offered for the best drilled and best equipped soldier in the regiment, and had taken great pride in his record, but this injury broke his spirit." Norris became a changed man. He was discharged due to wounds on January 24, 1865, and went to Pulaski, Tennessee, site of the regiment's camp the previous winter. There he "organized a squad of desperadoes which preyed on the community." Members of the Federal garrison of Pulaski chased and finally shot Norris.[39]

Aside from the Atlanta campaign, the pitched battles fought in the West saw only sporadic use of snipers during 1864. On the day after the battle of Tupelo, the opposing armies remained for a time on the field. Lt. Col. William H. Heath of the Thirty-third Missouri detailed three men from each company to snipe at Confederate skirmishers who were annoying his regiment. Isaac Shannon reported driving off a Federal battery at 450 yards within twenty minutes of opening on it with his Whitworth rifles at Spring Hill, just before the battle of Franklin. Shannon later opened fire at half a mile's distance on Battery A, Second Missouri Artillery before the Confederate attack at Franklin. He claimed to have shot four Federals with five rounds and to have driven off at least one gun, a Napoleon howitzer. In his report of the Tennessee campaign, Maj. Gen. William B. Bate credited Schell's "squad of sharpshooters (with Whitworth rifles)" with "marked gallantry on every occasion when brought into requisition."[40]

The war in Virginia also saw intense, continual contact between Union and Confederate armies in 1864. Some Federals took to the treetops at the Wilderness. Corps commander Richard S. Ewell was warned by a subordinate not to ride into an open space north of Orange Plank Road because "a sharpshooter who was posted in some high tree on the enemy's line had that spot covered and fired at everything seen there." Lee's army had by now acquired more Whitworths, which had been smuggled through the blockade early in the year. "Quite a scramble took place for these guns," recalled a veteran.[41]

A Whitworth was used in one of the most famous cases of sniper fire to occur during the war, the shooting of Sixth Corps commander Maj. Gen. John Sedgwick at Spotsylvania on May 9. Ben Powell of the Twelfth South Carolina used the telescopic sight to strike Sedgwick at a distance of more than half a mile across the open expanse of Spindle Field. Confederate snipers had shot several other men in this area, and Sedgwick was determined to shore up morale by sitting on his horse, remarking to others, "What are you dodging at? They can't hit an elephant at that distance." Just then Powell's bullet hit him under the left eye, causing almost instant death. Twenty-three-year-old Powell had used a Whitworth since before the battle of Gettysburg, receiving with it what he called "a roving commission as an Independent Sharpshooter and scout." Wounded badly in the shoulder at Gettysburg, he recovered in time to participate in the Overland campaign. Besides Powell, only one other man in McGowan's Sharpshooter Battalion carried a Whitworth. Both were allowed to "carry on war at their own sweet will," as comrade Berry Benson later put it.[42]

The First United States Sharpshooters detailed snipers to contend with the enemy at Spotsylvania. A Confederate signal party perched in a tree observed the Second Corps' approach to Po River on May 9, some 1,500 yards away from the snipers' position. Because the Sharps rifles carried by the unit had sights that went only to 1,000 yards, the men had to improvise "by cutting and fitting sticks to increase the elevation." They could not see the Rebels, only their flag flying from the treetop. With a Federal staff officer also up a tree to observe the effect, the Sharpshooters began to test their innovation. The observer guessed they were firing too low when the Confederates looked down, so the snipers cut new sticks and overshot their mark. By again cutting sticks, they finally were able to find the true range and soon forced the Rebel observers out of their tree.[43]

The First United States Sharpshooters had pushed the envelope of their guns' range in this case, but its members often dueled with Confederate marksmen at shorter ranges during the North Anna phase of the Overland campaign. A detail of forty Federals took shelter in log buildings 300 to 1,000 yards from the Confederates and silenced their fire. On another occasion, a squad used stumps and trees as shelter and fired on Confederate artillery at 400 to 600 yards. They claimed to be able to place bullets in the muzzles of the pieces and even shot away a flagstaff.[44]

The First Michigan Sharpshooters in the Ninth Corps had many Chippewa from the area of Saginaw in Company K. They proved to be good snipers. One, nicknamed "One Eye" because he had lost sight in his right eye, volunteered to deal with a Confederate sharpshooter who annoyed division headquarters on the Cold Harbor battlefield. One Eye observed his prey for half an hour, then left without a word. Later, Union pickets saw a Confederate soldier fall out of a tree. One Eye returned and reported, "Me got 'im." Other Chippewa of Company K were credited with shooting twenty-six horses at one location and thirty-seven at another as they harassed two Confederate batteries. Union skirmishers found and counted the carcasses after the positions were evacuated.[45]

As at Chickamauga, the Confederates sent snipers into the treetops at Cold Harbor. Several were shot in addition to the Rebel that One Eye accounted for. Second Corps headquarters detailed the sharpshooters of Maj. Gen. David B. Birney's Third Division for daily duty along the line. Because Birney did not need them, half the snipers were to cover the Second Division and the rest the First Division. Some Federals voluntarily performed the role of sniper at Cold Harbor. Pvt. Franklin M. Goff of the Fourteenth Connecticut initially tried a Springfield rifle musket, "using an ordinary charge of powder," but could not reach two Confederates estimated to be 800 yards behind the Rebel line. He then borrowed a Sharps rifle and hit both men. That the Confederates felt the sting of enemy sharpshooting at Cold Harbor is testified by the comment of Col. George M. Edgar, commander of the Twenty-sixth Virginia Battalion, who reported "an unusually galling fire from sharpshooters with long range rifles—who were reported to be posted in the tops of trees."[46]

The best opportunity for snipers to ply their trade took place along the Richmond-Petersburg lines, stretching for 35 miles by the end of the eleven-month-long campaign against Petersburg. Early in the campaign, the Ninth Corps lost 480 men in ten days to sniping and sporadic artillery fire. Zenas R. Bliss, who commanded a brigade in the corps, later recalled that on his part of the line the trenches ran almost north to south. The morning sun shone to his back, so the Federals were careful not to stand in front of a loophole early in the day as Rebel snipers automatically fired when the loophole blackened. In the afternoon, the Confederates had to take care.[47]

The commander of the Twentieth Maine in the Fifth Corps refused to take unending punishment from enemy snipers. He organized a detail of snipers, two men from each company, and obtained one rifle with telescopic sights for them. After just one day of practice on the skirmish line, these sharpshooters forced the Confederates into an informal truce wherein both sides agreed to stop shooting.[48]

Brigade leaders in the Ninth Corps soon learned the same lesson and began to organize sniper squads for self-defense. A circular dated July 9, 1864, required regimental commanders to report the names of their best marksmen. Twenty-seven men of the Second Pennsylvania Provisional Heavy Artillery and twenty-nine men of the Fourteenth New York Heavy Artillery were selected to operate only on their own regimental fronts. They were instructed to target Confederate snipers first, then artillery units, and finally all targets of opportunity. The snipers were excused from fatigue and picket duty.[49]

It appears that similar organizing efforts took place in all other corps of the Union forces opposing Petersburg. All along the line, snipers began to set up operations by sighting possible targets during the day and propping up their rifles with forked sticks so they could simply pull the trigger in the dark and hope to hit something. They constructed loopholes out of sandbags. Corps commanders shared their resources. "If you have not good sharpshooters," wrote Maj. Gen. Gouverneur Warren to the Second Corps commander, "General Griffin says he can send you ten that will keep every man's head down in the enemy's line in your front in the daytime." Maj. Gen. Winfield S. Hancock initially declined, noting that he had the services of the Second United States Sharpshooters, but two days later Griffin sent up to twenty men "armed with telescope rifles" to help his command deal with fierce Confederate skirmish fire near Fort Sedgwick.[50]

"We lose a few men in the trenches every day," reported Edgeworth Bird of the Third Georgia to his wife. "The sharp shooting on both sides is murderously active and accurate." Bird described a lucky shot by a Federal sniper that hit a Confederate captain and a private as they raised their heads a bit higher than they ought to have while passing each other in a narrow covered way: "a ball passed through the brain of each." Moreover, Federal sharpshooters killed the milk cows belonging to civilians in Petersburg even though they grazed as far as 700 yards to the rear of the works. Bird estimated the cattle were 1,100 yards from the Yankee snipers. When Confederate soldiers tried to cut up the carcasses for food, the Federals shot them down. "It would almost seem a Minnie ball never loses its force," Bird commented.[51]

The Federals who did this obviously were armed with telescopic rifles, although apparently there were relatively few of them available along the Union line. Most of the James rifles earlier issued to the First United States Sharpshooters had been replaced by Sharps rifles by this time, because they were heavy to carry and took more time and care to load. The few that remained were given to the best shots. James Ragin acquired one when another sharpshooter, James Heath, was killed on June 16, 1864. Ragin settled into a rifle pit 300 yards from the Confederate line a few days later and fought a duel with two Rebel sharpshooters on two consecutive days. His first opponent fired from a different spot about every half hour. Ragin observed until he found out the man's favorite place and nearly shot him, apparently only barking the tree that he used for cover. This was enough to scare the man away. The next day, an expert marksman armed with a telescopic rifle took his place and nearly killed Ragin. The two engaged in a deadly contest until both fired essentially simultaneously. Ragin's hair was clipped by the bullet, but the Rebel never fired again. When Ragin was wounded a month later, his gun went to Frederick H. Johnson. The telescopic rifle was so important to the work of the sniper, and so rare, that postwar writers often listed the names of otherwise obscure soldiers who carried them. Their possessor "was considered an independent character, used only for special service, with the privilege of going to any part of the line where in his own judgment he could do the most good."[52]

One telescopic rifle became available in Maj. Gen. William Mahone's division of Lee's army after July 30, 1864. To determine who had the right to use it, the division set up a target range 3 miles to the rear, and each man was allowed a shot at 100, 300, and 500 yards. The competition took place over several days because only a few men could be spared from the works at a time, and they worked in squads. "I was not interested because the fellow who got the gun was expected to pick off [enemy] officers from long range," wrote David Holt of the Sixteenth Mississippi. Nevertheless, he and a man in a Georgia regiment were the finalists. Both went to division headquarters, where a staff officer told the men to draw straws because there was no more time for them to compete, and Holt lost. In another brigade, Evan's Georgia unit, two soldiers were armed with Whitworth rifles by the end of the war.[53]

Although few in number, long-range target rifles spawned many stories among the Confederates who held the Petersburg line. Blackwood Benson, a member of McGowan's Sharpshooter Battalion, once fired Ben Powell's Whitworth at a Federal camp visible more than a mile away. Benson "raised the gun at its highest elevation and fired at random." He could but dimly see men scurry away, and later a New York newspaper reported two men were killed at a well in this vicinity on the day Benson tested the weapon. None among the Federals could hear the sound of the gun. Other stories of long-range shots included two from Archer's Tennessee brigade sharpshooters. One involved a Whitworth rifle and a shot that brought a Federal officer from his horse at the nearly unbelievable range of 2,250 yards. Francis Bass of the Seventh Tennessee reportedly shot a Union marksman out of a tree at 500 yards with an Enfield rifle and was in turn killed some time later by a Federal sharpshooter at long range.[54]

No one on either side of no-man's-land detailed so well the creation and activities of sniper units at Petersburg as did Daniel W. Sawtelle, a member of the Eighth Maine in the Eighteenth Corps of the Army of the James. The corps began to detail men for sharpshooting duties in late June. Fifteen men of the Eighth Maine volunteered, but only six were accepted. They began to assemble with men detailed from other units until 100 snipers were organized in John H. Martindale's First Division. Sawtelle was happy to be among the number. "At least we shall get rid of a great deal of night duty and shall not have to march in with large bodies of troops in the dust and there are a great many other things of which I cannot speak of here," he wrote his sister. Some of his comrades refused, saying "they would not set themselves up for marks to be shot at." Later, when writing his memoirs, Sawtelle gave a different reason for his enthusiasm about joining the sniper detachment. "I had seen so many of my comrades shot down, picked off by men in trees or any commanding position that they could find that when our officers called for volunteers to meet them, I had no hesitation although I knew it was very dangerous work." Sawtelle correctly pointed out that he would be a specific target of a very good marksman, whereas in battle, most bullets were fired at random in the general direction of the enemy.[55]

Sawtelle further remarked in his memoirs that nearly everyone considered sniping a more or less dirty business, but the Confederates started it first and pursued it with what he considered immoral relish. "They always took pride in killing every Yankee they could and boasted to the very last and even after the war closed," he argued. "I have heard a Reb exclaim when a gun was fired, 'There goes another Yank!' as though every Johnnie was a dead sure shot." Federal soldiers rarely gloated over the death of an enemy, Sawtelle believed. "There might not be much said but it was rather discouraged as barbarous and not an honorable way to fight." But they had to fight fire with fire.[56]

Maj. John Cooley of the 148th New York commanded the division sharpshooters, who were divided into two companies. Sawtelle and his comrades initially used their own muskets on sniping duty, but some of them managed to purchase target rifles and have them shipped to the front by family and friends. "If I had money I should get me a rifle to use here," Sawtelle wrote home on July 10. "We need some good ones to cope with the enemies sharpshooters successfully. Some of the boys have sent for globe sighted rifles. The others will get Sharp rifles." Sawtelle heard of a telescopic rifle available near his home in Maine and asked his sister to purchase it for $40 and send it to him, but he did not like the weapon when it arrived and sold it to another man in his unit.

Sawtelle's uncle later found another telescopic weapon, but it did not arrive before the end of the war. Other men in the sniper detachment were amazed at the long-range capabilities of target rifles. "With these sights one could see a tenpenny nail head half a mile away," reported Sawtelle. "One could distinguish a man or horse but could not shoot to be sure of the mark over half a mile unless he knew the exact distance or he could have time to get the range by firing a few shots and watching where the balls struck." Yet these telescopic rifles were too heavy and "clumsy," and many men found "they could not be handled quickly." Those armed with target rifles remained farther to the rear in more stationary positions, while the rest moved about in front with lighter weapons. Even though Sawtelle had expected Sharps rifles to replace the Enfields and Springfields already used by the detailed men, Spencer repeaters were issued before the end of the summer.[57]

The division snipers were given special training in estimating distance. "We worked day after day at it and became so proficient that we could tell the distance an object was from us within a few yards." Then they were put on duty during the daylight hours, given instructions to range along the division trenches and fire at targets of opportunity. The men received instructions about where to meet their officers at the end of their shift and return to the camp, which was hidden in a deep ravine behind the Union works. "We never fired without a mark," later reported Sawtelle, "and we could always find one. We often fired a hundred rounds in one day and seldom less than forty. We became so expert where we knew the distances that we seldom missed a loophole." The snipers were on duty for two days and then off for two days to rest, excused from fatigue and guard duty. "As long as we behaved and did our duty, we had a pleasant job in some ways."[58]

They sometimes worked in pairs, one man raising his hat on a ramrod to draw enemy fire and another returning it fast enough to catch the Rebel before he could duck away. Several snipers coordinated their efforts to neutralize one particularly good Confederate sharpshooter who had shot several Federals. Occupying different positions, everyone fired when they saw the Rebel loophole darken. Sawtelle and his comrades were never sure if they hit him, but after awhile he stopped making an appearance.[59]

The snipers took part in the division attack on Fort Harrison, part of Grant's Fifth Offensive at Petersburg, on September 29, 1864. They initially accompanied the division commander, Brig. Gen. George J. Stannard, but then joined the attacking column when it struck the fort. The snipers became scattered among the troops, but Stannard later ordered them to occupy captured Confederate huts and snipe at enemy gunners with their Spencers. A month later, during a small battle near Fair Oaks, the snipers were used as a reserve for the Federal skirmish line, but they were never called on to come forward and take part in the action.[60]

The snipers of the First Division, Eighteenth Corps remained on duty, detached from their parent regiments, until after the war ended. Not until June 3, 1865, after eleven months, were they relieved of that duty and returned to their original units.[61]

Assessing the Snipers

Everyone who commented on sniping during the Civil War marveled at the good marksmanship and long-range effect of their work. The snipers themselves, of course, sang their own praise after the war. Isaac Shannon thought his men never failed to silence a Federal battery as long as they had half a chance to do so. "It is due to Generals Cheatham, Brown, and Maney, who commanded our division, to say that our effectiveness was greatly enhanced by their good sense in letting us alone and leaving us unhampered with orders," Shannon wrote. "Not a man in the detachment but knew more about what to do and when and how to do it than any general officer in the army." Shannon admitted that the magazine-fed, bolt-action rifle that had become standard issue in all modern armies by the time he penned his Civil War recollections were "superior in point of general effectiveness to muzzle-loaders; but I do not believe a harder-shooting, harder-kicking, longer-range gun was ever made than the Whitworth rifle." Shannon loved his gun so much that when he gave it to a friend, Dr. W. E. Rogers of Memphis, at the end of the war, he extracted the promise that Rogers would return it if called for. "He is dead and I want the gun," Shannon wrote in 1907, "but have been unable to get it from his family or to hear from them in regard to it."[62]

Confederate snipers were so effective at Petersburg that the Union army sought ways to minimize officer casualties at their hands. Responding to reports from the trenches, chief engineer Richard Delafield suggested that officers dress less conspicuously so as to draw less attention. Halleck agreed and wanted to create a board to examine the problem. This resulted in a General Order from the Adjutant General's Office of the War Department, issued on

November 22, 1864, allowing officers in the field to dispense with wearing shoulder straps. They were allowed to wear overcoats "of the same color and shape as those of the enlisted men of their command. No ornaments will be required on the overcoats, hats, or forage caps; nor will sashes or epaulettes be required." The order specified, however, that some inconspicuous mark of rank had to be worn on the shoulder.[63]

Federal snipers were just as good as their Southern counterparts. A man of Pickett's Virginia division wrote an article for the *Baltimore Herald* in 1886 praising them. "I've seen them pick a man off who was a mile away," he wrote. "They could hit so far you couldn't hear the report of the gun. You wouldn't have any idea anybody was in sight of you, and all of a sudden, with every thing as silent as the grave and not a sound of a gun, here would come skipping along one of those 'forced' balls and cut a hole clear through you."[64]

Men who had a natural ability with weapons, armed with a long-range target rifle like the Whitworth, were Civil War precursors to the modern twentieth-century sniper. It is impossible to know exactly how many of them prowled Civil War battlefields, for information about the creation of sniping details is scanty and difficult to locate. If we take Sawtelle's experience as a rule of thumb and assume that 100 snipers were organized for each division in the major field armies of both sides during 1864–1865, the total amounts to 5,800. Even if that rough estimate were generously doubled, it would mean that no more than about 10,000 men out of some 3 million Federals and Confederates who served in the Civil War could be considered as operating in the mode of modern-day snipers. They could not have played more than a marginal role in military operations, even though they represented the epitome in the use of modern long range rifles. Sniping was a highly specialized, technical craft beyond the weapon capabilities or psychological makeup of average soldiers. If a small percentage of unusual soldiers hit targets that were beyond the comfort range of most men, which was at best about 200 yards, then that was not likely to produce a revolution in military affairs. Sniping made the battlefield a more dangerous place for men who had formerly felt comfortable several hundred yards behind the firing line, but it represented only a marginal change in military operations. How average soldiers used the standard rifle musket on the battle line affected Civil War combat the most.[65]

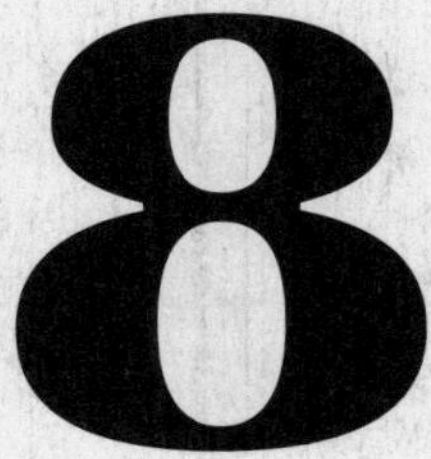

THE RIFLE'S IMPACT ON CIVIL WAR COMBAT

The impact of the rifle musket on Civil War combat has been exaggerated, misunderstood, and understudied ever since Union and Confederate volunteers shouldered firearms. It has been widely identified as the cause for a number of characteristics associated with the nature of warfare in the 1860s. The heavy battle casualties suffered by both armies, the apparent indecisiveness of engagements, the reduced role of cavalry and artillery in Civil War combat, and the increased role of field fortifications have all been ascribed to the use of rifle muskets by both armies. Furthermore, the increased role of skirmishing and the development of modern sniping have been linked to the new technology of the weapon. While sniping definitely grew out of the use of modern weapons then available, the other characteristics mentioned were at best marginally related to the use of rifle muskets. Many of them were not even distinctive features of Civil War combat, for they were characteristics shared with the mainstream of major battles fought in European and American wars before 1861, when smoothbore muskets ruled the battlefield.

Combat Losses

Of all the characteristics mentioned above, it is in the realm of battlefield losses that Civil War engagements have most in common with those of other wars. Historians who uphold the old interpretation of the rifle musket's impact on Civil War combat often cite the heavy battlefield casualties suffered by Union and Confederate armies to argue that the rifle had a devastating impact on warfare.[1]

A comparison of battlefield losses between Civil War engagements and major battles fought before 1861 is required to investigate the validity of this claim. McWhiney and Jamieson provided battlefield loss ratios for at least twenty-nine engagements of the Civil War, in both the East and West. For the Confederates, the ratio ranges from 1.4 percent at Chickasaw Bluffs to 30.2 percent at Gettysburg, but the average is 14.8 percent. For the Federals, the ratio ranges from 1.8 percent at Mine Run to 22.3 percent at Stones River, but the average is 9.8 percent.[2]

A survey of sixteen major battles fought by the French or the Prussians during the eighteenth and early nineteenth centuries indicates that the loss ratio was at least as high as that of the Civil War. The engagements range from Blenheim in 1704 to Waterloo in 1815 in which smoothbore muskets were universally used by both combatants. The loss ratio among French or Prussian armies in these engagements ranged from 8.9 percent to 61.1 percent, averaging 27.6 percent. Losses for the French or Prussian opponents in these battles range from 3.7 percent to 54.1 percent, averaging 28.3 percent.[3]

Most Civil War historians have denigrated the effectiveness of the smoothbore musket based on their assumption of the superiority of the new rifle. This is a modernist perspective that warps our understanding of the older weapon. Historians who specialize in eighteenth-century warfare give far more credit to the smoothbore than do Civil War scholars and students. The battle of Malplaquet on September 11, 1709, the bloodiest engagement of the wars of Louis XIV, saw 24.4 percent allied casualties and 14.6 percent French losses. During the Seven Years' War, the costly battle of Zorndorf on August 25, 1758, witnessed 37.5 percent losses among the Prussians and 50 percent among the Russians.

Table 8.1. Losses in selected Civil War battles

Battle	Confederate Losses (%)	Federal Losses (%)
First Bull Run, July 21, 1861	6.1	5.2
Pea Ridge, March 6–8, 1862	4.3	10.5
Shiloh, April 6–7, 1862	24.1	16.2
Williamsburg, May 5, 1862	4.9	4.6
Seven Pines, May 31–June 1, 1862	13.7	10.5
Seven Days, June 26–July 1, 1862	20.7	10.7
Second Bull Run, August 28–30, 1862	18.8	13.3
South Mountain, September 14, 1862	10.1	6.1
Antietam, September 17, 1862	22.6	15.5
Corinth, October 3–4, 1862	11.2	10.4
Perryville, October 8, 1862	19.7	10
Fredericksburg, December 13, 1862	6.4	10.9
Chickasaw Bluffs, December 26–29, 1862	1.4	3.9
Stones River, December 31, 1862–January 2, 1863	26.6	22.3
Chancellorsville, May 1–4, 1863	18.7	11.4
Champion's Hill, May 16, 1863	10.9	7.7
Gettysburg, July 1–3, 1863	30.2	21.2
Chickamauga, September 19–20, 1863	25.6	19.6
Chattanooga, November 23–25, 1863	5.5	9.7
Mine Run, November 26–December 2, 1863	1.5	1.8
Peachtree Creek, July 20, 1864	13.3	7.9
Atlanta, July 22, 1864	19	6.5
Ezra Church, July 28, 1864	22.2	4.2
Jonesboro, August 31–September 1, 1864	15	3.9
Third Winchester, September 19, 1864	12.3	12.4
Cedar Creek, October 19, 1864	10.1	13.2
Franklin, November 30, 1864	20.6	4.4
Nashville, December 15–16, 1864	25.9	5.9
Bentonville, March 19–21, 1865	8.9	5.8
Average	14.8	9.8

Source: Grady McWhiney and Perry D. Jamieson, *Attack and Die: Civil War Military Tactics and the Southern Heritage* (University: University of Alabama Press, 1982), 19–21.

At Kunersdorf, August 12, 1759, the Prussians lost 43.1 percent of their men and the Austro-Russian force lost 20 percent. The loss ratios at the battle of Mollwitz on April 10, 1741, were 22.4 percent (Prussian) and 23.1 percent (Austrian). Casualties amounting to 30 percent were not unusual in major smoothbore battles, according to B. P. Hughes, and the statistics bear out the general truth of that conclusion.[4]

There is some evidence that the loss ratio actually declined a bit in the wars of the French Revolution, primarily due to the increased size of field armies. One historian has charted loss ratios in European battles fought before 1789 as

Table 8.2. Losses in selected European battles before 1861

Battle	French or Prussian Losses (%)	Opponents' Losses
Blenheim, August 13, 1704 (War of the Spanish Succession)	33.3	23.2[a]
Oudenarde, July 11, 1708 (War of the Spanish Succession)	16.4	3.7[a]
Malplaquet, September 11, 1709 (War of the Spanish Succession)	14.6	24.4[a]
Mollwitz, April 10, 1741 (War of the Austrian Succession)	22.4	23.1[b]
Fontenoy, May 11, 1745 (War of the Austrian Succession)	Data not available	14[b]
Prague, May 6, 1757 (Seven Years' War)	21.8	21.2[b]
Zorndorf, August 25, 1758 (Seven Years' War)	37.5	50[b]
Kunersdorf, August 12, 1759 (Seven Years' War)	43.1	20[b]
Jemappes, November 6, 1792 (Wars of the French Revolution)	10	30.3[b]
Austerlitz, December 2, 1805 (Napoleonic Wars)	12.8	30.5[b]
Jena, October 14, 1806 (Napoleonic Wars)	8.9	54.1[b]
Auerstadt, October 14, 1806 (Napoleonic Wars)	30.7	23.8[b]
Eylau, February 8, 1807 (Napoleonic Wars)	44.4	Incomplete data[b]
Friedland, June 14, 1807 (Napoleonic Wars)	13.1	30[b]
Albuera, May 16, 1811 (Napoleonic Wars)	44	44[c]
Waterloo, June 18, 1815 (Napoleonic Wars)	61.1	32.3[c]
Average	27.6	28.3

[a] John A. Lynn, *The Wars of Louis XIV, 1667–1714* (London: Longman, 1999), 293, 320, 331–332.
[b] Steven Ross, *From Flintlock to Rifle: Infantry Tactics, 1740–1866* (Rutherford, NJ: Fairleigh Dickinson University Press, 1979), 31, 58, 108, 107, 108.
[c] Gunther E. Rothenburg, *The Art of Warfare in the Age of Napoleon* (Bloomington: Indiana University Press, 1980), 81.

"anywhere from 9 percent to 41 percent," while those of revolutionary engagements range "from 1 percent to 25 percent." Paddy Griffith believes that revolutionary battles resulted in an average of 15 percent losses for the French and 7.3 percent for the allies, but he notes that the ratio increased after 1804 to 20.6 percent for the French and 22.6 percent for the allies. At Austerlitz, December

2, 1805, the most decisive battle in modern history, the allies lost 30.5 percent of their manpower in a devastating defeat. The huge battles of the Napoleonic era could involve as many as half a million men, as at the battle of Leipzig. They show combined loss ratios for the antagonists of 32.5 percent (Borodino, September 7, 1812) and 28 percent (Leipzig, October 16–19, 1813).[5]

Loss ratios for units smaller than field armies remained high in the Napoleonic Wars. An English force of 14,000 men suffered 50 percent casualties in only one attack at the battle of Fontenoy (May 11, 1745). At Salamanca (July 22, 1812), the British Sixth Division lost 30 percent of its men, while a German brigade fighting for the British at Talavera (July 28, 1809) lost half its men. Losses among the five British brigades in action at Talavera amounted to 36 percent. Some units in the Napoleonic era recorded losses of up to 80 percent.[6]

The least that can be said about all this is that the loss ratios among armies armed entirely with smoothbore muskets before 1861 were on a par with antagonists armed with rifle muskets in the Civil War. In fact, it is possible to argue that the loss ratios for smoothbore battles was higher than for rifle battles, although a more thorough listing of pre–Civil War battles would have to be made to support that argument.[7]

The European smoothbore battles seem to have been very bloody affairs indeed, but a comparison of the Civil War with selected battles fought in North America before 1861 shows a rather similar relationship between the effect of smoothbore musketry and rifle musketry on the battlefield. Six battles, from Bunker Hill in 1775 to Buena Vista in 1847, involved loss ratios for the American army ranging from .09 percent to 36.6 percent, with an average of 19.5 percent. For their opponents, the British and Mexicans, the loss ratio ranges from 3.8 percent to 47.9 percent but averages 23.7 percent. Bunker Hill (June 17, 1775) seems to have been a particularly costly engagement, but battles such as Eutaw Springs (September 8, 1781) and Lundy's Lane (July 25, 1814) indicate that its loss ratio was not so unusual after all.[8]

It is possible, of course, to select abnormally costly engagements from every war, ignoring the fact that each conflict witnessed hundreds of small skirmishes that resulted in light casualties. But for our purposes, the large engagements are more important than the passing skirmishes, because here is where the real test of weaponry occurred. If it is possible for even one major battle fought with smoothbore muskets to have a loss ratio equal to that of a major battle fought with rifle muskets, then that calls into question the validity of the argument that the rifle had a revolutionary impact on the nature of combat.

Table 8.3. Losses in selected North American battles before 1861

	American Losses (%)	British or Mexican Battle Losses (%)
Bunker Hill, June 17, 1775 (American Revolution)	36.6	47.9[a]
Brandywine, September 11, 1777 (American Revolution)	.09	3.8[a]
Eutaw Springs, September 8, 1781 (American Revolution)	23	34.6[a]
Crysler's Farm, November 11, 1813 (War of 1812)	13	15[b]
Lundy's Lane, July 25, 1814 (War of 1812)	30.9	25[c]
Buena Vista, February 23, 1847 (Mexican War)	14	16[d]
Average	19.5	23.7

[a] Craig L. Symonds, *A Battlefield Atlas of the American Revolution* (Baltimore: Nautical & Aviation, 1991), 19, 53, 97.
[b] Donald E. Graves, *Field of Glory: The Battle of Crysler's Farm, 1813* (Toronto: Robin Brass Studio, 2000), 268, 270.
[c] Donald E. Graves, *Where Right and Glory Lead! The Battle of Lundy's Lane, 1814* (Toronto: Robin Brass Studio, 1999), 195–196.
[d] K. Jack Bauer, *The Mexican War, 1846–1848* (New York: Macmillan, 1974), 217.

Moreover, a brief look at loss ratios in post-1865 battle indicates that the Civil War was not bloodier than succeeding engagements that involved rifles used by armies employing modern tactics. Unit losses among the Germans in the Franco-Prussian War often reached 42 percent, 44 percent, even 73 percent of their strength. The German army lost 15 percent of its infantry strength in the first three months of its invasion of Russia in 1941. At Pelelieu (September 15–November 25, 1944), the First Marine Regiment suffered 56 percent casualties.[9]

Indeed, to make the point even further, the Roman army lost an estimated 64.7 percent to 75 percent of 80,000 men involved in the disastrous battle of Cannae on August 3, 216 BC, against a mixed army commanded by the Carthaginian general Hannibal. This Roman loss ratio took place 2,000 years before the introduction of the rifle musket.[10]

The point we should note is that Civil War combat did not produce more casualties than did other wars in different places and time periods. War by its nature produces casualties; a variety of factors determines whether the loss rate in any particular engagement is high or low, and the type of weapon used is not the most salient of those factors. No Civil War battle witnessed the killing of 75 percent of the men engaged, as took place at Cannae, because the unique tactical circumstances of that ancient battle and the martial culture of the era more readily supported the urge to slaughter even helpless enemy troops. Command decisions, the level of training, discipline, and motivation, the level of combat experience among the troops, and many other issues also determined how many men were lost in any given battle throughout world history.

The loss ratios for battles fought before 1861 indicate that the smoothbore musket was just as effective as the rifle musket of the Civil War when other conditions were equal. The American militia at Bunker Hill had literally no artillery support of any kind, yet these citizen-soldiers (who, at most, might have had some combat experience in the Lexington-Concord battle of April 19, 1775) shot down nearly half of the entire British force engaged in the battle with smoothbore muskets alone. It is impossible to ignore a fact like this and continue to assert that the rifle was a more effective weapon than the smoothbore.

Decisive Battle

Another major component of the old interpretation of the rifle musket in the Civil War is that its increased firepower robbed Civil War commanders of the possibility of producing a decisive battle that could win the war in one stroke. Typical of historians' views on this issue are the words of two respected military historians, Robert Doughty and Ira Gruber, who wrote that Civil War commanders "cherished the vision of a decisive victory over the enemy." This view has been echoed by nearly all Civil War historians, and there is ample evidence in the primary literature to prove that Union and Confederate soldiers of all ranks pined for a quick, decisive victory over the enemy. William B. Greene, a member of the Second United States Sharpshooters, expressed it well when he wrote his mother, "I think after one good battle that the rebels will be obliged to give in."[11]

Historians have been consistent in listing the reasons for the apparent indecisiveness of Civil War battles, usually starting with the effectiveness of the rifle musket. They cite the widespread use of field fortifications and argue that Civil War commanders usually had too few cavalry available to closely pursue a retreating enemy. The landscape of the Southern states, covered with thickets and rough ground, also hindered the ability of commanders to mount successful attacks or to effectively pursue a retreating foe.[12]

Francis Lippitt, a California lawyer who wrote *A Treatise on the Tactical Use of the Three Arms* right after the Civil War (based on his service in a California regiment that saw no combat), argued that battle was indecisive because the troops tended to fire too long "at a distance," rather than closing with the bayonet. He also thought Civil War commanders failed to provide adequate artillery preparation. Lippitt's thesis, based on examples drawn from both Civil War and European engagements, was that modern military leaders failed to coordinate the three arms of land warfare. "Artillery prepares the victory," he concluded. "Infantry achieves it; Cavalry completes it, and secures its fruits." In this, Lippitt echoed the argument of Henry W. Halleck, who had written in his *Elements of Military Art and Science* of a scientific approach to understanding the balance between the three arms in producing effective military action.[13]

William T. Sherman weighed in on the issue of decisiveness in his memoirs, noting that European observers often criticized American commanders "because we did not always take full advantage of a victory." Sherman argued that the wooded landscape he encountered during the Atlanta campaign too frequently "served as a screen, and we often did not realize the fact that our enemy had retreated till he was already miles away and was again intrenched, having left a mere skirmish-line to cover the movement, in turn to fall back to the new position." Gouverneur K. Warren made the same point about the Federal failure at Chancellorsville. In his report as the chief topographical engineer of the Army of the Potomac, Warren tried to counter the view that American commanders lacked skill "in handling troops in battle" by noting the effect of the tangled vegetation in the Wilderness. Here, engagements "had to be fought out hand to hand in forests, where artillery and cavalry could play no part; where the troops could not be seen by those controlling their movements; where the echoes and reverberations of sound from tree to tree were enough to appall the strongest hearts engaged, and yet the noise would often scarcely be heard beyond the immediate scene of strife." For Warren, there was no wonder "that such battles often terminated from the mutual exhaustion of both contending forces."[14]

The discussion about the indecisive nature of Civil War combat needs to be diverted to a new question—were Civil War battles really indecisive? And how did they compare in that regard with battles before 1861 and after 1865? Paddy Griffith has tried to measure the level of effectiveness in Civil War battles by comparing several of them with selected Napoleonic engagements and with offensives on the western front in World War I. He has concluded that "the overall efficacy of First World War attacks was almost half that of Napoleonic ones," and the effectiveness of Civil War attacks fell somewhere between those two marks. "If we express this in gambling terms," Griffith continued, "we could say that Napoleon knew he had a two in three chance of ending up on top if he launched a tactical attack; Foch and Hague had a chance of one in three, while Grant and Lee had evens."[15]

Doughty and Gruber also recognize that the supposed indecisiveness of Civil War combat begins to fade when compared with other wars. They admit that truly decisive battles were rare occurrences, perhaps taking place no more than six times every century since the Thirty Years' War of 1618 to 1648. Griffith further elaborates on the theme by arguing that Civil War commanders tended to miss many opportunities to make their battles more decisive. He believes "the individual personalities of the generals" tended to make them more cautious than was required for effective execution of tactical plans, especially in following up initial successes. In this, they were victims of "the predominant military culture," which "was not deeply rooted in rapid manoeuvre and crushing assault. A more tentative and sedate style of war seems to have been more generally preferred."[16]

Whether Griffith is right, and Civil War commanders deserve criticism for wanting to achieve a great deal quickly while employing a style of leadership that doomed them to long delays and frustrations, remains to be seen. But other historians have also noted that indecisiveness is far more common throughout Western military history than not. Russell Weigley has pointed out that it is in fact a consistent theme in the history of battles. In discussing the wars of Louis XIV, John Lynn writes of the modern concept of "war-as-event" versus the seventeenth-century concept of "war-as-process." The latter tended to consist of characteristics such as an emphasis on siege operations to capture fortified positions with minimal losses. This meant that conflicts often dragged on for years. The opposite style of warfare tends to consist of a willingness to risk all on one or a series of set-piece battles. In the dynastic wars of the seventeenth and eighteenth centuries, monarchs and generals alike tended to shy away from set-piece battles because they were costly and risky. Why take a chance of losing the game on one day of playing, losing thousands of peasant recruits carefully trained in the intricacies of linear tactics to boot? In ten years of campaigning in Flanders, Louis XIV's armies engaged in only four set-piece battles. "Sieges provided the most common form of combat during this era," according to Lynn, because they allowed for the capture of valuable resources on terms advantageous to the besieging army.[17]

Even among the various wars of the Napoleonic era, sieges tended to occur nearly as often as set-piece battles. Moreover, not all of the many pitched engagements of the Napoleonic era were decisive. In fact, Russell Weigley has noted that most of them were not.[18]

Within the American experience, it is easy to find engagements that failed to offer either side a decisive advantage before the Civil War. The battle of Crysler's Farm on November 11, 1813, was fought to a tactical draw, as was the fierce engagement at Lundy's Lane on July 25, 1814.[19]

Civil War contemporaries tended to be fooled into assuming that they could win the war with one decisive encounter, in part because citizens of democratic societies want to end wars quickly so they can get on with their peacetime lives. For the Americans, the tendency has always been to view armed conflict from the standpoint of Lynn's "war-as-event" rather than "war-as-process." This produces impatience among civilians and citizen-soldiers alike, which can be as easily noted among the men and women who fought World War II as among the Civil War generation.[20]

The Civil War generation also had the glittering example of Napoleon at his best to contemplate. The French emperor produced one of the world's most stunning examples of a decisive battle at Austerlitz, a large, costly engagement fought on December 2, 1805, which quickly led to the collapse of the allied coalition of Austria and Russia against the French. Austerlitz is the supreme example of a battlefield victory that led to political consequences and a swift end to a war.

But Civil War soldiers held themselves up to an impossibly high standard if they expected to duplicate the Austerlitz scenario, for that battle was unique in history. It was an unusually decisive engagement from a military perspective, with a crushing attack by 60,000 French troops on an unsuspecting allied army of 15,000 Austrians and 65,000 Russians. Napoleon used a hill to screen his army while maneuvering to strike the enemy as they prepared to attack his dimly understood positions. He waited until the allies committed their left wing before unleashing the main strength of his army against the allied center and right. The Russians and Austrians generally gave a good account of themselves, forcing the French to fight hard, but the end result of this great clash was the collapse of the allied position. The Russians retreated immediately to Hungary, leaving the Austrians to deal with the victorious French. Napoleon met Emperor Francis II on December 4, talked for less than one hour, and arranged an armistice to end the war the next day. The Treaty of Pressburg followed three weeks later.[21]

This great Napoleonic victory was both a military triumph and a political success for the French. Austerlitz was a severe drubbing for the allied army that led to an immediate collapse of political will on the part of the Austrian government to further endure the war. In effect, Emperor Francis II opted to quickly get out before worse disasters befell him, thereby saving his government from a possible takeover by the French.

Such a scenario would have been impossible to imagine in the Civil War, at least from the standpoint of hindsight. Civil War contemporaries could easily have assumed it was possible, given that Austerlitz was only half a century in the past and given that most Northerners tended to believe that the average, nonslaveholding Southerner could not possibly have his heart in the Confederate war effort. The belief in a large proportion of loyalists inhabiting the Confederate states gave support to the notion that the Southern effort might collapse if the hotheads who ran the government in Richmond could have the starch taken out of their will by one or two big battlefield defeats. Once the Confederacy proved it had more staying power than expected, however, it is increasingly difficult to excuse Northerners for continuing to believe that they could pull off an American version of Austerlitz.

The truth is, there were militarily decisive battles in the Civil War; not every engagement ended in a tactical draw. The battle of Richmond, Kentucky, fought on August 30, 1862, resulted in the capture of 4,303 out of 6,500 Federal troops engaged. Most of those Yankees served in newly raised regiments. They had very little training and no combat experience, and the opposition, consisting of battle-hardened Confederates, gave them one of the most impressive drubbings ever witnessed on a Civil War battlefield. Chickamauga was another impressive Confederate victory, although bought at dear cost in life and without a follow-through that could have resulted in the destruction of the Union army. For their part, the Confederates suffered decisive military defeat at Missionary Ridge and at Nashville. The point is that none of these militarily decisive battles resulted in political activity that ended the war. The governments in Washington, DC, and in Richmond continued to fight after each of them. The Civil War came to have some important aspects of a total war, with enough political determination on the part of civilians and government alike to see the endeavor through to its bitter conclusion. In a situation like that, it would have been impossible to duplicate another Austerlitz. With political will to pursue the war, with armies that were armed, trained, and disciplined on a level of parity with each other, the Civil War was characterized by many battles, decisive and indecisive alike, that taken together all had a slow cumulative effect on the outcome of the war. The Confederacy disappeared under the weight of this effect, slowly and over time, rather than coming to an end in a spectacular moment of history.

The rifle musket seems to have had little if any impact on making Civil War battles tactically indecisive, for there were many such battles in the age of the smoothbore as well. Moreover, a few tactically decisive engagements did occur in the Civil War, too. Many other factors besides the type of weaponry used influenced the nature of Civil War combat. As long as there was a willingness on both sides to fight to the bitter end, there would be no decisive battle that quickly led to an end of the war.

Artillery

Historians who adhere to the traditional view of the rifle's impact on Civil War combat argue that it altered the role of field artillery by robbing it of offensive power on the battlefield. The increased range of infantry small arms meant that batteries could not deploy close to the enemy and prepare the way for an infantry assault. This meant that Civil War field artillery could be effective only as a defensive weapon, deployed at a distance from the enemy but able to fire on attacking columns. Spectacular examples of its efficacy on the defense included the repulse of John C. Breckinridge's division on January 2, 1863, at Stones River. Its failures to prepare the way for offensive strikes include Pickett's Charge at Gettysburg.[22]

There is no doubt that artillerists were in greater danger during the Civil War because of the rifle musket. Marcus Woodcock of the Ninth Kentucky recalled that his comrades commenced firing at a section of Confederate artillery on the second day at Chickamauga at a distance of 150 yards and compelled the guns to retire. Other contemporaries noted that field artillery had comparatively little effect on skirmish lines because the skirmishers had the opportunity to take cover behind natural obstacles while creeping closer to battery positions.[23]

The argument that Civil War field artillery lost an offensive capability previously held by its predecessor in eighteenth- and early nineteenth-century European wars is weakened by closer study of the artillery arm in those earlier conflicts. It is true that Prussian doctrine of the 1750s advocated that field artillery accompany the advancing infantry, occupying the intervals between battalions. The gunners were to open fire with round shot when 500 paces from the defender and continue to fire, rolling the guns forward between rounds. They were to switch to canister at 100 paces. While the longer-range firing is assumed to have been for psychological effect, the closer-range canister fire presumably was more physically damaging.[24]

It is not easy to determine if this artillery fire was effective in breaking up a defender's position and opening the way for the attacking infantry to succeed. Moreover, the guns employed were 3 and 6 pounders, light enough to be manhandled across easy terrain. Civil War field artillery typically consisted of 12 pounders, or even heavier ordnance, making it quite difficult to handle in this fashion. The heavier pieces meant that infantry more often had to coordinate its movements to protect the guns rather than the gunners coordinating their movements to support the foot soldiers. Moreover, the heavier ordnance fired by these heavier pieces did not necessary have a greater effect on the enemy, because little advance had been made in projectile design and detonation. Fuses often failed to explode shells properly, and the cast-iron fragmentation projectiles broke into irregular, often large pieces that flew at slow speeds for short distances when the shell exploded.

In short, even though some artillerists such as Edward Porter Alexander dreamed of doing so, it was impractical to push Civil War field guns up in close support of advancing infantry for reasons other than the effect of rifle fire. In a time before improvements in the technical aspects of ammunition and shells, one cannot expect too much of Civil War field artillery in the way of softening up the objective for a successful infantry attack. The devastating impact of World War I artillery proved that technical advances in these areas were the key to allowing artillery to dominate a static battlefield, and these improvements were unavailable to Civil War gunners.[25]

As Paddy Griffith has rightly argued, the effectiveness of Civil War field artillery depended on a number of factors in every given battlefield situation. It is dangerous to make generic conclusions about this complex issue based on only one factor, such as the kind of weapons used by defending troops. The quality of decision making by battery commanders, the nature of the terrain, the kind of tactical mission given the artillerists, and many other issues came into play to determine whether the guns were effective or not. Ironically, there are several case studies in Civil War combat to show that field artillery actually could be effective as offensive weapons. Franz Sigel coordinated the fire of several batteries at the battle of Pea Ridge on March 8, 1862, to support a successful infantry attack. The Confederates assembled fifty-three guns to fire at the Hornet's Nest at Shiloh on April 6, 1862, at a range of about 500 yards, contributing to the reduction of that Union strongpoint. If artillery could contribute to offensive victories such as these, it questions the old notion that the rifle completely altered its effectiveness on the battlefield. The guns were more effective at defense than offense because of limitations in technology. Griffith has also argued that even Napoleonic artillery was better at defense than offense for the same reason. Its effectiveness on the attack has been overblown by Napoleonic observers and modern historians. Griffith also correctly points out that Civil War batteries suffered on average fewer casualties than infantry units, typically amounting to about 10 percent.[26]

Cavalry

Historians have made arguments for the rifle's impact on cavalry operations similar to those for its impact on field artillery, asserting that troops armed with the new weapon robbed the mounted arm of its effectiveness in attacking infantry. Thus Civil War troopers were forced to fight on foot alongside the infantry, forswearing dashing mounted charges designed to break up enemy infantry formations as in the days of Napoleon.[27]

It is true that Civil War cavalry often dismounted to fight on foot, but there is no reason to believe this was due to the effect of rifle musketry. Samuel B. Barron of the Third Texas Cavalry wrote in his memoirs that his comrades "always fought on foot. Sometimes behind breastworks, sometimes not, sometimes confronting infantry and sometimes cavalry." Barron never mentioned the fire of rifle musketry as the cause of this; in fact, he implies it was a habitual practice against all types of opposing troops and in all situations, regardless of the weapons used by the enemy. In other words, dismounted fighting occurred in the Civil War for reasons other than the rifle musket. It is important to note that the Civil War did not witness the first experiments in mixing the role of cavalry and infantry. Europeans dating back to the early seventeenth century had tried mounting infantry troops for greater mobility. It may well be that the Civil War witnessed the first experiment with dismounting cavalry troops to fight as infantry, but this was a phenomenon much larger than the widespread use of rifle muskets.[28]

Even more to the point, historians who adhere to the traditional interpretation of the impact of the rifle musket on Civil War combat argue that cavalry had the ability to attack infantry formations and break them up on the battlefield in the smoothbore era. Like the idea that pre–Civil War field artillery was effective on the offensive, this is an erroneous notion. Even Arthur L. Wagner, one of the originators of the interpretation that the rifle musket had a revolutionary impact on warfare, admitted that infantry armed with smoothbores had always been able to defend itself against cavalry attacks. As long as the infantrymen remained calm and maintained their formations, they were safe even if there were no rifles available. A thoughtful reading of battle accounts of pre–Civil War engagements bears this out.[29]

Rifle advocates of the 1850s wrestled with this issue, trying to determine whether the use of rifle muskets would negate any battlefield role for the mounted arm. Cadmus Wilcox argued that cavalry could still break up infantry formations if it attacked them at a vulnerable point, but that is merely to say that any mounted or dismounted assault that hit a weak defensive spot had a chance of success, whether the defender was armed with rifles or not.[30]

European commanders had typically used cavalry and infantry as separate arms on the battlefield during the previous two centuries. Normally they placed their mounted troops so as to cover the flanks of their infantry formations. When on the offensive, both infantry and cavalry advanced against whatever opponents were in front of them. Mounted attacks against infantry units that were still in good formation and ready for a fight were rare. When they occurred, they invariably failed.[31]

The same was true of the Civil War. Griffith notes that during the first three years of the conflict, cavalry attached to the Army of the Potomac made but five mounted attacks against Confederate infantry. All were done on a small scale at the end of a major battle, after the decision had been reached, and all were unsuccessful. During the battle of Froeschwiller, August 6, 1870, 1,200 French cavalry conducted a mounted attack on Prussian infantry. The Germans opened fire with breech-loading rifles at a range of 400 yards. Initially, they volleyed, then fired at will. The attack was stopped, with 800 French casualties, after the French had advanced to no closer than 50 yards from the German formation.[32]

It is easy enough to cite examples like the above for the Army of the Potomac and the French at Froeschwiller to conclude that the rifle made it impossible for cavalry to play a decisive role on the battlefield. But these examples occurred because of bad decisions by cavalry commanders, who acted either out of desperation or based on a misguided understanding of the proper role and history of cavalry operations. The same was true of cavalry commanders in the smoothbore age. Those Confederate and German infantry formations most likely would have held their ground and blunted the cavalry attacks even if they had been armed with smoothbores, if the history of pre–Civil War battle in Europe is any guide.

Field Fortifications

Historians have assumed that the full-scale use of the rifle musket by the midpoint of the Civil War accounts for the surprising use of field fortifications in the spring of 1864. They argue that the rifle gave defenders a much greater chance of repelling attacks and view the use of fieldworks as part of a growing ascendancy of the defensive over the offensive. Interestingly, prewar rifle enthusiasts did not predict increased use of field fortifications, and not all Civil War contemporaries believed it was impossible to attack and carry fieldworks on the battlefield. Postwar commentators were greatly impressed by the widespread, almost habitual use of fieldworks among some troops during the last year of the war. Arthur L. Wagner was particularly impressed by what his contemporaries called "hasty intrenchments," earthworks dug quickly under fire or only minutes before meeting the enemy. This was "the most marked tactical feature" of the Civil War and one of the true innovations to arise from the conflict. When modern historians delve a bit more deeply into the cause-and-effect relationship between the rifle and field fortifications, they usually note that skirmishing and sniping, which utilized the new weapon more effectively than the operations of the battle line, were the true causes of the growing sophistication and strength of fieldworks in Virginia and Georgia during 1864–1865. This is a view supported by the opinion of Horace William Shaler Cleveland in 1862.[33]

William B. Hazen was a strong advocate of cover on the battlefield based on his experience as a brigade commander in the Army of the Cumberland and as a division leader in the Army of the Tennessee during the Atlanta campaign. He argued in his memoirs that Americans paid little attention to field fortifications early in the war due to ignorance of military matters in general. Although assuming the smoothbore musket was "random-firing," Hazen still thought some sort of cover was necessary even when armies used the older weapon. With the rifle, Hazen thought fieldworks were essential and pointed as proof to the results of the battle of Ezra Church and the skirmishing experienced during the so-called siege of Atlanta in August 1864. At Ezra Church, the Second Division of the Fifteenth Corps repelled repeated Confederate attacks with only slight breastworks, suffering 12 killed while the Confederates left 320 dead in its front. In August, Union and Confederate units were locked for weeks in static positions that they heavily fortified, delivering volumes of skirmish fire across no-man's-land. Hazen not only had faith in the accuracy of the rifle musket, but he also thought that fieldworks offered the soldier a deeper sense of "composure and deliberation. He aims carefully, and fires at his mark; while without cover he is excited, seldom aims at all, and often fires high in the air."[34]

Paddy Griffith discounts the role of the rifle musket in fostering the greater use of fieldworks but offers no good explanation for their existence. He believes, in fact, that they were overused by both Union and Confederate armies and assumes that a variety of causes accounted for the unnatural tendency of commanders and men alike to dig them.[35]

The notion that the rifle caused Northern and Southern armies to entrench is not convincing. Field fortifications were constructed from the very beginning of the war, long before most regiments were armed with rifles. Entrenchment can be seen at the battle of Big Bethel, on June 10, 1861. The Confederates had dug earthwork defenses before and especially right after the battle ended. The same pattern applied at First Bull Run, where the Confederates had only light defenses before the battle but more heavily dug in after they repulsed the first Union advance toward Richmond. Earthwork use was sporadic from 1861 to 1863 but occurred much more frequently than historians have normally admitted. Both sides dug in a lot during the Peninsula campaign. Lee's army fortified its entire line right after the battles of Fredericksburg and Gettysburg, and used earthworks well during the Chancellorsville campaign. Lee also dug immensely strong earthworks to protect his crossing of the Potomac River on his retreat from Pennsylvania and to hold his position along Mine Run in late 1863. In the West, the Confederates consistently fortified important towns and river sites with fieldworks, and the Federals dug in very extensively during their advance on Corinth, Mississippi, in May 1862. Union and Confederate soldiers built field defenses at Stones River and Chickamauga.[36]

The most intensive period of fortification use took place in the spring of 1864. The fact that most Union and Confederate units were armed with rifles by this time has given rise to the assumption that the two factors are related. But the intensive use of fieldworks is more closely related to another factor: the employment of continuous campaigning by the Federals. Grant introduced this new concept when he became general in chief of the Union armies and insisted it be implemented by all field commanders beginning in the first week of May 1864. Essentially, only the Army of the Potomac in Virginia and Sherman's army group in Georgia fulfilled that directive, the former maintaining continuous contact until the fall of Petersburg on April 2, 1865, but the latter only until the fall of Atlanta on September 2, 1864. It is not surprising that the most intensive use of field defenses exactly coincides with the duration of continuous contact. Operations in the West reverted to more mobile campaigning and a reversion to the sporadic use of field defenses witnessed before 1864 by the fall of that year.

It is true that the armies that maintained continuous contact were widely armed with rifle muskets, but it is wrong to ignore the fact that they were continually within striking distance of each other during that time. In the earlier part of the war, armies often fought battles with little use of field defenses, only to see one or both sides dig in immediately after the fighting ended and while the opponents remained within striking distance of each other, before one commander or the other decided to retire. Fredericksburg and Gettysburg are among the best examples of this phenomenon, which resulted from the psychological shock of combat and a fear that the fighting might be renewed. With continuous contact, the shock of combat and the fear of another engagement remained present for weeks, even months, at a time. Even if the armies had been armed with smoothbores, it seems likely one would see more extensive use of earthworks than had been true of the early part of the war as long as the opponents remained within attacking range of each other.

Because continuous contact was a new concept, we cannot compare its effects with pre–Civil War battles. But although the Franco-Prussian War of 1870–1871 saw the use of much more sophisticated and powerful small arms than the Civil War rifle musket, nothing like the fieldworks of Virginia or Georgia were ever used. That war was characterized by fluid movements, engagements that resulted in intense, costly rifle and artillery battles lasting one or two days before contact was broken. Only the French sometimes employed light fieldworks on a few occasions. In short, the use of heavy earthworks was not an inevitable result of fighting wars with rifles.[37]

Even in the smoothbore era, heavy earthworks were used whenever opposing armies engaged in static warfare within close range of each other—in other words, during siege operations. In the twentieth century, when armies were universally armed with powerful rifled small arms, warfare generally remained mobile in character except when armies became locked in static positions such as on the western front during World War I and during the Italian campaign of World War II. The desire to seek protection made no distinction between weapon types. Soldiers were tempted to dig in as long as the enemy was close enough to fire at or attack them, whether that enemy was armed with a smoothbore musket or an M-16.

While the use of the rifle by the battle line does not seem to have played a large role in spurring the use of sophisticated field fortifications, even in the gory campaigns of 1864, there is every reason to believe that skirmishers and snipers and even artillerists forced the battle line to dig in ever deeper while the armies remained in static positions during the last year of the war. Hazen noted that the Fifteenth Corps troops needed only slight breastworks to shoot down their opponents at Ezra Church, an engagement closer to an open field fight than an attack on a fortified line. But harassing the enemy during the siege of Atlanta required much more extensive field protection. There were no attacks here, only the daily grinding away of enemy morale, stamina, and personnel across a month of deadly trench warfare.[38]

9

AFTER THE RIFLE MUSKET

The Civil War was the only major conflict in history in which both sides were fully armed with the rifle musket. Breech-loading rifles superseded that weapon very soon after 1865. The small arms used in the Franco-Prussian War made the Springfield and Enfield seem antiquated only five years after Appomattox. Lessons regarding the rifle musket's role in warfare are therefore muted, for the rifle musket era was a brief episode in the fast-paced development of modern weaponry in the nineteenth and twentieth centuries.

Of course, when the Civil War came to an end, few could have predicted that the rifle musket would have such a short day in the sun. Francis Lippitt, the San Francisco lawyer who based his treatise on the use of combined arms on examples derived from European warfare and the tactical manuals of the day, had a decidedly conservative vision of infantry warfare. He still believed in the effectiveness of bayonet attacks. Although Lippitt recognized that sharpshooters could severely harass artillery crews, he doubted that breechloaders were superior to muskets because they encouraged troops to waste ammunition. Lippitt understood that rifle muskets were important for effective skirmishing, but he discouraged long-range firing by the battle line (which he defined as a range of 300 yards), for it was wasteful and unproductive. He recommended that infantrymen fire volleys at 50 paces; independent fire was not only evidence of poor discipline but was less effective than controlled firing. Lippitt firmly believed that soldiers tended to fire too rapidly to be effective. It is true that Lippitt's study was based on theory rather than experience, and that it represented some old ideas predating the rifle musket, but he accurately pointed to the effect that weapon had on skirmishing and was generally correct in assuming that infantry fire, even with the rifle musket, would be at short range.[1]

But Lippitt had no impact on U.S. Army doctrine in the postwar era. Emory Upton, author of *Infantry Tactics,* caught the imagination of many army officers with a manual that the army adopted in 1867. It had the appearance of significantly breaking from tradition, in that Upton did not merely translate a French manual for American publication. He rewrote the basic European tactical manual in an American idiom, lending a new twist to an old tradition in military literature in this country. But it is wrong to assume that he revolutionized the way Americans thought about tactics. His only innovation was to incorporate the concept of fours, taken from prewar skirmish drill, into the operations of the battle line. This had the effect of loosening up the battle line a bit without changing its essential nature. Upton's book was still a manual of linear tactics.[2]

In his School of the Battalion, Upton kept faith with previous manual authors in laying out the process of firing by company, wing, battalion, file, and rank. He recommended that fire by subunits be conducted alternately, like the old platoon-firing system of the previous century, and argued that firing by file was most commonly used on the battlefield. Also like previous manual authors, Upton took the recruit through the process of firing and reloading while kneeling and while lying on the ground. Little had changed in the manual literature of the immediate postwar era in terms of teaching soldiers how to handle their weapons.[3]

A major change, however, took place in training men to use the rifle, for a widespread system of rifle practice was finally created after the Civil War. Edward O. C. Ord, commander of the Department of California, initially inspired this development. He established guidelines for all aspects of target practice in May 1869, and the entire army adopted his scheme. The idea was spread by a number of army officers who believed that their experience fighting Indians in the West justified it. Civilians became involved in the process, too. The National Rifle Association, created in 1871, established a target range at Creedmoor,

New York, where the army and National Guard sent teams of marksmen for regular competitions. The craze for target shooting reached a peak in the decade of the 1880s, including among army units the need to estimate distances and practice in real-world settings like woods and hilly terrain.[4]

Advances in weapons technology continued at a rapid pace after the Civil War. Smokeless powder was initially developed in the 1860s, but it was crude, "often unstable and dangerous in storage." A French chemist improved the powder in 1884, and it was widely adopted by national armies. Smokeless powder had many advantages; it cleared up the battlefield, allowing soldiers to enjoy greater concealment because the smoke of their guns no longer revealed their location. While black powder burned fast, fouling the rifling of small arms, the new smokeless powder "burned more evenly and produced a prolonged power impulse with a relatively low chamber pressure." The sharpshooter Hiram Berdan and Englishman Edward Boxer independently developed an improved cartridge for breech-loading rifles. It was encased in light metal and had an internal primer. Both men worked in the late 1860s, so the metallic cartridge was ready for the new powder when the latter became commercially feasible.[5]

National armies began a process of adopting new rifles to take advantage of the improvements in powder and cartridge. The British converted their Enfield rifle muskets into breechloaders and called them Snider-Enfields, using a center-fired cartridge encased in brass. The Americans converted their Springfield rifle muskets into breechloaders by opening up a "trapdoor" for the insertion of the new cartridge. In 1871, the British replaced the Snider-Enfield with the Martini-Henry, the first general-issue small arm designed from the start as a breech-loader in their army. By 1888, the British were ready to replace the Martini-Henry with the Lee-Metford, a bolt-action weapon with a magazine of eight rounds. This weapon gave way to the Short Magazine Lee-Enfield in 1907, which the British used until after World War I. Even the Spanish army adopted a modern bolt-action weapon, the Mauser, in 1893.[6]

The Americans quickly fell behind European nations in weapons development. They relied on their single-shot, breech-loading Springfield for twenty years before finally adopting a Danish weapon, the Krag-Jorgenson, in 1892. It already was beginning to be outdated by new developments in weapons technology, most notably the magazine clip, invented in 1885, but at least the American army had a magazine-fed, bolt-action rifle. It was the first repeating small arms adopted as standard issue by the U.S. Army. At .30 caliber, it was consistent with a major trend in weapons development to create small arms of much smaller caliber than the .58-caliber Springfield of the Civil War. The Americans replaced the Krag-Jorgenson with the Model 1903 Springfield, representing a resurgence in American-made small arms, and retained guns with this traditional American stamp for many decades to come.[7]

Enthusiasm for programs designed to produce well-trained marksmen suffered reverses by the 1890s. The new magazine weapons gave soldiers the opportunity to deliver rapid fire without much aiming, and many American and European officers felt this was the wave of the future. Advocates of rifle training and target competitions had always emphasized the need to develop an individualistic element among infantrymen, encouraging them to take pride in their shooting. Now the German army was tending toward a firing doctrine that discounted individualism in favor of officers who estimated the range and directed the fire of their men. "The era of an army of marksmen is over," proclaimed Lt. George B. Davis of the U.S. Army. "The era of mutual support and cooperation is with us."[8]

The period from 1865 to the onset of World War I was a time of theory and expectation, reinforced to one degree or another by a series of short wars that only tentatively indicated which way tactical doctrine and firing systems were likely to go in the modern rifle age. Arthur L. Wagner stressed the need for intensive training to achieve fire discipline, for he did not think that long-range small-arms fire was very effective. But Wagner argued that American armies had to respond to long-range fire or suffer demoralization. He had great faith in the long-range capabilities of modern rifles, arguing that they could annoy troops at a distance of 2,500 yards. At 1,700 yards, rifle fire "becomes serious," and it was "really effective" at 1,000 yards. Wagner termed its effect as "decisive" at 500 yards and "practically annihilating" at 300 yards. What he based these estimates on is not clear, for Americans engaged in no major battles against troops armed with modern rifles before he published these views in 1895. Wagner probably based his observations on minimal information derived from the battle of Plevna during the Russo-Turkish War of 1873 and on his own suppositions regarding the probable performance of the new weapons in battle. From whatever source they came, these estimates of rifle range and effectiveness seem very unrealistic. Even if the ranges achieved at Plevna were long, that does not mean they could be duplicated on other battlefields. The notion that soldiers could see any targets, much less deliver "serious" fire at 1,700 yards, the equivalent of a mile away, is hard to believe.[9]

Wagner preferred to open fire at 500 yards but feared that modern weapons would encourage opponents to fire at an advancing party 700 or 800 yards away. At this range, he preferred volley fire over individual fire, as it allowed officers more opportunity to control and limit the exchange. Moreover, the sudden falling of many enemy soldiers would demoralize the survivors. If the volleys became ragged, Wagner wanted officers to order individual fire instead. To cover the last 500 yards of ground before closing with the enemy, the range at which Wagner described rifle fire as "decisive" and "practically annihilating," he advocated successive "bounds or rushes" of about 50 yards each. This could be done by the entire attacking force or only parts of it in alternating fashion.[10]

The wars of the late nineteenth century did not necessarily prove Wagner's predictions regarding range and effectiveness of rifle fire. The Austro-Prussian War of 1866 offered little in the way of direction. While the Prussians used their old needle gun, a breech-loading rifle developed in the 1830s, the Austrians used a single-shot muzzle-loader. The war, for reasons not necessarily related to the imbalance of firearms, lasted only seven weeks.[11]

The Franco-Prussian War of 1870–1871 was the first conflict in which both opponents were armed with breech-loading rifles. In fact, the older German Dreyse gun was outclassed by a new French weapon, the chassepot, developed by 1866. The French gun had an effective range of 1,000 yards while Dreyse's invention had an effective range of only 400 yards. The chassepot could fire at an estimated rate of eight to fifteen rounds per minute, compared to the Dreyse's four to five rounds per minute, and it was lighter in weight. The French weapon also had improved powder and cartridge, which gave its ammunition more punching power.[12]

Despite their superiority in small arms, the French lost the war due to a number of factors. The Germans often suffered heavy losses from French small-arms fire while conducting frontal attacks, but they compensated for this by skillful use of their field artillery to dominate the French guns. Another great advantage to the Germans was consistently poor generalship on the part of French commanders, who usually preferred to act strictly on the defensive, refusing to follow up bloody repulses of German attacks with offensive action of their own. French commanders consistently yielded the strategic and grand tactical initiative to their opponents, negating most of their advantage in small-arms capability.[13]

Albrecht von Boguslawski, a German battalion commander in the Franco-Prussian War, reported receiving long-range small-arms fire when the terrain offered French troops the opportunity to do so. Yet, "long shots from 800 to

1,500 paces . . . really did us little harm." Only if the Germans were massed in large formations and the vegetation offered a view of them could such long-range fire be more than annoying. While Boguslawski tentatively supported German calls for a longer-range infantry weapon than the Dreyse after the war, he also wondered if it was truly necessary because he thought it was extremely difficult to hit an individual beyond 300 yards anyway.[14]

The Second Boer War of 1899–1902 has often been cited as a good illustration of rifle effectiveness. The men of the South African Republic had a heritage of pioneering use of firearms, similar to that of the American West, and their government possessed 50,000 of the best rifles in the world at the start of the war, in addition to 150,000 lesser but still modern weapons. British troops were armed with the Lee-Metford rifle. The traditional view of historians is that the Boers fought as individuals, sniping at the British army from long range, while their opponents trained and operated in traditional tactical formations. Individual marksmanship, similar in traditional lore to the American view of the Revolution, rules the mythic image of this draining colonial war in South Africa. But at least one recent historian has argued that this mythic image is misleading. M. A. Ramsay believes that most Boers were not expert marksmen, but did their best execution "with volume of fire . . . at long ranges, and good snap shooting at 300 yards." This was a performance level probably similar to what the Turks accomplished at Plevna.[15]

There was enough evidence from the wars of the late nineteenth century to encourage advocates of individual marksmanship among American officers, and enough doubt about the effectiveness of long-range rifle fire to encourage their opponents. These two groups continued to argue with each other until John J. Pershing insisted that individual marksmanship remain a key feature of American training for World War I. In part he wanted to distinguish American training from that of the European allies, and in part he truly believed this might be a solution to trench stalemate on the battlefield. Despite the celebration of Sgt. Alvin York's remarkable achievement in almost single-handedly killing 25 and capturing 132 Germans on October 8, 1918, in the Meuse-Argonne Offensive, there is no reason to believe Pershing was right in arguing that adroit use of the rifle could compensate for successive lines of entrenched machine guns, artillery, and infantry weapons. In fact, exponents of the machine gun in the early twentieth century rightly argued that the new weapon was the equivalent of fifty rifle-firing infantrymen.[16]

Nevertheless, the Pershing model of an individual rifleman carefully picking out his targets and dispatching them one by one remained an ideal, if not a reality, in important army circles for decades to come. Between the world wars the U.S. Army "developed a paradigm centered on the primacy of the infantry rifleman and steadfastly clung to that idea." Although a variety of new weapons came to the fore, "the army attempted to fit them into the framework of the existing paradigm rather than develop a new approach." A combined arms doctrine, which de-emphasized the rifle-firing infantryman in favor of coordinated power exerted by foot soldiers, armored units, and air power, began to be implemented in the 1930s, against great resistance. The Americans retained Pershing's square division of World War I, a huge configuration giving each division some 25,000 troops to maximize the power of rifle-firing infantrymen, well into the late 1930s. Not until the German successes of 1939–1942, and the demands of American participation in the war, did conservative elements in the army finally cave in. The square division had to give way to the triangular division, which gave more flexibility for the use of armored units and crew-served weapons rather than the individual infantryman.[17]

The Americans introduced the first general-issue rifle capable of semiautomatic firing in World War II. In fact, they were the only combatants to do so in that conflict. The M-1 Garand channeled the gases from the fired cartridge into a mechanism to reload the chamber with a fresh round. It had an eight-round magazine clip and a range of 511 yards. The M-16 had exactly the same range, but a twenty-round magazine. It was capable of firing either semiautomatic or fully automatic. Initially issued in 1965 for service in Vietnam, the M-16 officially replaced the M-14 four years later as standard issue in the U.S. Army. By the 1990s, an improved version of the original M-16 used a thirty-round magazine and had an effective range of 611 yards.[18]

The M-16 was of little use as a target rifle. Its strength lay in the soldier's ability to spray bullets quickly over a given area a short distance away. The effective range of rifles and the typical range of infantry fire during actual engagements had not changed much by the mid-twentieth century compared to the Civil War. In the Second World War and the Korean conflict, range of fire normally was about 100 yards. In Vietnam, it was even less.[19]

Succeeding generations of new small-arms weapons improved dramatically in power and efficiency, but by the twentieth century their effective ranges were limited to little more than that of the Springfield and Enfield of the Civil

War era. Ironically, soldiers from the age of Marlborough to the cold war continued to fire their weapons more often than not at ranges of roughly 50 to 125 yards. Whether soldiers were armed with a Brown Bess or an M-16, most infantry combat for the past 300 years has taken place at very close range. Never has the long-range firing capability of any kind of rifle issued generally to the rank and file of any national army been fully realized. It is impractical to expect otherwise, due to the varying levels of proficiency with firearms among soldiers and the natural tendency to fire at ranges close enough so the shooter can more easily see his target and have a greater chance of hitting it.

Skirmishing and Sniping

The rifle continued to have an important impact on skirmishing and sniping after 1865. Both were, to different degrees, specialized activities offering much opportunity for long-range firing. Moreover, both activities demanded certain skills and personality traits. Although skirmishing could be adequately performed by any given line unit, the best skirmishers were men who preferred the opportunity to exercise individual initiative and were comfortable with taking risks. Sniping was skirmishing boiled down to one of its fundamental traits, demanding the most specialized characteristics of the men who engaged in it, preferably with specialized weapons to match.

Francis Lippitt knew that skirmishing would continue to be important in warfare, but he argued against a trend evident in the Civil War, which was to strengthen the skirmish line. The line had to be highly mobile and flexible to do its job properly. Skirmishers effected their aim through sharpshooting, not attacking the enemy as if they were a minibattle line. "Skirmishers are only auxiliary to the main force," Lippitt correctly concluded, "and are not capable, by themselves, of effecting any decisive result."[20]

Emory Upton, however, placed more faith on the skirmish line as an increasingly valuable asset on the battlefield. He wanted officers to "constantly aim to impress each man with the idea of his individuality, and the responsibility that rests upon him." Upton wanted skirmishers who possessed "the feeling that they cannot be whipped."[21]

Many post–Civil War observers went too far in predicting that skirmishing would soon become more important on the battlefield than the actions of the battle line. Arthur L. Wagner argued that skirmishers "have become the most important element in modern tactics, and now not only begin the action, but fight it out to the end." Yet neither Upton's tactical manual nor any other adopted by the U.S. Army gave precedence to skirmishing over the operations of the main force. In this case, the conservative Lippitt was more prescient than the progressive Wagner. Skirmishers could never be as numerous as the men who made up the battle line, or else they would lose the very qualities that made them effective on the battlefield. Precisely because they were a small force, they could take advantage of cover, move about in order to locate enemy weak spots, snipe at targets of opportunity, and serve as a warning line for enemy attacks.[22]

Modern sniping originated in the Civil War, where the role of the sniper and the role of the skirmisher often blended. While skirmishers often spent time taking potshots at individual targets during the Civil War, modern sniping was a different affair. Engaged in by individuals who were not only naturally gifted marksman but who did not mind working alone or in small teams, sometimes waiting for a long time before getting an opportunity to fire at a man, modern snipers appeared for the first time in the Civil War. They were essentially hunters of men rather than typical soldiers, and their role in warfare increased with each major conflict to come.

The Germans took the lead in developing a force of snipers early in World War I by issuing six specialized rifles to each company and allowing commanders to decide which men should use them. They established sniper schools to give quick training to the troops selected for the task. The Germans created a battalion sniper section consisting of twenty-four men who were usually allowed to work along the lines as they saw fit. The snipers carried portable steel plates with loopholes that they could easily fit atop the parapet. Historians believe the Germans continued to hold the edge in the sniper war on the western front, using them both offensively to take out observers and machine gunners, and defensively as rear guards during a retreat.[23]

The Germans also seem to have had a greater supply of high-quality long-range small arms available, while the British had to scramble to find appropriate arms. They duplicated the German system of selection, training, and deployment of snipers, while improving on the Germans by using their snipers as scouts and intelligence gatherers. British snipers took a ten-day course at sniper school, the most difficult and important part of which was estimating distances and setting the sights properly. Commonwealth troops and the Americans attended the British schools, the latter becoming eager pupils. Like the

Commonwealth armies, the Americans preferred to pair snipers to work as teams. The French created their own system similar to that of the Germans and British, although never deploying as many snipers as either of them.[24]

The history of sniping in World War II was similar to that in the First World War. In both conflicts, captured snipers often were killed in cold blood, their captors shocked and angry at unseen enemies who stalked individuals. Snipers tended to be treated as strange, unapproachable figures even by their own countrymen. In part, it was because they literally were not part of the normal routine of army life. Absolved from daily duties, they were also told not to talk about their exploits. Most snipers tended to be loners, quiet and self-sufficient men who would have been thoroughly lost in the crowd anyway even if it were not for their solitary occupation.[25]

Finnish snipers are credited with a high level of effectiveness during their Winter War against the Soviets in 1939–1940, working in two-man teams protected by camouflage. Some Finnish snipers chalked up 400 to 500 confirmed hits. On the eastern front of World War II, the Germans deployed 22 snipers for every battalion after sending candidates to sniper school. Matthias Hetzenaur racked up the highest number of hits on the eastern front: 345 confirmed kills and another 150 probable hits. His longest shot was 1,000 meters, about three quarters of a mile. The Russians trained their sniper candidates for three weeks, often allowing women to train and deploy as snipers. The highest tally by a Russian sniper was 309 confirmed kills. It has been estimated that a sample group of 1,061 Soviet snipers accounted for 12,000 German casualties on the eastern front. U.S. Marine forces operating in the Pacific war often created teams of three snipers armed with a mixture of scoped rifles for long-range work and semiautomatic guns for short-range fire.[26]

The U.S. Army paid little attention to sniper training between the major wars of the twentieth century, reconstructing the system of sniper schools at the start of each conflict and trying to find appropriate weapons all over again. Not until after Vietnam did the Marines create a more effective, permanent system of sniper training, using specialized weapons. One sees a similar history of neglect until recent times in the British army as well.[27]

Some areas of weapons technology have seen important developments for snipers. Equipment designed to suppress sound and provide the capability for night vision became available after World War II. Also, the development of armor-piercing ammunition has given the sniper the opportunity to damage machines at long range, and improvements in optical equipment have increased his ability to hit distant targets. While custom-made rifles are currently the norm for snipers in all national armies, most prefer bolt-action rifles over semiautomatic weapons.[28]

While sniping increased in importance, skirmishing soon declined, despite the optimistic predictions of Wagner and his colleagues. Battle lines became thinner in response to changed conditions on the battlefield, but they did not disappear. As long as there were battle lines, there were skirmishers, although late nineteenth-century tactical manuals gave equal importance to scouts. Acting individually, scouts were able to locate enemy positions and provide a flow of information to the battle line more effectively than skirmishers. Lee's sharpshooter battalions of 1864–1865 gave rise to the first recorded use of scouts in American military history.

Skirmishers became all but irrelevant by World War I because of the prolonged, close-range contact of the opposing armies along the western front. When the Americans finally converted to a decentralized tactical organization, abandoning linear formations altogether, there was no longer any need for skirmishers. Small teams armed with a mixture of weapons and trained to take advantage of cover while approaching enemy positions replaced the age-old linear system by the 1940s, and the skirmish line disappeared, too. Close-order drill, however, remains a fundamental component of training in all services up to today. It is an important reminder of the continued utility of what Civil War historians have all too often erroneously termed "outdated" concepts in organizing, moving, and fighting troops. Completely useful in fighting units armed with smoothbore muskets and rifle muskets alike, linear tactics became outdated only with the introduction of modern artillery, machine guns, semiautomatic and automatic small arms, and tactical nuclear explosives in the twentieth century.

Notes

Abbreviations

CHS	Chicago Historical Society
FSNMP	Fredericksburg-Spotsylvania National Military Park, Fredericksburg, VA
LC	Manuscripts Division, Library of Congress, Washington, DC
Mary-HS	Maryland Historical Society, Baltimore
NARA	National Archives and Records Administration, Washington, DC
OR	*The War of the Rebellion: A Compilation of the Official Records of the Union and Confederate Armies,* 70 vols. (Washington, DC: Government Printing Office, 1880–1901)
PNB	Petersburg National Battlefield, Petersburg, VA
RBMD-NYPL	Rare Books and Manuscripts Division, New York Public Library, New York
SC-CWM	Special Collections, College of William and Mary, Williamsburg, VA
SC-E	Special Collections, Emory University, Atlanta, GA
SCL-DU	Special Collections Library, Duke University, Durham, NC
SC-UTK	Special Collections, University of Tennessee, Knoxville
SC-WLU	Special Collections, Washington and Lee University, Lexington, VA
SHC-UNC	Southern Historical Collection, University of North Carolina, Chapel Hill
USAMHI	United States Army Military History Institute, Carlisle, PA

Introduction

1. Jack Coggins, *Arms and Equipment of the Civil War* (Garden City, NY: Doubleday, 1962), 38–39; Brent Nosworthy, *The Bloody Crucible of Courage: Fighting Methods and Combat Experience of the Civil War* (New York: Carroll & Graf, 2003), 42, 45, 50–53, discusses the European and American writers on rifle issues before 1861.

2. Review of *Field Tactics for Infantry,* by William H. Morris, *North American Review* 99 (July 1864): 291.

3. Stephen Vincent Benet to William T. Sherman, January 29, 1883, in John C. Tidball, "Rifle Target Practice in the Army," *Ordnance Notes,* January 30, 1883, 1; John K. Mahon, "Civil War Infantry Assault Tactics," in *Military Analysis of the Civil War: An Anthology by the Editors of Military Affairs* (Millwood, NY: KTO, 1977), 253; Edward Hagerman, *The American Civil War and the Origins of Modern Warfare: Ideas, Organization, and Field Command* (Bloomington: Indiana University Press, 1988), 10; Margaret E. Wagner, Gary W. Gallagher, and Paul Finkelman, eds., *Civil War Desk Reference* (New York: Simon & Schuster, 2002), 486; Robert A. Doughty and Ira D. Gruber, *American Military History and the Evolution of Warfare in the Western World* (Lexington, MA: D. C. Heath, 1996), 96; Allan R. Millett and Peter Maslowski, *For the Common Defense: A Military History of the United States of America* (New York: Free

Press, 1984), 123; Russell F. Weigley, *A Great Civil War: A Military and Political History, 1861–1865* (Bloomington: Indiana University Press, 2000), 33; Thomas Vernon Moseley, "Evolution of the American Civil War Infantry Tactics" (Ph.D. diss., University of North Carolina, Chapel Hill, 1967), 1–2.

4. James M. McPherson, *Battle Cry of Freedom: The Civil War Era* (New York: Oxford University Press, 1988), 477; McPherson, *Ordeal by Fire: The Civil War and Reconstruction,* 3rd ed. (New York: McGraw-Hill, 2001), 216–217; Weigley, *Great Civil War,* xix, 131; Russell F. Weigley, *History of the United States Army,* rev. ed. (Bloomington: Indiana University Press, 1984), 235; Perry D. Jamieson, *Crossing the Deadly Ground: United States Army Tactics, 1865–1899* (Tuscaloosa: University of Alabama Press, 1994), 106; Edward Hagerman, *The American Civil War and the Origins of Modern Warfare: Ideas, Organization, and Field Command* (Bloomington: Indiana University Press, 1988), xi; Herman Hattaway, "The Changing Face of Battle," *North and South* 4, 6 (2001): 36; James M. Morris, *America's Armed Forces: A History,* 2nd ed. (Upper Saddle River, NJ: Prentice Hall, 1996), 84; John Morgan Dederer, "Side Arms, Standard Infantry," in *The Oxford Companion to American Military History,* ed. John Whiteclay Chambers II (New York: Oxford University Press, 1999), 659–660; Jack Coggins, "The Engineers Played a Key Role in Both Armies," *Civil War Times Illustrated* 3, 9 (1965): 44.

5. McPherson, *Battle Cry of Freedom,* 475; McPherson, *Ordeal by Fire,* 215.

6. Paddy Griffith, *Battle Tactics of the Civil War* (New Haven, CT: Yale University Press, 1989), 74–75, 81, 90, 146–149.

7. Ibid., 29, 67.

8. Ibid., 17, 20, 27, 191–192.

9. Earl J. Hess, *The Union Soldier in Battle: Enduring the Ordeal of Combat* (Lawrence: University Press of Kansas, 1997), 56–57; Hess, *Field Armies and Fortifications in the Civil War: The Eastern Campaigns, 1861–1864* (Chapel Hill: University of North Carolina Press, 2005), 308–314. I continued to adhere to the standard interpretation for some time after the publication of Griffith's work, as witness my "Tactics, Trenches, and Men in the Civil War," in Stig Forster and Jorg Nagler, eds., *On the Road to Total War: The American Civil War and the German Wars of Unification, 1861–1871* (New York: Cambridge University Press, 1997), 483–484, originally written as a conference paper in 1992.

10. Mark Grimsley, "Surviving Military Revolution: The U.S. Civil War," in *The Dynamics of Military Revolution, 1300–2050,* ed. Macgregor Knox and Williamson Murray (Cambridge: Cambridge University Press, 2001), 76; Grimsley, "Review Essay: The Continuing Battle of Gettysburg," *Civil War History* 49, 2 (2003): 186. Herman Hattaway noted Griffith's thesis and my support of it in "Changing Face of Battle," 40.

11. Nosworthy, *Bloody Crucible,* 278, 573–577, 592. Joseph G. Bilby, in a little-known book entitled *Civil War Firearms: Their Historical Background and Tactical Use* (Conshohocken, PA: Combined Books, 1996), 12, agreed with Griffith about the short range of Civil War firing, the similarities in combat losses in smoothbore battles of the Napoleonic era, and the assertion that breech-loading rifles were the true revolutionary factor in warfare. Bilby also argued for the continued effectiveness of smoothbore muskets on Civil War battlefields.

12. Hess, *Field Armies and Fortifications,* 311–312; Earl J. Hess, *Trench Warfare under Grant and Lee: Field Fortifications in the Overland Campaign* (Chapel Hill: University of North Carolina Press, 2007), 216.

13. Wagner, Gallagher, and Finkelman, *Civil War Desk Reference,* 518.

Chapter One. The Smoothbore and Rifle Heritage

1. Brent Nosworthy, *The Anatomy of Victory: Battle Tactics, 1689–1763* (New York: Hippocrene, 1992), 3, 11–12, 18.

2. Ibid., 16, 22; B. P. Hughes, *Firepower: Weapons Effectiveness on the Battlefield, 1630–1850* (New York: Sarpedon, 1997), 10–11.

3. Hughes, *Firepower,* 11; Nosworthy, *Anatomy of Victory,* 4–5, 12, 33–35, 39, 184, 209; Steven Ross, *From Flintlock to Rifle: Infantry Tactics, 1740–1866* (Rutherford, NJ: Fairleigh Dickinson University Press, 1979), 25; John A. Lynn, *Giant of the Grand Siecle: The French Army, 1610–1715* (New York: Cambridge University Press, 1997), 457, 459–461, 463.

4. Hughes, *Firepower,* 11; Nosworthy, *Anatomy of Victory,* 12.

5. Nosworthy, *Anatomy of Victory,* xv–xvi, 346.

6. Ibid., 14, 22, 49–51; Christopher Duffy, *The Army of Frederick the Great* (London: David & Charles, 1974), 89–90.

7. Nosworthy, *Anatomy of Victory,* 47–48.

8. Ibid., 55–57, 344; Lynn, *Giant of the Grand Siecle,* 486.

9. Nosworthy, *Anatomy of Victory,* 57–58, 61, 189, 92, 276–277; Ross, *Flintlock to Rifle,* 26.

10. Nosworthy, *Anatomy of Victory,* 208, 210.

11. Hughes, *Firepower,* 119, 164; Nosworthy, *Anatomy of Victory,* 6, 112–114. Hughes defined effective range as the distance "at which a trained marksman could expect no more than 30% of his shots to hit a target" (*Firepower,* 119).

12. Hughes, *Firepower,* 27, 81, 83, 85; Nosworthy, *Anatomy of Victory,* 40.

13. Nosworthy, *Anatomy of Victory,* 157–158, 214–215; Lynn, *Giant of the Grand Siecle,* 455.

14. Nosworthy, *Anatomy of Victory,* 214, 263; Ross, *Flintlock to Rifle,* 28–30.

15. Nosworthy, *Anatomy of Victory,* 211–212, 216; Ross, *Flintlock to Rifle,* 29–30.

16. Ross, *Flintlock to Rifle,* 34–35, 37–40, 53.

17. Ibid., 56, 67; Paddy Griffith, *The Art of War of Revolutionary France, 1789–1802* (London: Greenhill, 1998), 210–212.

18. Griffith, *Art of War of Revolutionary France,* 211; Ross, *Flintlock to Rifle,* 127–128.

19. Hughes, *Firepower,* 10, 26, 29.

20. Ibid., 114–116, 129–130, 137–138, 140–148, 151.

21. Rory Muir, *Tactics and the Experience of Battle in the Age of Napoleon* (New Haven, CT: Yale University Press, 1998), 77–78.

22. Ibid., 83.

23. Ibid.

24. Muir, *Tactics and the Experience of Battle,* 81–82. Griffith's findings are reported in Muir's work.

25. Hughes, *Firepower,* 27, 29.

26. Ibid., 29.

27. Ibid., 124, 133.

28. Muir, *Tactics and the Experience of Battle,* 46, 78, 84, 148; Hughes, *Firepower,* 164–165; Lynn, *Giant of the Grand Siecle,* 489.

29. Muir, *Tactics and the Experience of Battle,* 53, 55–56; Gunther E. Rothenburg, *The Art of Warfare in the Age of Napoleon* (Bloomington: Indiana University Press, 1980), 138.

30. Muir, *Tactics and the Experience of Battle,* 53–54, 60–62.

31. Ross, *Flintlock to Rifle,* 137, 149–150; Muir, *Tactics and the Experience of Battle,* 51–52.

32. Ross, *Flintlock to Rifle,* 158, 173–176.

33. Ibid., 162–163; Nosworthy, *Bloody Crucible,* 53–59.

34. Ross, *Flintlock to Rifle,* 163–164.

35. Ibid., 176.

36. John W. Wright, "The Rifle in the American Revolution," *American Historical Review* 29 (1923–1924): 294.

37. Lewis Nicola, *A Treatise of Military Exercise, Calculated for the Use of the Americans* (Philadelphia: Stymer & Cist, 1776), 1–2, 2n, 4; Howard R. Marraro, "Unpublished Letters of Colonel Nicola, Revolutionary Soldier," *Pennsylvania History* 12, 4 (1946): 274.

38. Robert Kenneth Wright Jr., "Organization and Doctrine in the Continental Army, 1774 to 1784" (Ph.D. diss., College of William and Mary, 1980), 45–47; Wright, "Rifle in the American Revolution," 293–294.

39. John W. Wright, "The Corps of Light Infantry in the Continental Army," *American Historical Review* 31 (1925–1926): 454–459, 461.

40. Wright, "Rifle in the American Revolution," 297–298.

41. Ibid., 295; William Duane, *The American Military Library; Or, Compendium of the Modern Tactics,* 2 vols. (Philadelphia: William Duane, 1809), 2: 1–2.

42. Nicola, *Treatise of Military Exercise,* 32–39.

43. Ibid., 40–41, 44, 46.

44. Baron von Steuben, *Revolutionary War Drill Manual* (New York: Dover, 1985), 63–65; James Kirby Martin, ed., *Ordinary Courage: The Revolutionary War Adventures of Joseph Plumb Martin* (St. James, NY: Brandywine, 1993), 19.

45. Brent Nosworthy, *The Bloody Crucible of Courage: Fighting Methods and Combat Experience of the Civil War* (New York: Carroll & Graf, 2003), 25–26.

46. Ibid., 26–28.

47. Ibid., 35–36; Ross, *Flintlock to Rifle,* 175–176.

48. Allan R. Millett and Peter Maslowski, *For the Common Defense: A Military History of the United States of America* (New York: Free Press, 1984), 123; Ross, *Flintlock to Rifle,* 161–162; Nosworthy, *Bloody Crucible,* 29; John Morgan Dederer, "Side Arms, Standard Infantry," in *The Oxford Companion to American Military History,* ed. John Whiteclay Chambers II (New York: Oxford University Press, 1999), 659.

49. Ross, *Flintlock to Rifle,* 164, 170; Nosworthy, *Bloody Crucible,* 29, 84–86, 96; Millett and Maslowski, *For the Common Defense,* 123.

50. Nosworthy, *Bloody Crucible,* 41–42, 45.

51. Ibid., 50, 52.

52. Ibid., 30–32, 53.

53. Ibid., 78–79.

54. Grady McWhiney and Perry D. Jamieson, *Attack and Die: Civil War Military Tactics and the Southern Heritage* (University: University of Alabama Press, 1982), 29, 40; Robert A. Doughty and Ira D. Gruber, *American Military History and the Evolution of Warfare in the Western World* (Lexington, MA: D. C. Heath, 1996), 95; Jamieson, "Background to Bloodshed: The Tactics of the U.S.–Mexican War and the 1850s," *North and South* 4, 6 (2001): 28.

55. Winfried Baumgart, *The Crimean War, 1853–1856* (London: Arnold, 1999), 64, 70, 79, 86; entry of February 22, 1855, in Colin Robins, ed., *Captain Dunscombe's Diary* (Bowdon, UK: Withycut House, 2003), 81, 81n.

56. A number of books have been published on the Sebastopol campaign, but Baumgart, *Crimean War,* provides the best short description and analysis of it.

57. Entries of December 3–4, 1854, and February 18, 1855, in Robins, *Captain Dunscombe's Diary,* 51–52, 77; Earl J. Hess, *Field Armies and Fortifications in the Civil War: The Eastern Campaigns, 1861–1864* (Chapel Hill: University of North Carolina Press, 2005), 333.

58. Baumgart, *Crimean War,* 145–162.

59. Hans Busk, *The Rifle: And How to Use It,* 8th ed. (London: Routledge, Warne, & Routledge, 1862), 85–166; C. M. Wilcox, *Rifles and Rifle Practice: An Elementary Treatise upon the Theory of Rifle Firing* (New York: D. Van Nostrand, 1859), 175–207; Horace William Shaler Cleveland, "The Use of the Rifle," *Atlantic Monthly,* March 1862, 301–303, 305; Arthur Walker, *The Rifle: Its Theory and Practice* (Westminster, UK: J. B. Nichols and Sons, 1864), 10.

60. Wilcox, *Rifles and Rifle Practice,* 1–146, 167–207, 222–229, 248.

61. Ibid., 65–67, 179; Walker, *The Rifle,* 84–85, 142; Nosworthy, *Bloody Crucible,* 30–32.

62. Nosworthy, *Bloody Crucible,* 32.

63. Ibid., 34–35; Walker, *The Rifle,* 13–25; Busk, *The Rifle,* 201–215; Capt. Binney, "Muskets and Musketry," *Aide Memoir to the Military Sciences* 2 (1860): 442–467.

64. John C. Tidball, "Rifle Target Practice in the Army," *Ordnance Notes,* January 30, 1883, 1–2.

65. Ibid., 2–3.

66. Bell Irvin Wiley, *The Life of Billy Yank: The Common Soldier of the Union* (Baton Rouge: Louisiana State University Press, 1983), 53–54.

67. Wilcox, *Rifles and Rifle Practice,* 243; Busk, *The Rifle,* 22–23.

68. Nosworthy, *Bloody Crucible,* 58–59, 89–90.

69. Ibid., 48, 50, 90–91.

70. Wilcox, *Rifles and Rifle Practice,* 173.

71. Ibid., 237–238.

72. Nosworthy, *Bloody Crucible,* 43–44; Busk, *The Rifle,* 18–19.

73. Nosworthy, *Bloody Crucible,* 49, 93.

Chapter Two. The First Rifle War

1. Russell F. Weigley, *A Great Civil War: A Military and Political History, 1861–1865* (Bloomington: Indiana University Press, 2000), 32–33; George H. Thomas to William F. Patterson, November 4, 1861, William Franklin Patterson Papers, LC; *The Story of the Fifty-fifth Regiment Illinois Volunteer Infantry in the Civil War, 1861–1865* (Clinton, MA: W. J. Coulter, 1887), 41; Camp Inspection Return, 3rd Kentucky, March 19, 1863, United States Sanitary Commission Records, series 7, RBMD-NYPL.

2. Entry of March 26, 1862, in Mildred Throne, ed., *The Civil War Diary of Cyrus F. Boyd, Fifteenth Iowa Infantry, 1861–1863* (Baton Rouge: Louisiana State University Press, 1998), 24; entry of June 1, 1862, in Mark Grimsley and Todd D. Miller, eds., *The Union Must Stand: The Civil War Diary of John Quincy Adams Campbell, Fifth Iowa Volunteer Infantry* (Knoxville: University of Tennessee Press, 2000), 45, 239n.

3. Joseph J. Crute Jr., *Units of the Confederate States Army* (Midlothian, VA: Derwent, 1987), 42–43, 124–126, 128–130, 247, 250, 320; Charles E. Dornbusch, *Military Bibliography of the Civil War,* 3 vols. (New York: New York Public Library, 1994), vol. 1, pt. 2, 73; *Military Service*

Records: A Select Catalog of National Archives Microfilm Publications (Washington, DC: National Archives and Service Administration, 1985), 77–78.

4. Michael A. Mullins, *The Fremont Rifles: A History of the 37th Illinois Veteran Volunteer Infantry* (Wilmington, NC: Broadfoot, 1990), 1–2, 11–12, 16, 75.

5. Robert V. Bruce, *Lincoln and the Tools of War* (Indianapolis: Bobbs-Merrill, 1956), 40, 42.

6. Carl L. Davis, *Arming the Union: Small Arms in the Civil War* (Port Washington, NY: Kennikat, 1973), v–viii.

7. Ibid., 64–67, 167–168.

8. Ken Baumann, *Arming the Suckers, 1861–1865: A Compilation of Illinois Civil War Weapons* (Dayton, OH: Morningside Bookshop, 1989), vii, ix, 70–152, 154–222.

9. Robert Beecham reminiscences, in Michael E. Stevens, ed., *As if It Were Glory: Robert Beecham's Civil War from the Iron Brigade to the Black Regiments* (Madison, WI: Madison House, 1998), 3; Leander Stillwell, *The Story of a Common Soldier of Army Life in the Civil War, 1861–1865,* 2nd ed. (Kansas City, MO: Franklin Hudson, 1920), 28; entry of June 18, 1861, in Mary Ann Andersen, ed., *The Civil War Diary of Allen Morgan Geer, Twentieth Regiment, Illinois Volunteers* (New York: Cosmos, 1977), 3–4; Henry C. Matrau to parents, July 22, 1861, and Matrau to father and mother, July 23, 1865, in Marcia Reid-Green, ed., *Letters Home: Henry Matrau of the Iron Brigade* (Lincoln: University of Nebraska Press, 1993), 7, 124; Lyman Richard Comey, ed., *A Legacy of Valor: The Memoirs and Letters of Captain Henry Newton Comey, 2nd Massachusetts Infantry* (Knoxville: University of Tennessee Press, 2004), 6; Grant testimony before House Select Committee on Government Contracts, October 31, 1861, in John Y. Simon, ed., *The Papers of Ulysses S. Grant,* 24 vols. (Carbondale: Southern Illinois University Press, 1967–2000), 3: 90, 98n.

10. Isaac H. Elliott, *History of the Thirty-third Regiment Illinois Veteran Volunteer Infantry in the Civil War, 22nd August, 1861, to 7th December, 1865* (Gibson City, IL: Gibson Courier, 1902), 254; W. A. Neal, ed., *An Illustrated History of the Missouri Engineer and the Twenty-fifth Infantry Regiments* (Chicago: Donohue & Henneberry, 1889), 22–14.

11. Benjamin T. Smith journal, February 16–17, July 20, 1862, quoted in Baumann, *Arming the Suckers,* 134.

12. *Story of the Fifty-fifth Regiment,* 41–42; Baumann, *Arming the Suckers,* 140.

13. Philip B. Fouke to John A. McClernand, November 9, 1861, *OR,* vol. 3, 287; Baumann, *Arming the Suckers,* 80.

14. Stillwell, *Story of a Common Soldier,* 28; F. Y. Hedley, *Marching through Georgia: Pen-Pictures of Every-Day Life in General Sherman's Army, from the Beginning of the Atlanta Campaign until the Close of the War* (Chicago: Donohue, Henneberry, 1890), 29; George P. Woodruff reminiscences, quoted in Baumann, *Arming the Suckers,* 89.

15. Samuel Beardsley to Freddy, September 12, 1861, in Annette Tapert, ed., *The Brothers' War: Civil War Letters to Their Loved Ones from the Blue and Gray* (New York: Vintage, 1988), 20; Joseph R. Reinhart, *A History of the 6th Kentucky Volunteer Infantry, U.S.: The Boys Who Feared No Noise* (Louisville, KY: Beargrass, 2000), 22; entries of summer 1861 and December 19, 1861; February 26 and March 5, 1862, in "Diary of Colonel William Camm, 1861 to 1865," *Journal of the Illinois State Historical Society* 18 (1925–1926): 802, 834, 836; Elliott, *History of the Thirty-third Regiment,* 179; Thomas H. Parker, *History of the 51st Regiment of P.V. and V.V* (Philadelphia: King & Baird, 1869), 93.

16. Larry J. Daniel, *Soldiering in the Army of Tennessee: A Portrait of Life in a Confederate*

Army (Chapel Hill: University of North Carolina Press, 1991), 39–41; Inspection Report, Provisional Army of Tennessee, July 31, 1861, *OR,* vol. 52, pt. 2, 123.

17. Nathaniel C. Hughes, ed., *Liddell's Record: St. John Richardson Liddell, Brigadier General, C.S.A.* (Dayton, OH: Morningside Bookshop, 1985), 41; Davis to Leonidas Polk, September 2, 1861, in Lynda Lasswell Crist, ed., *The Papers of Jefferson Davis,* 10 vols. (Baton Rouge: Louisiana State University Press, 1971–1999), 7: 318; Inspection Report, Provisional Army of Tennessee, July 31, 1861, *OR,* vol. 52, pt. 2, 122–123.

18. I. G. Bradwell, "Soldier Life in the Confederate Army," *Confederate Veteran* 24 (1916): 21; Samuel W. Hankins, *Simple Story of a Soldier: Life and Service in the 2d Mississippi Infantry* (Tuscaloosa: University of Alabama Press, 2004), 8–9.

19. William T. Alderson, ed., "The Civil War Reminiscences of John Johnston, 1861–1865," *Tennessee Historical Quarterly* 13 (1954): 79; Alex Spence to Tom, May 30, 1861, in Mark K. Christ, ed., *Getting Used to Being Shot At: The Spence Family Civil War Letters* (Fayetteville: University of Arkansas Press, 2002), 7, 148n; G. W. Nichols, *A Soldier's Story of His Regiment* (Kennesaw, GA: Continental, 1961), 24, 27; Arthur W. Bergeron Jr., ed., *The Civil War Reminiscences of Major Silas T. Grisamore, C.S.A.* (Baton Rouge: Louisiana State University Press, 1993), 2; R. Lockwood Tower, ed., *A Carolinian Goes to War: The Civil War Narrative of Arthur Middleton Manigault, Brigadier General, C.S.A.* (Columbia: University of South Carolina Press, 1988), 5.

20. Finley P. Curtis, "The Black Shadow of the Sixties," *Confederate Veteran* 24 (1916): 353–354; Winchester Hall, *The Story of the 26th Louisiana Infantry, in the Service of the Confederate States* (N.p., n.d.), 31–32.

21. Judah P. Benjamin to Nathan Ross, December 20, 1861; George B. Crittenden to Samuel Cooper, December 20, 1861, *OR,* vol. 7, 780; Grant Taylor to wife and children, May 16, 1862, in Ann K. Blomquist and Robert A. Taylor, eds., *This Cruel War: The Civil War Letters of Grant and Malinda Taylor, 1862–1865* (Macon, GA: Mercer University Press, 2000), 20.

22. Daniel, *Soldiering in the Army of Tennessee,* 42; D. Leib Ambrose, *From Shiloh to Savannah: The Seventh Illinois Infantry in the Civil War* (DeKalb: Northern Illinois University Press, 2003), 58; Kenneth W. Noe, ed., *A Southern Boy in Blue: The Memoirs of Marcus Woodcock, 9th Kentucky Infantry (U.S.A.)* (Knoxville: University of Tennessee Press, 1996), 74. The Thirty-second Tennessee used "inferior flint-lock muskets" at Fort Donelson and many of them became inoperable due to "the priming being wet." Edward C. Cook to John C. Brown, February 16, 1862, *OR,* vol. 7, 356.

23. George P. Woodruff quoted in Baumann, *Arming the Suckers,* 176; J. W. Gaskill, *Footprints through Dixie: Everyday Life of the Man under a Musket on the Firing Line and in the Trenches, 1862–1865* (Alliance, OH: Bradshaw, 1919), 22.

24. Theodore F. Upson letter, undated, but ca. August 1862, in Oscar Osburn Winther, ed., *With Sherman to the Sea: The Civil War Letters, Diaries and Reminiscences of Theodore F. Upson* (Bloomington: Indiana University Press, 1958), 27–28; K. Jack Bauer, ed., *Soldiering: The Civil War Diary of Rice C. Bull, 123rd New York Volunteer Infantry* (San Rafael, CA: Presidio, 1978), 4; David Gould and James B. Kennedy, eds., *Memoirs of a Dutch Mudsill: The "War Memories" of John Henry Otto, Captain, Company D, 21st Regiment Wisconsin Volunteer Infantry* (Kent, OH: Kent State University Press, 2004), ix, 10, 18.

25. Dan McCook to Julius Garesche, December 6, 1862; McCook to Capt. Wiseman, December 17, 1862, 14th Army Corps, Letters Sent (3rd Brig., 2nd Div.), vol. 42/109, RG393, pt. 2,

NARA; Baumann, *Arming the Suckers,* 221; William H. Bentley and Richard L. Howard, quoted in Baumann, *Arming the Suckers,* 161, 215.

26. Fredrick Pettit to parents, brothers, and sisters, September 21, 1862, in Tapert, *Brothers' War,* 90; William Henry Harrison Clayton to father and mother, August 31, 1862, in Donald C. Elder III, ed., *A Damned Iowa Greyhound: The Civil War Letters of William Henry Harrison Clayton* (Iowa City: University of Iowa Press, 1998), 8; Edward Hastings Ripley to William, July 21, 1862, in Otto Eisenschiml, ed., *Vermont General: The Unusual War Experiences of Edward Hastings Ripley, 1861–1865* (New York: Devin-Adair, 1960), 9.

27. A. F. Sperry, *History of the 33d Iowa Infantry Volunteer Regiment, 1863–6* (Fayetteville: University of Arkansas Press, 1999), 7–8; Baumann, *Arming the Suckers,* 72, 90–91.

28. Entry of March 19, 1863, in Andersen, *Civil War Diary of Allen Morgan Geer,* 83; Mullins, *Fremont Rifles,* 224; Record of Events, May 1863, 41st Ohio, in *Supplement to the Official Records of the Union and Confederate Armies,* 100 vols. (Wilmington, NC: Broadfoot, 1993–2000), pt. 2, vol. 52, 420; Henry Roback, *The Veteran Volunteers of Herkimer and Otsego Counties in the War of the Rebellion: Being a History of the 152nd N.Y.V.* (Utica, NY: L. C. Childs, 1888), 15, 26; William Bircher, *A Drummer-Boy's Diary: Comprising Four Years of Service with the Second Regiment Minnesota Veteran Volunteers, 1861 to 1865* (St. Paul, MN: St. Paul Book and Stationery, 1889), 61; Stillwell, *Story of a Common Soldier,* 28–29; Baumann, *Arming the Suckers,* 223–229.

29. Louis A. Garavaglia and Charles G. Worman, "Arms and the Man," *North and South* 4, 6 (2001): 53; Francis A. Lord, "Springfield Armory Produced 793,000 Muskets." *Civil War Times Illustrated* 2, 7 (1963): 19; Margaret E. Wagner, Gary W. Gallagher, and Paul Finkelman, eds., *Civil War Desk Reference* (New York: Simon & Schuster, 2002), 486.

30. Daniel, *Soldiering in the Army of Tennessee,* 45–47.

31. Richard S. Bostwick to Charles P. Hatch, December 19, 1862, *OR,* vol. 21, 255–256; Robert Goodyear to Sarah, February 14, 1863, in Tapert, *Brothers' War,* 132; Harvey Reid to Libbie, June 21, 1863, in Frank L. Byrne, ed., *Uncommon Soldiers: Harvey Reid and the 22nd Wisconsin March with Sherman* (Knoxville: University of Tennessee Press, 2001), 73.

32. John P. Hawkins to W. T. Clark, November 15, 1863, *OR,* vol. 26, pt. 1, 800; John C. Chadwick to Lorenzo Thomas, June 13, 1864, ibid., vol. 34, pt. 1, 444.

33. Richard Taylor to William R. Boggs, February 27, 1864; William L. Cabell to John S. Marmaduke, February 27, 1864, *OR,* vol. 34, pt. 2, 1000–1002.

34. John C. Moore, "Battle of Lookout Mountain," *Confederate Veteran* 6 (1898): 426; Joseph E. Chance, *The Second Texas Infantry: From Shiloh to Vicksburg* (Austin, TX: Eakin, 1984), 148.

35. Entry of November 17, 1862, in "William J. Rogers' Memorandum Book," *West Tennessee Historical Society Papers* 9 (1955): 75; Willis H. Claiborne diary, April 6, 1864, in *Supplement to the Official Records,* pt. 1, vol. 6, 209.

36. General Orders No. 22, Headquarters, 18th Corps, May 22, 1864, *OR,* vol. 36, pt. 2, 108; Nathan B. Forrest to P. Ellis, July 1, 1864, ibid., vol. 39, pt. 1, 225; Daniel, *Soldiering in the Army of Tennessee,* 46; Thomas Jordan, "Notes of a Confederate Staff-Officer at Shiloh," in Robert Underwood Johnson and Clarence Clough Buel, eds., *Battles and Leaders of the Civil War* (New York: Thomas Yoseloff, 1956), 1: 603; Report of Inspection, Arms and Accoutrements, Lost at Shiloh, 3rd Corps, Army of the Mississippi, June 1, 1862, Confederate Inspection Reports, roll 18, M935, NARA. It was possible for the craftsmen at armories to rehabilitate even guns that had been deliberately wrecked by the enemy and harvested from battlefields. The master armorer at the Richmond Armory managed to make 400 carbines out of damaged

Springfield rifle muskets even though the Federals had broken them "over gun carriage wheels, and around trees." Samuel Adams to Jefferson Davis, December 2, 1862, in Crist, *Papers of Jefferson Davis,* 8: 526.

37. Gould and Kennedy, *Memoirs of a Dutch Mudsill,* 242; entry of April 6, 1862, in "Diary of Colonel William Camm," 847.

38. Charles I. Switzer, ed., *Ohio Volunteer: The Childhood and Civil War Memoirs of Captain John Calvin Hartzell, OVI* (Athens: Ohio University Press, 2005), 97; Noe, *Southern Boy in Blue,* 142; Tower, *A Carolinian Goes to War,* 100.

39. Thomas D. Cockrell and Michael B. Ballard, eds., *A Mississippi Rebel in the Army of Northern Virginia: The Civil War Memoirs of Private David Holt* (Baton Rouge: Louisiana State University Press, 1995), 90–91, 272, 274.

40. John A. Cheatham to Hippolyte Oladowski, October 20, 1863; A. J. Paine to not stated, October 18, 1863; Lewis T. Woodruff to J. M. Macon, September 18, 1863, *OR,* vol. 30, pt. 2, 81–82, 122, 408; Daniel, *Soldiering in the Army of Tennessee,* 43.

41. O. T. Gibbes, statement of captured stores, n.d.; William B. Bate to R. A. Hatcher, October 9, 1863, *OR,* vol. 30, pt. 2, 42–43, 386.

42. Briscoe G. Baldwin to not stated, January 20, 1863, *OR,* vol. 21, 568; William Allan to G. C. Brown, December 28, 1864, ibid., vol. 36, pt. 1, 1076.

43. James C. Bates to Ma, December 29, 1862, in Richard Lowe, ed., *A Texas Cavalry Officer's Civil War: The Diary and Letters of James C. Bates* (Baton Rouge: Louisiana State University Press, 1999), 220.

44. Dwight Morris to J. W. Plume, September 19, 1862, *OR,* vol. 19, pt. 1, 333; Charles H. Morgan account, n.d., in David L. Ladd and Audrey J. Ladd, eds., *The Bachelder Papers: Gettysburg in Their Own Words,* 3 vols. (Dayton, OH: Morningside Bookshop, 1994–1995), 3: 1367; Earl J. Hess, *Pickett's Charge—The Last Attack at Gettysburg* (Chapel Hill: University of North Carolina Press, 2001), 345–346; Hankins, *Simple Story of a Soldier,* 54, 57; Eric A. Campbell, "The Aftermath and Recovery of Gettysburg," pt. 2, *Gettysburg Magazine* 12 (January 1995): 108.

45. Winfield S. Hancock to Andrew A. Humphreys, May 12, 1864; Seth Williams to Hancock, May 14, 1864; Circular, Headquarters, 2nd Corps, May 14, 1864, *OR,* vol. 36, pt. 2, 661, 749, 754; Robert McAllister to Ellen and family, May 15, 1864, in James I. Robertson Jr., ed., *The Civil War Letters of General Robert McAllister* (New Brunswick, NJ: Rutgers University Press, 1965), 420.

46. John A. Logan to John A. McClernand, November 11, 1861, *OR,* vol. 3, 289; William S. Morris, *History 31st Regiment Volunteers, Organized by John A. Logan* (Evansville, IN: Keller, 1902), 19; Baumann, *Arming the Suckers,* 72, 104, 185; Morgan L. Smith to J. H. Hammond, May 19, 1862, *OR,* vol. 10, pt. 1, 841; Elliott, *History of the Thirty-third Regiment,* 233.

47. A. Mordecai to George H. Thomas, February 5, 1865, *OR,* vol. 45, pt. 1, 48–49; Circular, Headquarters 2nd Division, 15th Corps, February 18, 1865; T. J. Goodwyn affidavit, February 19, 1865, ibid., vol. 47, pt. 2, 477, 488; Ulysses S. Grant, *Personal Memoirs,* 2 vols. (New York: Viking, 1990), 1: 384–385; Thomas G. Baylor to William T. Sherman, April 7, 1865, *OR,* vol. 47, pt. 1, 180–182, 186.

48. Robert M. Rogers, *The 125th Regiment Illinois Volunteer Infantry* (Champaign, IL: Gazette Steam, 1882), 99.

49. P. H. Watson to O. F. Winchester, August 19, 1862, *OR,* ser. 3, vol. 2, 412–413.

50. Winther, *With Sherman to the Sea,* 107, 157–158; Trueman S. Powell letter, published in

Ottawa Republican, July 9, 1864, and quoted in Baumann, *Arming the Suckers,* 150; entry of April 9, 1865, in Janet Correll Ellison, ed., *On to Atlanta: The Civil War Diaries of John Hill Ferguson, Illinois Tenth Regiment of Volunteers* (Lincoln: University of Nebraska Press, 2001), 117.

51. Ambrose, *From Shiloh to Savannah,* 175; Hedley, *Marching through Georgia,* 27; Robert L. Mountjoy to not stated, July 29, 1864, Atlanta (Illinois) Public Library and Museum, quoted in Baumann, *Arming the Suckers,* 71; Baumann, *Arming the Suckers,* 70, 75, 115, 122, 150, 153, 179, 187, 192, 213; Thomas G. Baylor to William T. Sherman, April 7, 1865, *OR,* vol. 47, pt. 1, 185.

52. Eisenschiml, *Vermont General,* 249.

53. James L. Bowen, *History of the Thirty-seventh Regiment Mass. Volunteers, in the Civil War of 1861–1865* (Holyoke, MA: Clark W. Bryan, 1884), 354–355.

54. Samuel C. Williams, *General John T. Wilder: Commander of the Lightning Brigade* (Bloomington: Indiana University Press, 1936), 1–4, 6–10; B. F. McGee, *History of the 72d Indiana Volunteer Infantry of the Mounted Lightning Brigade* (Lafayette, IN: S. Vater, 1882), 119.

55. Williams, *General John T. Wilder,* 10–12.

56. George S. Wilson, "Wilder's Brigade of Mounted Infantry in the Tullahoma-Chickamauga Campaigns," in *War Talks in Kansas: A Series of Papers Read before the Kansas Commandery of the Military Order of the Loyal Legion of the United States* (Kansas City, MO: Franklin Hudson, 1906), 6, 49; Williams, *General John T. Wilder,* 74; McGee, *History of the 72d Indiana,* 120–121.

57. Davis, *Arming the Union,* 169; armament and ammunition report, Army of Tennessee, week ending June 19, 1864, *OR,* vol. 38, pt. 4, 782.

58. Brent Nosworthy, *The Bloody Crucible of Courage: Fighting Methods and Combat Experience of the Civil War* (New York: Carroll & Graf, 2003), 180, 624–628; Williams, *General John T. Wilder,* 18–19, 79; Wilson, "Wilder's Brigade," 53; James A. Connolly to wife, July 5, 1863, in Paul M. Angle, ed., *Three Years in the Army of the Cumberland: The Letters and Diary of Major James A. Connolly* (Bloomington: Indiana University Press, 1959), 93.

59. Virginia Kepler, "'My God, We Thought You Had a Division Here!'" *Civil War Times Illustrated* 5, 9 (1967): 10; James A. Connolly to wife, October 21, 1863, in Angle, *Three Years in the Army of the Cumberland,* 127.

60. Hedley, *Marching through Georgia,* 217; Winther, *With Sherman to the Sea,* 137–138; W. L. Truman to editor, May 17, 1908, *Confederate Veteran* Papers, SCL-DU.

61. Bowen, *Thirty-seventh Regiment,* 372–373, 417; entry of July 18–19, 1864, in Robert Hunt Rhodes, ed., *All for the Union: The Civil War Diary and Letters of Elisha Hunt Rhodes* (New York: Orion, 1985), 172–173; Peter H. Buckingham, ed., *All's for the Best: The Civil War Reminiscences and Letters of Daniel W. Sawtelle, Eighth Maine Volunteer Infantry* (Knoxville: University of Tennessee Press, 2001), 132, 136.

62. Davis, *Arming the Union,* 90, 130, 142, 153–161, 175–176; Bruce, *Lincoln and the Tools of War,* 284–288; Nosworthy, *Bloody Crucible,* 626–628.

63. T. T. S. Laidley, "Breech-Loading Musket," *United States Service Magazine* 3 (January 1865): 67, 69; C. A. Stevens, *Berdan's United States Sharpshooters in the Army of the Potomac, 1861–1865* (Dayton, OH: Morningside Bookshop, 1972), 236.

64. Horace William Shaler Cleveland, "Rifle-Clubs," *Atlantic Monthly,* September 1862, 306; Thomas W. Hyde, *Following the Greek Cross; Or, Memories of the Sixth Army Corps* (Columbia: University of South Carolina Press, 2005), 205.

65. Robert Stiles, *Four Years under Marse Robert* (New York: Neale, 1904), 274.

66. Gordon C. Rhea, *Cold Harbor: Grant and Lee, May 26–June 3, 1864* (Baton Rouge: Louisiana State University Press, 2002), 197–202; Williams, *General John T. Wilder,* 15.

67. Davis, *Arming the Union,* 146; Garavaglia and Worman, "Arms and the Man," 47. Joseph Bilby, in *Civil War Firearms: Their Historical Background and Tactical Use* (Conshohocken, PA: Combined Books, 1996), 12, expresses his viewpoint as a modern historian who firmly believes that the smoothbore musket still had an important role to play in Civil War combat.

68. George H. Thomas to William F. Patterson, November 4, 1861, William Franklin Patterson Papers, LC; Nosworthy, *Bloody Crucible,* 375; Davis, *Arming the Union,* 147–148.

69. William E. Baldwin to George B. Cosby, March 12, 1862, *OR,* vol. 7, 341.

70. Nosworthy, *Bloody Crucible,* 374; Hess, *Pickett's Charge,* 210.

Chapter Three. The Gun Culture of Civil War Soldiers

1. Anywhere from 0.68 percent to 1.5 percent of the rounds fired by Civil War soldiers hit a human target, according to Nosworthy. In comparison, 0.3 percent to 0.5 percent of the rounds fired by British soldiers during the Peninsula campaigns were effective. His cite for the American army during the Mexican War is 0.80 percent. At the battle of Solferino (June 24, 1859), the Austrians fired 700 rounds to hit a human target. Brent Nosworthy, *The Bloody Crucible of Courage: Fighting Methods and Combat Experience of the Civil War* (New York: Carroll & Graf, 2003), 182, 587–592. The notion that a small proportion of soldiers accounted for the greatest share of fire effectiveness is supported by a recent study of Eighth Air Force fighter pilots in World War II and American fighter pilots in the Korean conflict, which indicates that 5 percent of the pilots accounted for 40 percent of enemy planes shot down. Thomas A. Horner, "Killers, Fillers, and Fodder," *Parameters* 12, 3 (1982): 27–34, 30; Fred L. Ray, *Shock Troops of the Confederacy: The Sharpshooter Battalions of the Army of Northern Virginia* (Asheville, NC: CFS, 2006), 328.

2. Nosworthy, *Bloody Crucible,* 588, 590, 592.

3. Charles I. Switzer, ed., *Ohio Volunteer: The Childhood and Civil War Memoirs of Captain John Calvin Hartzell, OVI* (Athens: Ohio University Press, 2005), 10.

4. Switzer, *Ohio Volunteer,* 10–12.

5. Ibid., xvii, 12.

6. Richard W. Johnson, *A Soldier's Reminiscences in Peace and War* (Philadelphia: J. B. Lippincott, 1886), 17; E. F. Ware, *The Lyon Campaign in Missouri: Being a History of the First Iowa Infantry* (Iowa City: Press of the Camp Pope Bookshop, 1991), 86.

7. W. C. Whitthorne to Col. Claiborne, February 16, 1862, *OR,* vol. 7, 887; W. J. Tancig, ed., *Confederate Military Land Units, 1861–1865* (New York: Thomas Yoseloff, 1967), 13–109.

8. Ware, *Lyon Campaign,* 85, 87, 292–293.

9. Leander Stillwell, *The Story of a Common Soldier of Army Life in the Civil War, 1861–1865,* 2nd ed. (Kansas City, MO: Franklin Hudson, 1920), 55, 252.

10. Sam R. Watkins, *"Co. Aytch": A Side Show of the Big Show* (New York: Collier, 1962), 161.

11. Thomas D. Cockrell and Michael B. Ballard, eds., *A Mississippi Rebel in the Army of Northern Virginia: The Civil War Memoirs of Private David Holt* (Baton Rouge: Louisiana State University Press, 1995), 298; Peter H. Buckingham, ed., *All's for the Best: The Civil War Reminiscences and Letters of Daniel W. Sawtelle, Eighth Maine Volunteer Infantry* (Knoxville: University of Tennessee Press, 2001), 99; Robert J. Burdette, *The Drums of the 47th* (Urbana: Univer-

sity of Illinois Press, 2000), 45–46. Historian Carl Davis has written, "The regard for good muzzle-loaders among the men who carried them remained high throughout the war and for many years thereafter." Carl L. Davis, *Arming the Union: Small Arms in the Civil War* (Port Washington, NY: Kennikat, 1973), x.

12. Lee Kennett and James LaVerne Anderson, *The Gun in America: The Origins of a National Dilemma* (Westport, CT: Greenwood, 1975), 33–63, 71–77.

13. Ibid., 86–91, 111–112; Michael A. Bellesiles, *Arming America: The Origins of a National Gun Culture* (New York: Alfred A. Knopf, 2000), 384–386, 431. After the Civil War, the proliferation of guns, especially pistols, brought the gun culture to urban areas of the East as well.

14. Theodore F. Upson journal, [May] 12, 1865, in Oscar Osburn Winther, ed., *With Sherman to the Sea: The Civil War Letters, Diaries and Reminiscences of Theodore F. Upson* (Bloomington: Indiana University Press, 1958), 173. Confederate veteran F. S. Harris claimed a comrade of his "could successively hit the bottom of a pint tin cup 1,000 yards with an army rifle." This was more than half a mile. If true, he was an astoundingly good shot. F. S. Harris, "Fine Shots in the Virginia Army," *Confederate Veteran* 4 (1896): 74.

15. Samuel W. Hankins, *Simple Story of a Soldier: Life and Service in the 2d Mississippi Infantry* (Tuscaloosa: University of Alabama Press, 2004), 12; Hugh C. Bailey, ed., "An Alabamian at Shiloh: The Diary of Liberty Independence Nixon," *Alabama Review* 11 (1958): 152; Stillwell, *Story of a Common Soldier,* 54–55.

16. Burdette, *Drums of the 47th,* 25, 27.

17. Entry of December 28, 1863, in Terrence J. Winschel, ed., *The Civil War Diary of a Common Soldier: William Wiley of the 77th Illinois Infantry* (Baton Rouge: Louisiana State University Press, 2001), 29.

18. John William DeForest, "Sheridan's Victory of the Opequon," in James H. Croushore, ed., *A Volunteer's Adventures: A Union Captain's Record of the Civil War* (New Haven, CT: Yale University Press, 1946), 184–185.

19. Cockrell and Ballard, *Mississippi Rebel,* 282.

20. Janet B. Hewett, ed., *The Roster of Confederate Soldiers, 1861–1865,* 16 vols. (Wilmington, NC: Broadfoot, 1995–1996), 9: 32, 12: 384, 393; J. B. Polley, "J. B. Polley to 'Charming Nellie,'" *Confederate Veteran* 5 (1897): 425; J. B. Polley, "Texas in the Battle of the Wilderness," *Confederate Veteran* 5 (1897): 292–293.

21. J. B. Polley, *A Soldier's Letters to Charming Nellie* (New York: Neale, 1908), 242–244.

22. J. W. Gaskill, *Footprints through Dixie: Everyday Life of the Man under a Musket on the Firing Line and in the Trenches, 1862–1865* (Alliance, OH: Bradshaw, 1919), 23.

23. Paddy Griffith, *Battle Tactics of the Civil War* (New Haven, CT: Yale University Press, 1989), 86–88.

24. Stephen Vincent Benet to William T. Sherman, January 29, 1883, in John C. Tidball, "Rifle Target Practice in the Army," *Ordnance Notes,* January 30, 1883, 3.

25. Circular, Headquarters, Army of the Potomac, April 19, 1864, *OR,* vol. 33, 907–908; Andrew Haughton, *Training, Tactics and Leadership in the Confederate Army of Tennessee: Seeds of Failure* (Portland, OR: Frank Cass, 2000), 116–117.

26. William J. Hardee, *Rifle and Light Infantry Tactics, for the Instruction, Exercises and Manoeuvres of Riflemen and Light Infantry, Including School of the Soldier and School of the Company* (New York: J. O. Kane, 1862), 88.

27. Daniel E. Burbank to parents, October 17, 1861, quoted in Bell Irvin Wiley, *The Life of Billy Yank: The Common Soldier of the Union* (Baton Rouge: Louisiana State University Press,

1983), 63; entry for June 22, 1862, in Harold B. Simpson, ed., *The Bugle Softly Blows: The Confederate Diary of Benjamin M. Seaton* (Waco, TX: Texian Press, 1965), 14; Joseph K. Taylor to father, October 2, 1862, in Kevin C. Murphy, ed., *The Civil War Letters of Joseph K. Taylor of the Thirty-seventh Massachusetts Volunteer Infantry* (Lewiston, NY: Edwin Mellen, 1998), 45, 45n.

28. Dan McCook to Capt. Nevin, June 13, 1863; Alex. Marshall to Henry M. Cist, June 18, 1863, 14th Army Corps, Letters Sent (3rd Brig., 2nd Div.), vol. 42/109, RG393, pt. 2, NARA; Circular, Headquarters, 2nd Brigade, 2nd Division, Reserve Corps, August 5, 1863: General Orders No. 3, Headquarters, 3rd Brig., 2nd Division, 14th Corps, March 2, 1864, 14th Army Corps Orders, vol. 43/112, RG393, pt. 2, NARA.

29. Camp Inspection Return, 4th New Hampshire, March 12, 1863, Document 1203, United States Sanitary Commission Records, series 7, roll 25, RBMD-NYPL.

30. General Orders No. 3, Headquarters, Army of the Mississippi, March 14, 1862, *OR,* vol. 10, pt. 2, 325–326.

31. Haughton, *Training, Tactics and Leadership,* 115.

32. Thomas Jefferson Newberry to Marshall C. Newberry, June 27, 1863, in Enoch L. Mitchell, ed., "The Civil War Letters of Thomas Jefferson Newberry," *Journal of Mississippi History* 10, 1 (1948): 73; Edwin H. Rennolds diary, May 5, 1863, SC-UTK; Haughton, *Training, Tactics and Leadership,* 146–147; entries of April 29–30, 1863, in "William J. Rogers' Memorandum Book," *West Tennessee Historical Society Papers* 9 (1955): 87. Similar target training took place sporadically among Federal units. See Michael A. Mullins, *The Fremont Rifles: A History of the 37th Illinois Veteran Volunteer Infantry* (Wilmington, NC: Broadfoot, 1990), 285.

33. Circular, Headquarters, Army of Tennessee, February 16, 1864, *OR,* vol. 32, pt. 2, 751; Joseph E. Johnston to Jefferson Davis, August 3, 1861, in Lynda Lasswell Crist, ed., *The Papers of Jefferson Davis,* 10 vols. (Baton Rouge: Louisiana State University Press, 1971–1999), 7: 272.

34. Entry of November 27, 1862, in Christopher Looby, ed., *The Complete Civil War Journal and Selected Letters of Thomas Wentworth Higginson* (Chicago: University of Chicago Press, 2000), 48, 50.

35. Entry of February 28, 1862, in "Diary of Colonel William Camm, 1861 to 1865," *Journal of the Illinois State Historical Society* 18 (1925–1926): 835.

36. "The Battle at Fort Gregg," *Southern Historical Society Papers* 28 (1900): 266–267; David Gould and James B. Kennedy, eds., *Memoirs of a Dutch Mudsill: The "War Memories" of John Henry Otto, Captain, Company D, 21st Regiment Wisconsin Volunteer Infantry* (Kent, OH: Kent State University Press, 2004), 365; entry of April 11, 1865, in Arnold Gates, ed., *The Rough Side of War: The Civil War Journal of Chesley A. Mosman, 1st Lieutenant, Company D, 59th Illinois Volunteer Infantry Regiment* (Garden City, NY: Basin, 1987), 350.

37. J. G. Benton, *Ordnance and Gunnery* (New York: D. Van Nostrand, 1867), 341; Orders No. 20, Headquarters, 5th Division, Army of the Tennessee, April 12, 1862, *OR,* vol. 10, pt. 2, 103–104; William Gould to Marve, December 31, 1863, in Robert F. Harris and John Niflot, eds., *Dear Sister: The Civil War Letters of the Brothers Gould* (Westport, CT: Praeger, 1998), 115; Mitchell Thompson letter, n.d., quoted in Ken Baumann, *Arming the Suckers, 1861–1865: A Compilation of Illinois Civil War Weapons* (Dayton, OH: Morningside Bookshop, 1989), 168; Stillwell, *Story of a Common Soldier,* 90–91. The author of the official manual for taking care of the Model 1855 Springfield rifle musket recommended using water to loosen caked powder residue in the barrel, then swabbing it with a dry cloth. The same applied for the Model 1863 Springfield. E. S. Allin, *Rules for the Management and Cleaning of the Rifle Musket, Model 1855* (Washington, DC: Government Printing Office, 1862), 22; *Rules for the Management and Clean-*

ing of the Rifle Musket, Model 1863 (Washington, DC: Government Printing Office, 1863), 19–20.

38. James M. Williams to Lizzie, January 4, 1862, in John Kent Folmar, ed., *From That Terrible Field: Civil War Letters of James M. Williams, Twenty-first Alabama Infantry Volunteers* (Tuscaloosa: University of Alabama Press, 1981), 20; George F. Cram to mother, November 3, 1863, in Jennifer Cain Bohrnstedt, ed., *Soldiering with Sherman: Civil War Letters of George F. Cram* (DeKalb: Northern Illinois University Press, 2000), 61; entry of June 7, 1862, in William C. Davis, ed., *Diary of a Confederate Soldier: John S. Jackman of the Orphan Brigade* (Columbia: University of South Carolina Press, 1990), 45.

39. Joseph K. Taylor to father, June 20, 1863, in Murphy, *Civil War Letters,* 129; Hankins, *Simple Story of a Soldier,* 48.

40. Entry of July 3, 1863, in George A. Bowen, ed., "The Diary of Captain George D. Bowen, 12th Regiment New Jersey Volunteers," *Valley Forge Journal* 2 (1984): 133–134; Susan Williams Benson, ed., *Berry Benson's Civil War Book: Memoirs of a Confederate Scout and Sharpshooter* (Athens: University of Georgia Press, 1992), 23; Aquila Wiley to John Crowell Jr., September 25, 1863, *OR,* vol. 33, pt. 1, 773; Robert McAllister to Gershom Mott, January 24, 1882, in James I. Robertson Jr., ed., *The Civil War Letters of General Robert McAllister* (New Brunswick, NJ: Rutgers University Press, 1965), 419; J. Catlett Gibson and William W. Smith, "The Battle of Spotsylvania Courthouse, May 12, 1864," *Southern Historical Society Papers* 32 (1904): 211, 213.

41. D. Leib Ambrose, *From Shiloh to Savannah: The Seventh Illinois Infantry in the Civil War* (DeKalb: Northern Illinois University Press, 2003), 85; Gould and Kennedy, *Memoirs of a Dutch Mudsill,* 96; Henry C. Matrau to Cousin Rusha, May 30, 1864, in Marcia Reid-Green, ed., *Letters Home: Henry Matrau of the Iron Brigade* (Lincoln: University of Nebraska Press, 1993), 77–78.

42. William D. Chadick to C. D. Anderson, April 12, 1862, *OR,* vol. 10, pt. 1, 546; entry of April 8, 1862, in Mildred Throne, ed., *The Civil War Diary of Cyrus F. Boyd, Fifteenth Iowa Infantry, 1861–1863* (Baton Rouge: Louisiana State University Press, 1998), 40; Dan McCook to Capt. Wiseman, September 9, 1863, 14th Army Corps, Letters Sent (3rd Brig., 2nd Div.), vol. 42/109, RG393, pt. 2, NARA; Circulars, Headquarters, 3rd Brigade, 4th Division, 17th Corps, October 3, 22, 1864, 17th Army Corps, Letters Sent, vol. 20/79, RG393, pt. 2, NARA.

43. John William DeForest to wife, August 8, 1864, in Croushore, *A Volunteer's Adventures,* 165.

44. J. T. Scott, inspection report, December 1, 1864, in Joseph E. Chance, *The Second Texas Infantry: From Shiloh to Vicksburg* (Austin, TX: Eakin, 1984), 142–143.

45. Donald R. Collins, "How Bullets Were Made," *Civil War Times Illustrated* 4, 9 (1966): 22–23.

46. Ibid., 22–24.

47. Ware, *Lyon Campaign,* 135–136.

48. Ibid., 136.

49. Robert Hunt Rhodes, ed., *All for the Union: The Civil War Diary and Letters of Elisha Hunt Rhodes* (New York: Orion, 1985), 20; entry of April 6, 1862, in Throne, *Civil War Diary of Cyrus F. Boyd,* 29.

50. Wayne Austerman, "Abhorrent to Civilization," *Civil War Times Illustrated* 24, 5 (1985): 37–38; Robert V. Bruce, *Lincoln and the Tools of War* (Indianapolis: Bobbs-Merrill, 1956), 191.

51. Bruce, *Lincoln and the Tools of War,* 191–192; Robert R. Hardy, "Explosive Bullets," *Civil War Times Illustrated* 5, 6 (1966): 43.

52. Horace Edwin Hayden, "Explosive or Poisoned Musket or Rifle Balls," *Southern Historical Society Papers* 8 (1880): 158; Austerman, "Abhorrent to Civilization," 38, 40; Bruce, *Lincoln and the Tools of War,* 282.

53. Hayden, "Explosive or Poisoned Musket or Rifle Balls," 22; Donald R. Collins, "More on Bullets," *Civil War Times Illustrated* 5, 4 (1966): 36–37; Albert Kern, "Bullets Used in the Civil War," *Confederate Veteran* 24 (1916): 311; George Brown, "Explosive Bullets," *Confederate Veteran* 24 (1916): 95; Horace Edwin Hayden, "Explosive and Poisoned Bullets," *Confederate Veteran* 7 (1899): 156–157.

54. Austerman, "Abhorrent to Civilization," 37–40; Hayden, "Explosive or Poisoned Musket or Rifle Balls," 26.

55. Ulysses S. Grant, "The Vicksburg Campaign," in Robert Underwood Johnson and Clarence Clough Buel, eds., *Battles and Leaders of the Civil War* (New York: Thomas Yoseloff, 1956), 3: 522; Samuel H. Lockett, "The Defense of Vicksburg," in Johnson and Buel, *Battles and Leaders of the Civil War,* 491; Austerman, "Abhorrent to Civilization," 40.

56. Austerman, "Abhorrent to Civilization," 38; Benson J. Lossing, *Pictorial Field Book of the Civil War: Journeys through the Battlefields in the Wake of Conflict,* 3 vols. (Baltimore: Johns Hopkins University Press, 1997), 3: 78.

57. Austerman, "Abhorrent to Civilization," 40; *The Medical and Surgical History of the Civil War,* 12 vols. (Wilmington, NC: Broadfoot, 1991), 12: 701–703, 702n; Hardy, "Explosive Bullets," 44.

58. Henry H. Cook, "Explosive Bullets," *Confederate Veteran* 7 (1899): 27; Thomas G. J. Doughman reminiscences, in Matt Spruill, ed., *Guide to the Battle of Chickamauga* (Lawrence: University Press of Kansas, 1993), 258, 260.

59. Bruce, *Lincoln and the Tools of War,* 282; *Medical and Surgical History,* 12: 701; Hayden, "Explosive and Poisoned Bullets," 158.

60. Frederic Shriver Klein, "The Civil War as a Pitchman's Paradise," *Civil War Times Illustrated* 2, 3 (1963): 30; *St. Louis Daily Missouri Democrat,* September 26, 1862.

61. Bruce, *Lincoln and the Tools of War,* 135–136.

62. "Steel Breast Plates," *Southern Historical Society Papers* 32 (1904): 221; Bruce, *Lincoln and the Tools of War,* 136.

63. Entry of March 24, 1862, in Throne, *Civil War Diary of Cyrus F. Boyd,* 23; entry of May 22, 1862, in Gates, *Rough Side of War,* 15.

64. William Vermilion to Mary, January 23, 1863, in Donald C. Elder III, ed., *Love amid the Turmoil: The Civil War Letters of William and Mary Vermilion* (Iowa City: University of Iowa Press, 2003), 46; entry of January 19, 1863, in Throne, *Civil War Diary of Cyrus F. Boyd,* 110; Gould and Kennedy, *Memoirs of a Dutch Mudsill,* 25.

65. Bruce, *Lincoln and the Tools of War,* 135–136; Chance, *Second Texas Infantry,* 76, 133; Peter Cozzens, *The Darkest Days of the War: The Battles of Iuka and Corinth* (Chapel Hill: University of North Carolina Press, 1997), 253–254, 265.

66. Charles Beneulyn Johnson, *Muskets and Medicine; Or, Army Life in the Sixties* (Philadelphia: F. A. Davis, 1917), 133; Nosworthy, *Bloody Crucible,* 597.

67. Johnson, *Muskets and Medicine,* 132.

68. Stillwell, *Story of a Common Soldier,* 64–65.

69. Buckingham, *All's for the Best,* 89.

70. Johnson, *Muskets and Medicine,* 133; entry of June 26, 1861, in Mary Ann Andersen, ed., *The Civil War Diary of Allen Morgan Geer, Twentieth Regiment, Illinois Volunteers* (New York:

Cosmos, 1977), 5; entry of October 10, 1864, in W. Springer Menge and J. August Shimrak, eds., *The Civil War Notebook of Daniel Chisholm: A Chronicle of Daily Life in the Union Army, 1864–1865* (New York: Ballantine, 1989), 42.

71. W. A. Neal, ed., *An Illustrated History of the Missouri Engineer and the Twenty-fifth Infantry Regiments* (Chicago: Donohue & Henneberry, 1889), 135; General Orders No. 9, 2nd Brig., 2nd Div., Reserve Corps, June 14, 1863, 14th Army Corps Orders, vol. 43/112, RG393, NARA.

72. Gould and Kennedy, *Memoirs of a Dutch Mudsill,* 359, 375–376.

73. Ulysses S. Grant to Edwin M. Stanton, May 29, 1865, in John Y. Simon, ed., *The Papers of Ulysses S. Grant,* 24 vols. (Carbondale: Southern Illinois University Press, 1967–2000), 15: 105–106; Henry Roback, *The Veteran Volunteers of Herkimer and Otsego Counties in the War of the Rebellion: Being a History of the 152nd N.Y.V.* (Utica, NY: L. C. Childs, 1888), 162.

74. William E. Elwood, Thomas Goodman, James Guyer, Lawrence Guyer, Jacob F. Hahn, Benjamin F. Hamilton, Isaac R. Hays, service records, 1st Missouri Engineers, M405, Compiled Service Records of Volunteer Union Soldiers Who Served in Organizations from the State of Missouri, RG94, NARA; Francis A. Lord, "Disposal of Post-war Surplus," *Civil War Times Illustrated* 6, 10 (1968): 36.

75. William Hartsuff to J. A. Campbell, May 12, 1865; C. F. Vanderford to Archer Anderson, May 2, 1865; Circular, Headquarters, Army of Tennessee, May 2, 1865; Edmund W. Pettus to Archer Anderson, May 3, 1865; Joseph E. Johnston to Mansfield Lovell, May 5, 1865, *OR,* vol. 47, pt. 3, 483, 862, 864, 869, 872; entry of April 29, 1865, in Roger S. Durham, ed., *The Blues in Gray: The Civil War Journal of William Daniel Dixon and the Republican Blues Daybook* (Knoxville: University of Tennessee Press, 2000), 281–282.

76. Gould and Kennedy, *Memoirs of a Dutch Mudsill,* 83.

Chapter Four. The Rifle Musket in Battle

1. E. F. Ware, *The Lyon Campaign in Missouri: Being a History of the First Iowa Infantry* (Iowa City: Press of the Camp Pope Bookshop, 1991), 319, 327; Sam R. Watkins, *"Co. Aytch": A Side Show of the Big Show* (New York: Collier, 1962), 159–160; W. B. Smith, *On Wheels and How I Came There* (New York: Hunt & Eaton, 1892), 55.

2. Smith, *On Wheels,* 55; Ware, *Lyon Campaign,* 319.

3. Robert Hunt Rhodes, ed., *All for the Union: The Civil War Diary and Letters of Elisha Hunt Rhodes* (New York: Orion, 1985), 26; *Adjutant-General's Report, Illinois,* 4: 67, noted in Ken Baumann, *Arming the Suckers, 1861–1865: A Compilation of Illinois Civil War Weapons* (Dayton, OH: Morningside Bookshop, 1989), 142.

4. Leander Stillwell, *The Story of a Common Soldier of Army Life in the Civil War, 1861–1865,* 2nd ed. (Kansas City, MO: Franklin Hudson, 1920), 45.

5. Nathaniel Cheairs Hughes Jr., ed., *Sir Henry Morton Stanley, Confederate* (Baton Rouge: Louisiana State University Press, 2000), 121, 123.

6. Kenneth W. Noe, ed., *A Southern Boy in Blue: The Memoirs of Marcus Woodcock, 9th Kentucky Infantry (U.S.A.)* (Knoxville: University of Tennessee Press, 1996), 132–133.

7. Susan Williams Benson, ed., *Berry Benson's Civil War Book: Memoirs of a Confederate Scout and Sharpshooter* (Athens: University of Georgia Press, 1992), 23; entry of July 3, 1863, in

George A. Bowen, ed., "The Diary of Captain George D. Bowen, 12th Regiment New Jersey Volunteers," *Valley Forge Journal* 2 (1984): 133–134.

8. J. Catlett Gibson, and William W. Smith, "The Battle of Spotsylvania Courthouse, May 12, 1864," *Southern Historical Society Papers* 32 (1904): 211; Horace William Shaler Cleveland, *Hints to Riflemen* (New York: D. Appleton, 1864), 171, as noted in Thomas Vernon Moseley, "Evolution of the American Civil War Infantry Tactics" (Ph.D. diss., University of North Carolina, Chapel Hill, 1967), 136–137; Edwin C. Bennett memoir, noted in Steve Fratt, "American Civil War Tactics: The Theory of W. J. Hardee and the Experience of E. C. Bennett," *Indiana Military History Journal* 10, 1 (1985): 13.

9. William J. Hardee, *Rifle and Light Infantry Tactics, for the Instruction, Exercises and Manoeuvres of Riflemen and Light Infantry, Including School of the Soldier and School of the Company* (New York: J. O. Kane, 1862), 57, 59; Marcus M. Spiegel to wife and children, March 28, 1862, in Frank L. Byrne and Jean Powers Soman, eds., *Your True Marcus: The Civil War Letters of a Jewish Colonel* (Kent, OH: Kent State University Press, 1985), 85; Horatio Newhall to mother, December 22, 1862, in Annette Tapert, ed., *The Brothers' War: Civil War Letters to Their Loved Ones from the Blue and Gray* (New York: Vintage, 1988), 113; Francis C. Barlow to not stated, n.d., in Christian G. Samito, ed., *"Fear Was Not in Him": The Civil War Letters of Major General Francis C. Barlow, U.S.A.* (New York: Fordham University Press, 2004), 70–71; John William DeForest, "Sheridan's Victory of the Opequon," in James H. Croushore, ed., *A Volunteer's Adventures: A Union Captain's Record of the Civil War* (New Haven, CT: Yale University Press, 1946), 183.

10. Henry Pearson to friend, September 5, 1862, in Tapert, *Brothers' War,* 81; DeForest, "First Time under Fire," in Croushore, *A Volunteer's Adventures,* 70.

11. M. F. Bonham to E. A. O'Neal, May 9, 1863, *OR,* vol. 25, pt. 1, 956; Michael E. Stevens, ed., *As if It Were Glory: Robert Beecham's Civil War from the Iron Brigade to the Black Regiments* (Madison, WI: Madison House, 1998), 71–72; Henry C. Matrau to parents, July 29, 1863, in Marcia Reid-Green, ed., *Letters Home: Henry Matrau of the Iron Brigade* (Lincoln: University of Nebraska Press, 1993), 59; Francis Adams Donaldson to Auntie, July 21, 1863, in J. Gregory Acken, ed., *Inside the Army of the Potomac: The Civil War Experience of Captain Francis Adams Donaldson* (Mechanicsburg, PA: Stackpole, 1998), 303; entry of March 21, 1865, in Janet Correll Ellison, ed., *On to Atlanta: The Civil War Diaries of John Hill Ferguson, Illinois Tenth Regiment of Volunteers* (Lincoln: University of Nebraska Press, 2001), 115.

12. Paddy Griffith, *Battle Tactics of the Civil War* (New Haven, CT: Yale University Press, 1989), 86; *Army and Navy Journal,* December 24, 1864, noted in Moseley, "Evolution of the American Civil War Infantry Tactics," 167; Hardee, *Rifle and Light Infantry Tactics,* 89.

13. Edmund Newsome, *Experience in the War of the Great Rebellion* (Carbondale, IL: E. Newsome, 1879), 5–6, 15; Francis Adams Donaldson to brother, September 23, 1862, in Acken, *Inside the Army of the Potomac,* 133.

14. T. T. S. Laidley, "Breech-Loading Musket," *United States Service Magazine* 3 (January 1865): 69.

15. J. G. Benton, *Ordnance and Gunnery* (New York: D. Van Nostrand, 1867), 340; Stillwell, *Story of a Common Soldier,* 252.

16. James W. A. Wright, "An Eyewitness Account of General Bragg's Chattanooga Campaign," in William Stanley Hoole, *A Historical Sketch of the Thirty-sixth Alabama Infantry Regiment, 1862–1865* (University, AL: Confederate Publishing, 1986), 24–25.

17. John A. Cheatham to Hippolyte Oladowski, October 20, 1863; Frederick B. Dallas to Oladowski, October 24, 1863, *OR,* vol. 30, pt. 2, 82–83, 333.

18. Thomas D. Cockrell and Michael B. Ballard, eds., *A Mississippi Rebel in the Army of Northern Virginia: The Civil War Memoirs of Private David Holt* (Baton Rouge: Louisiana State University Press, 1995), 256.

19. Special Orders No. 8, Headquarters, Army of the Mississippi, April 3, 1862, *OR,* vol. 10, pt. 1, 394–395; Orders No. 20, Headquarters, 5th Division, April 12, 1862, ibid., vol. 10, pt. 2, 104; Thomas Hindman address, December 4, 1862, in Nannie M. Tilley, ed., *Federals on the Frontier: The Diary of Benjamin F. McIntyre, 1862–1864* (Austin: University of Texas Press, 1963), 77; George F. Cram to uncle, November 7, 1862, in Jennifer Cain Bohrnstedt, ed., *Soldiering with Sherman: Civil War Letters of George F. Cram* (DeKalb: Northern Illinois University Press, 2000), 14.

20. Entry of August 25, 1864, in W. Springer Menge and J. August Shimrak, eds., *The Civil War Notebook of Daniel Chisholm: A Chronicle of Daily Life in the Union Army, 1864–1865* (New York: Ballantine, 1989), 36.

21. R. E. McBride, *In the Ranks: From the Wilderness to Appomattox Court-House* (Cincinnati: Walden & Stowe, 1881), 97.

22. Marion D. Joyce, "Tactical Lessons of the War," *Civil War Times Illustrated* 2, 10 (1964): 42–43.

23. Stillwell, *Story of a Common Soldier,* 55–56; Benson J. Lossing, *Pictorial Field Book of the Civil War: Journeys through the Battlefields in the Wake of Conflict,* 3 vols. (Baltimore: Johns Hopkins University Press, 1997), 3: 78n; Special Orders No. 93, Headquarters, Bate's Division, May 6, 1864, *OR,* vol. 38, pt. 4, 671; Robert J. Burdette, *The Drums of the 47th* (Urbana: University of Illinois Press, 2000), 26–27; J. F. J. Caldwell, *The History of a Brigade of South Carolinians, Known First as "Gregg's," and Subsequently as "McGowan's Brigade"* (Philadelphia: King & Baird, 1866), 46.

24. William Watson, *Life in the Confederate Army* (New York: Scribner & Welford, 1888), 343.

25. Stewart Bennett and Barbara Tillery, eds., *The Struggle for the Life of the Republic: A Civil War Narrative by Brevet Major Charles Dana Miller, 76th Ohio Volunteer Infantry* (Kent, OH: Kent State University Press, 2004), 18.

26. Silas Casey, *Infantry Tactics, for the Instruction, Exercise, and Manoeuvres of the Soldier, a Company, Line of Skirmishers, Battalion, Brigade, or Corps d'Armee,* 3 vols. (New York: D. Van Nostrand, 1862), 2: 12, 18, and 3: 12; Hardee, *Rifle and Light Infantry Tactics,* 55–56.

27. Charles F. Manderson to W. H. H. Sheets, January 6, 1863; Bernard F. Mullen to Samuel W. Price, January 5, 1863, *OR,* vol. 20, pt. 1, 594, 611.

28. General Orders No. 3, Headquarters, Army of the Mississippi, March 14, 1862, *OR,* vol. 10, pt. 2, 326; Special Orders No. 93, Headquarters, Bate's Division, May 6, 1864, ibid., vol. 38, pt. 4, 671.

29. Hardee, *Rifle and Light Infantry Tactics,* 54, 56, 88.

30. Robert McAllister to LeGrand Benedict, August 3, 1863, *OR,* vol. 27, pt. 1, 553; letter from unidentified member of Colquitt's brigade, June 7, 1864, *Augusta Telegraph,* quoted in Gordon C. Rhea, *Cold Harbor: Grant and Lee, May 26–June 3, 1864* (Baton Rouge: Louisiana State University Press, 2002), 332; Charles G. Elliott, "Kirkland's Brigade, Hoke's Division, 1864–65," *Southern Historical Society Papers* 23 (1895): 173–174.

31. Emerson Opdycke to wife, April 29, 1862, in Glenn V. Longacre and John E. Haas, eds.,

To Battle for God and the Right: The Civil War Letterbooks of Emerson Opdycke (Urbana: University of Illinois Press, 2003), 33.

32. Hardee, *Rifle and Light Infantry Tactics,* 54–55.

33. Erasmus L. Mottley to Samuel Beatty, January 6, 1863, *OR,* vol. 20, pt. 1, 592; John A. Davis to Capt. Fox, April 8, 1862, ibid., vol. 10, pt. 1, 228.

34. F. Y. Hedley, *Marching through Georgia: Pen-Pictures of Every-Day Life in General Sherman's Army, from the Beginning of the Atlanta Campaign until the Close of the War* (Chicago: Donohue, Henneberry, 1890), 227.

35. Aquila Wiley to John Crowell Jr., September 25, 1863; George H. Cram to O. O. Miller, September 26, 1863, *OR,* vol. 30, pt. 1, 773–774, 813; DeForest, "First Time under Fire," 64.

36. William B. Hazen to D. W. Norton, September 28, 1863, *OR,* vol. 30, pt. 1, 764; Griffith, *Battle Tactics of the Civil War,* 112.

37. Noe, *Southern Boy in Blue,* 123.

38. John Logan to N. G. Williams, April 12, 1862, *OR,* vol. 10, pt. 1, 214.

39. Moseley, "Evolution of the American Civil War Infantry Tactics," 383; Louis D. Kelley to James C. Veatch, April 10, 1862, *OR,* vol. 10, pt. 1, 226.

40. Cockrell and Ballard, *Mississippi Rebel,* 256, 259; Buxton Reives Conerly reminiscences, Robert G. Evans, ed., *The 16th Mississippi Infantry: Civil War Letters and Reminiscences* (Jackson: University Press of Mississippi, 2002), 257.

41. John L. Brady to John Bachelder, May 24, 1886, in David L. Ladd and Audrey J. Ladd, eds., *The Bachelder Papers: Gettysburg in Their Own Words,* 3 vols. (Dayton, OH: Morningside Bookshop, 1994–1995), 3: 1398.

42. Stephen W. Sears, *Landscape Turned Red: The Battle of Antietam* (New Haven, CT: Ticknor & Fields, 1983), 240; entry of September 19, 1862, in Mark Grimsley and Todd D. Miller, eds., *The Union Must Stand: The Civil War Diary of John Quincy Adams Campbell, Fifth Iowa Volunteer Infantry* (Knoxville: University of Tennessee Press, 2000), 60; Francis Adams Donaldson to Auntie, July 21, 1863, in Acken, *Inside the Army of the Potomac,* 303; entry of May 12, 1864, in George A. Bowen, ed., "The Diary of Captain George D. Bowen, 12th Regiment New Jersey Volunteers," *Valley Forge Journal* 2 (1985): 185.

43. Entry of April 6, 1862, in "Diary of Colonel William Camm, 1861 to 1865," *Journal of the Illinois State Historical Society* 18 (1925–1926): 847; K. Jack Bauer, ed., *Soldiering: The Civil War Diary of Rice C. Bull, 123rd New York Volunteer Infantry* (San Rafael, CA: Presidio, 1978), 149; Watkins, *"Co. Aytch,"* 160; James I. Hall, "History of Company C," in James R. Fleming, *Band of Brothers: Company C, 9th Tennessee Infantry* (Shippensburg, PA: White Mane, 1996), 70–71.

44. Griffith, *Battle Tactics of the Civil War,* 83; Josiah Gorgas, "Contributions to the History of the Confederate Ordnance Department," *Southern Historical Society Papers* 12 (1884): 92; Thomas Ward Osborne to A. C. Osborne, December 26, 1864, in Richard Harwell and Philip N. Racine, eds., *The Fiery Trail: A Union Officer's Account of Sherman's Last Campaigns* (Knoxville: University of Tennessee Press, 1986), 52. An ordnance officer estimated that each musket was typically not fired more than 500 times per year during the Civil War and cited French tests that concluded that a good musket could stand up to 25,000 discharges during its lifetime. He therefore concluded that a musket ought to last fifty years under the worst conditions, an estimate that does not ring true when one considers the high rate of fire, the fouling with powder residue, the fact that many soldiers did not take care of their weapons properly,

and the damages that occurred to many muskets on the battlefield (Benton, *Ordnance and Gunnery,* 339).

45. Isaac O. Best, *History of the 121st New York State Infantry* (Chicago: W. B. Conkey, 1921), 157.

46. Larry J. Daniel, *Soldiering in the Army of Tennessee: A Portrait of Life in a Confederate Army* (Chapel Hill: University of North Carolina Press, 1991), 48.

47. Mason Whiting Tyler, *Recollections of the Civil War, with Many Original Diary Entries and Letters Written from the Seat of War, and with Annotated References* (New York: G. P. Putnam's Sons, 1912), 191; James L. Bowen, "General Edwards's Brigade at the Bloody Angle," in Robert Underwood Johnson and Clarence Clough Buel, eds., *Battles and Leaders of the Civil War* (New York: Thomas Yoseloff, 1956), 4: 177; entry of May 12, 1864, in Rhodes, *All for the Union,* 152; G. Norton Galloway, "Hand-to-Hand Fighting at Spotsylvania," in Johnson and Buel, *Battles and Leaders of the Civil War,* 4: 174, 174n.

48. Galloway, "Hand-to-Hand Fighting at Spotsylvania," 173–174; Varina D. Brown, *A Colonel at Gettysburg and Spotsylvania* (Columbia, SC: State Company, 1931), 259; Cadwallader Jones, "Tree Cut Down by Bullets," *Confederate Veteran* 34 (1926): 8; Tyler, *Recollections of the Civil War,* 195; entry of May 12, 1864, in Mark DeWolfe Howe, ed., *Touched with Fire: Civil War Letters and Diary of Oliver Wendell Holmes, Jr., 1861–1864* (Cambridge, MA: Harvard University Press, 1946), 116–117; Gordon C. Rhea, *The Battles for Spotsylvania Court House and the Road to Yellow Tavern, May 7–12, 1864* (Baton Rouge: Louisiana State University Press, 1997), 311–312.

49. Ulysses S. Grant, *Personal Memoirs,* 2 vols. (New York: Viking, 1990), 1: 205.

50. John Logan to N. G. Williams, April 12, 1862, *OR,* vol. 10, pt. 1, 215.

51. John M. Thayer to Fred Knefler, April 10, 1862, ibid., 194; Daniel D. T. Cowen to J. A. Mallory, October 9, 1862, ibid., vol. 16, pt. 1, 1086; Daniel R. Collier to Edmund R. Kerstetter, January 5, 1863; Charles F. Manderson to W. H. H. Sheets, January 6, 1863, ibid., vol. 20, pt. 1, 489, 594.

52. Samuel Davis to not stated, January 7, 1863, *OR,* vol. 20, pt. 1, 891; Horace Rice to R. M. Harwell, September 29, 1863, ibid., vol. 30, pt. 2, 114.

53. Francis W. Dawson to Jacob H. Manning, n.d., in *Supplement to the Official Records of the Union and Confederate Armies,* 100 vols. (Wilmington, NC: Broadfoot, 1993–2000), pt. 1, vol. 5, 687.

54. Myron Baker to Charles V. Ray, September 25, 1863; Henry V. N. Boynton to J. R. Beatty, September 24, 1863, in ibid., pt. 1, vol. 30, 419–421, 436.

55. John A. Cheatham to Hippolyte Oladowski, October 20, 1863; A. J. Paine to not stated, October 18, 1863, *OR,* vol. 30, pt. 2, 83, 122.

56. Cadmus M. Wilcox to G. Moxley Sorrel, May 25, 1862, ibid., vol. 11, pt. 1, 592; Robert McAllister to John Hancock, August 11, 1864; James R. Hagood to A. C. Sorrel, December 20, 1864, ibid., vol. 36, pt. 1, 489, 1068; Galloway, "Hand-to-Hand Fighting at Spotsylvania," 173; Cockrell and Ballard, *Mississippi Rebel,* 261; Rhea, *Battles for Spotsylvania Court House,* 280.

57. General Orders No. 36, Headquarters, Hood's Corps, March 7, 1864, *OR,* vol. 32, pt. 3, 593; Circular, Headquarters, Army of the Potomac, May 6, 1864; Walter H. Taylor to Richard S. Ewell, May 7, 1864, ibid., vol. 36, pt. 2, 439, 969; Taylor to John C. Breckinridge, June 3, 1864, ibid., vol. 36, pt. 3, 870.

58. D. H. Hamilton to R. H. Finney, May 9, 1863, *OR,* vol. 25, pt. 1, 903.

59. See Perry D. Jamieson's review of Hess, *Field Armies and Fortifications,* in *America's Civil War* 18, 5 (2005): 66, 68.

60. Griffith, *Battle Tactics of the Civil War,* 146–147; Mark Grimsley, "Surviving Military Revolution: The U.S. Civil War," in *The Dynamics of Military Revolution, 1300–2050,* ed. Macgregor Knox and Williamson Murray (Cambridge: Cambridge University Press, 2001), 176; Brent Nosworthy, *The Bloody Crucible of Courage: Fighting Methods and Combat Experience of the Civil War* (New York: Carroll & Graf, 2003), 278, 573–577, 592.

61. General Orders No. 36, Headquarters, Hood's Corps, March 7, 1864, *OR,* vol. 32, pt. 3, 593; T. B. Roy to Patrick Cleburne, May 14, 1864; Special Orders No. 93, Headquarters, Bate's Division, May 6, 1864, ibid., vol. 38, pt. 4, 671, 709; David Gould and James B. Kennedy, eds., *Memoirs of a Dutch Mudsill: The "War Memories" of John Henry Otto, Captain, Company D, 21st Regiment Wisconsin Volunteer Infantry* (Kent, OH: Kent State University Press, 2004), 210.

62. John William DeForest to not stated, n.d., in Croushore, *A Volunteer's Adventures,* 153–154. Lt. Edward Mitchell Whaley of the 1st South Carolina described a remarkable random shot at extremely long range that killed one of his comrades after the fighting ended at Averasboro during the Carolinas campaign. He saw Lt. Oscar LaBorde "collapse and fall to the ground," killed instantly by a Federal bullet that entered the top of his head and came out under his chin. LaBorde was standing up when hit. Whaley could account for this only by assuming that the Yankees, over a mile and half away, had fired rounds directly into the air and one of them came down exactly where LaBorde was standing. Edward Mitchell Whaley account, in Sion H. Harrington III and John Hairr, eds., *Eyewitnesses to Averasboro* (Erwin, NC: Averasboro Press, 2001), 1: 84.

63. Hendrick E. Paine to P. Sidney Post, January 10, 1863, *OR,* vol. 20, pt. 1, 273; Maris R. Vernon to J. S. Wilson, March 27, 1865, ibid., vol. 47, pt. 1, 516–517; Gould and Kennedy, *Memoirs of a Dutch Mudsill,* 83; Grant, *Personal Memoirs,* 1: 235–236; William T. Shaw to Samuel J. Kirkwood, October 26, 1862; Benjamin H. Bristow to Jacob Lauman, April 9, 1862; Frederick C. Jones to Jacob Ammen, April 8, 1862, *OR,* vol. 10, pt. 1, 153, 243, 340; Francis T. Sherman to sister, January 12, 1863, in C. Knight Aldrich, ed., *Quest for a Star: The Civil War Letters and Diaries of Colonel Francis T. Sherman of the 88th Illinois* (Knoxville: University of Tennessee Press, 1999), 22.

64. Hagerman Tripp to P. P. Baldwin, January 4, 1863, *OR,* vol. 20, pt. 1, 339; Myron Baker to Charles V. Ray, September 25, 1863; Granville A. Frambes to Charles F. King, September 26, 1863, ibid., vol. 30, pt. 1, 420, 832; Bennett and Tillery, *Struggle for the Life of the Republic,* 164; William R. Marshall to Henry Hoover, July 22, 1864, *OR,* vol. 39, pt. 2, 273; Benjamin H. Bristow to Jacob Lauman, April 9, 1862, ibid., vol. 10, pt. 1, 243.

65. Andrew Moon to Sade, December 4, 1864, in Tapert, *Brothers' War,* 227; W. W. Gist, "The Battle of Franklin," *Tennessee Historical Magazine* 6, 3 (1920): 233.

66. B. P. Hughes, *Firepower: Weapons Effectiveness on the Battlefield, 1630–1850* (New York: Sarpedon, 1997).

67. William S. Rosecrans to Lorenzo Thomas, February 12, 1863, *OR,* vol. 20, pt. 1, 187, 197.

68. Earl J. Hess, *Field Armies and Fortifications in the Civil War: The Eastern Campaigns, 1861–1864* (Chapel Hill: University of North Carolina Press, 2005), 163–164, 171; Charles Powell reminiscences, 17, SCL-DU.

69. Archibald K. Jones to Nathaniel Harris, April 12, 1878, in N. H. Harris, "Defence of Battery Gregg," *Southern Historical Society Papers* 8 (1880): 477, 483; A. K. Jones, "The Battle of Fort Gregg," *Southern Historical Society Papers* 31 (1903): 58.

70. A. Wilson Greene, *Breaking the Backbone of the Rebellion: The Final Battles of the Petersburg Campaign* (Mason City, IA: Savas, 2000), 393–399, 406; Jones, "Battle of Fort Gregg,"

58–60; Archibald K. Jones to Nathaniel Harris, April 12, 1878, in Harris, "Defence of Battery Gregg," 483; John G. Barnard to wife, April 2, 1865, John Gross Barnard Papers, SCL-DU; Noah Andre Trudeau, *The Last Citadel: Petersburg, Virginia, June 1864–April 1865* (Baton Rouge: Louisiana State University Press, 1991), 389. Archibald Jones later claimed that most Confederate casualties were shot after they surrendered to the Federals, due to rage, frustration, and the influence of whiskey, but there is no supporting evidence for this.

71. Louis D. Kelley to James C. Veatch, April 10, 1862; Thomas T. Crittenden to not stated, n.d., *OR,* vol. 10, pt. 1, 226–227, 311.

72. Charles R. Woods to F. M. Crandal, January 12, 1863, ibid., vol. 17, pt. 1, 768–769.

73. Bennett and Tillery, *Struggle for the Life of the Republic,* 65, 76–80.

74. Thomas J. Churchill to Theophilus H. Holmes, May 6, 1863; James Deshler to B. S. Johnson, March 25, 1863, *OR,* vol. 17, pt. 1, 781, 791–792, 795; Record of Events, Company D, 24th Texas Cavalry (dismounted), in *Supplement to the Official Records,* pt. 2, vol. 68, 159; Carlos W. Colby to not stated, January 2, 1863 (section written January 15, 1863), in John S. Painter, ed., "Bullets, Hardtack and Mud: A Soldier's View of the Vicksburg Campaign," *Journal of the West* 4, 2 (1965): 138.

75. Return of casualties: Deshler to Johnson, March 25, 1863, *OR,* vol. 17, pt. 1, 718, 793, 795.

Chapter Five. The Art of Skirmishing

1. Lew Wallace, "The Capture of Fort Donelson," in Robert Underwood Johnson and Clarence Clough Buel, eds., *Battles and Leaders of the Civil War* (New York: Thomas Yoseloff, 1956), 1: 407; the reports of both Union and Confederate commanders for the Atlanta campaign in *OR,* vol. 38, indicate frequent interchange of the terms *pickets* and *skirmishers.*

2. Martin Pegler, *Out of Nowhere: A History of the Military Sniper* (Oxford: Osprey, 2004), is one of the few general histories of sniping.

3. Brent Nosworthy, *The Anatomy of Victory: Battle Tactics, 1689–1763* (New York: Hippocrene, 1992), 339.

4. Steven Ross, *From Flintlock to Rifle: Infantry Tactics, 1740–1866* (Rutherford, NJ: Fairleigh Dickinson University Press, 1979), 38; William Duane, *The American Military Library; Or, Compendium of the Modern Tactics,* 2 vols. (Philadelphia: William Duane, 1809), 2: 1, 3–5, 20.

5. Duane, *American Military Library,* 2:v, 17.

6. Paddy Griffith, *Battle Tactics of the Civil War* (New Haven, CT: Yale University Press, 1989), 111; Silas Casey, *Infantry Tactics, for the Instruction, Exercise, and Manoeuvres of the Soldier, a Company, Line of Skirmishers, Battalion, Brigade, or Corps d'Armee,* 3 vols. (New York: D. Van Nostrand, 1862), 1: 184–187; Earl J. Hess, *Lee's Tar Heels: The Pettigrew-Kirkland-MacRae Brigade* (Chapel Hill: University of North Carolina Press, 2002), 84–86.

7. Casey, *Infantry Tactics,* 1: 181–182, 184; Thomas Vernon Moseley, "Evolution of the American Civil War Infantry Tactics" (Ph.D. diss., University of North Carolina, Chapel Hill, 1967), 267–268.

8. Casey, *Infantry Tactics,* 1: 182, 202–203, 205.

9. J. Monroe, *The Company Drill of the Infantry of the Line, Together with the Skirmishing Drill of the Company and Battalion, after the Method of Gen. LeLouteril, Bayonet Fencing, with a Supplement on the Handling and Service of Light Infantry* (New York: D. Van Nostrand, 1863), 144–145.

10. Marcus M. Spiegel to wife and children, March 2, 1862, in Frank L. Byrne and Jean Powers Soman, eds., *Your True Marcus: The Civil War Letters of a Jewish Colonel* (Kent, OH: Kent State University Press, 1985), 53; Gerald J. Prokopowicz, *All for the Regiment: The Army of the Ohio, 1861–1862* (Chapel Hill: University of North Carolina Press, 2001), 50.

11. W. H. Chamberlin, "The Skirmish Line in the Atlanta Campaign," in *Sketches of War History, 1861–1865: Papers Prepared for the Ohio Commandery of the Military Order of the Loyal Legion of the United States,* vol. 3. (Cincinnati: Robert Clarke, 1890), 182–183; John W. DeForest, "Port Hudson," in James H. Croushore, ed., *A Volunteer's Adventures: A Union Captain's Record of the Civil War* (New Haven, CT: Yale University Press, 1946), 111–112.

12. Wilbur Fisk to *Green Mountain Freeman,* September 24, 1864, in Emil Rosenblatt and Ruth Rosenblatt, eds., *Hard Marching Every Day: The Civil War Letters of Private Wilbur Fisk, 1861–1865* (Lawrence: University Press of Kansas, 1992), 254–255.

13. Francis A. Walker, *History of the Second Army Corps in the Army of the Potomac* (New York: Charles Scribner's Sons, 1887), 450–451.

14. Casey, *Infantry Tactics,* 1: 184; Grady McWhiney and Perry D. Jamieson, *Attack and Die: Civil War Military Tactics and the Southern Heritage* (University: University of Alabama Press, 1982), 56, 99; Marion D. Joyce, "Tactical Lessons of the War," *Civil War Times Illustrated* 2, 10 (1964): 46.

15. Ken Baumann, *Arming the Suckers, 1861–1865: A Compilation of Illinois Civil War Weapons* (Dayton, OH: Morningside Bookshop, 1989), 150–153.

16. John A. Andrew to Edwin M. Stanton, June 7, 1862; George D. Ruggles to Andrew, June 12, 1862; Stanton to William Sprague, September 19, 1862; General Orders No. 149, Adjutant General's Office, War Department, October 2, 1862, *OR,* ser. 3, vol. 2, 113–114, 147, 574, 643–644.

17. *Military Service Records: A Select Catalog of National Archives Microfilm Publications* (Washington, DC: National Archives and Service Administration, 1985), 77–80, 83; Record of Events, 1st Michigan Sharpshooters, sometime after August 25, 1864, in *Supplement to the Official Records of the Union and Confederate Armies,* 100 vols. (Wilmington, NC: Broadfoot, 1993–2000), pt. 2, vol. 30, 423; Raymond J. Herek, *These Men Have Seen Hard Service: The First Michigan Sharpshooters in the Civil War* (Detroit: Wayne State University Press, 1998), 9, 42, 90, 97, 103.

18. C. A. Stevens, *Berdan's United States Sharpshooters in the Army of the Potomac, 1861–1865* (Dayton, OH: Morningside Bookshop, 1972), 2–3; Ronald H. Bailey, *Forward to Richmond* (Alexandria, VA: Time-Life, 1983), 101.

19. Bailey, *Forward to Richmond,* 4–7, 9–12, 27–28, 75, 526; William B. Greene to mother, February 1, 1862, in William H. Hastings, ed., *Letters from a Sharpshooter: The Civil War Letters of Private William B. Greene, Co. G, 2nd United States Sharpshooters (Berdan's), Army of the Potomac, 1861–1865* (Belleville, WI: Historic Publications, 1993), 64, 70.

20. Russell C. White, ed., *The Civil War Diary of Wyman S. White* (Baltimore: Butternut & Blue, 1991), 96; Stevens, *Berdan's United States Sharpshooters,* 17–19.

21. Stevens, *Berdan's United States Sharpshooters,* 83–159, 217, 220, 232–233, 398–399, 486, 494, 499; John L. Parker, *Henry Wilson's Regiment: History of the Twenty-second Massachusetts Infantry, the Second Company Sharpshooters, and the Third Light Battery, in the War of the Rebellion* (Baltimore: Butternut & Blue, 1997), 20–21.

22. Detachment Record of Events, 5th USCT, June to December 1864; Detachment Record of Events, 6th USCT, June 1864; Detachment Record of Events, 8th USCT, January 1865; De-

tachment Record of Events, 22nd USCT, June 1864, in *Supplement to the Official Records,* pt. 2, vol. 77, 337–338, 354, 410, 556–557; Records of Events, Company E, 107th USCT, December 1864; Detachment Record of Events, 116th USCT, January 1865; Detachment Record of Events, 127th USCT, February 1865, in *Supplement to the Official Records,* pt. 2, vol. 79, 160, 231–232, 275.

23. Jefferson Davis to Leonidas Polk, September 2, 1861, in Lynda Lasswell Crist, ed., *The Papers of Jefferson Davis,* 10 vols. (Baton Rouge: Louisiana State University Press, 1971–1999), 7: 318; General Orders No. 34, Adjutant and Inspector General's Office, War Department, May 3, 1862, *OR,* ser. 4, vol. 1, 1110.

24. General Orders No. 34, Adjutant and Inspector General's Office, War Department, May 3, 1862, *OR,* ser. 4, vol. 1, 1110. It is possible that G. T. Beauregard was the originator of the idea for dedicated skirmish units. Russell K. Brown, *"Our Connection with Savannah": History of the First Battalion Georgia Sharpshooters, 1861–1865* (Macon, GA: Mercer University Press, 2004), 1.

25. Joseph J. Crute Jr., *Units of the Confederate States Army* (Midlothian, VA: Derwent, 1987), 15–16, 19, 41–42, 49, 80, 82, 84–85, 149, 165, 171, 175, 200, 211, 247, 270, 297, 321, 377; "Col. Erasmus I. Stirman," *Confederate Veteran* 22 (1914): 226; Ras Stirman to sister, August 10, 1862, in Pat Carr, ed., *In Fine Spirits: The Civil War Letters of Ras Stirman* (Fayetteville, AR: Washington County Historical Society, 1986), 46, 48; Brown, *"Our Connection with Savannah,"* 1–5, 12, 27; Record of Events, Field and Staff, 14th Battalion Louisiana Sharpshooters, in *Supplement to the Official Records,* pt. 2, vol. 24, 329; Kenneth A. Hafendorfer, ed., *Civil War Journal of William L. Trask: Confederate Sailor and Soldier* (Louisville, KY: KH, 2003), 44; Record of Events, Company A, 9th Battalion Missouri Sharpshooters, November 11, 1862, in *Supplement to the Official Records,* pt. 2, vol. 38, 594; Record of Events, Field and Staff, Company A, B, and C, 24th Battalion Tennessee Sharpshooters, in *Supplement to the Official Records,* pt. 2, vol. 67, 21–26; Record of Events, Company A, B, C, D, and F, 30th Battalion Virginia Sharpshooters, in *Supplement to the Official Records,* pt. 2, vol. 72, 153–155.

26. Record of Events, Company A, 15th Battalion Mississippi Sharpshooters, June 7, 1862, in *Supplement to the Official Records,* pt. 2, vol. 33, 386–387.

27. Record of Events, Company A, 9th Battalion Mississippi Sharpshooters, July 1864, in *Supplement to the Official Records,* pt. 2, vol. 33, 169; Crute, *Units of the Confederate States Army,* 171.

28. Irving A. Buck, *Cleburne and His Command* (Jackson, TN: McCowat-Mercer, 1959), 128; Edwin H. Rennolds diary, April 27, 1863, SC-UTK; William B. Bate to R. A. Hatcher, October 9, 1863, *OR,* vol. 30, pt. 2, 387; Andrew Haughton, *Training, Tactics and Leadership in the Confederate Army of Tennessee: Seeds of Failure* (Portland, OR: Frank Cass, 2000), 147, 184; R. A. Jarman, "History of Company K, 27th Mississippi, *Aberdeen Examiner,* March 14, 1890; Fred L. Ray, *Shock Troops of the Confederacy: The Sharpshooter Battalions of the Army of Northern Virginia* (Asheville, NC: CFS, 2006), 291.

29. Ray, *Shock Troops,* 45, 47, 49, 51, 340–342; "Sharpshooting in Lee's Army," *Confederate Veteran* 3 (1895): 98; Frederick L. Ray, "Shock Troops of the South," *America's Civil War* 15, 3 (2002): 38; Circular, Headquarters Rodes's Brigade, January 12, 1863, Orders, Rodes's and Battle's Brigades, Army of Northern Virginia, 1861–65, chap. 2, vol. 66, RG109, NARA.

30. Eugene Blackford to mother, January 15, 25, 1863, Gordon-Blackford Papers, Mary-HS.

31. Samuel A. Burney to wife, April 24 and 26, 1863, in Nat S. Turner III, ed., *A Southern Sol-*

dier's Letters Home: The Civil War Letters of Samuel A. Burney, Cobb's Georgia Legion, Army of Northern Virginia (Macon, GA: Mercer University Press, 2002), 246, 248.

32. General Orders No. 9, May 12, 1863, and Circular, May 19, 1863, Headquarters Rodes's Brigade, Orders, Rodes's and Battle's Brigades, Army of Northern Virginia, 1861–65, chap. 2, vol. 66, RG109, NARA; Ray, *Shock Troops,* 343; W. S. Dunlop, *Lee's Sharpshooters; Or, The Forefront of Battle* (Dayton, OH: Morningside Bookshop, 1982), 11; Susan Williams Benson, ed., *Berry Benson's Civil War Book: Memoirs of a Confederate Scout and Sharpshooter* (Athens: University of Georgia Press, 1992), 45, 57, 61–65; George J. Winter, "A Battalion of Sharpshooters," *Transactions of the Huguenot Society of South Carolina* 79 (1974): 90; Louis Leon, *Diary of a Tar Heel Confederate Soldier* (Charlotte, NC: Stone, 1913), 30.

33. Cadmus M. Wilcox to William S. Dunlop, November 4, 1887, in Dunlop, *Lee's Sharpshooters,* x–xi; Ray, *Shock Troops,* 93, 95, 97; Ray, "Shock Troops of the South," 38.

34. Jerome B. Yates to Ma, April 20, 1864, in Robert G. Evans, ed., *The 16th Mississippi Infantry: Civil War Letters and Reminiscences* (Jackson: University Press of Mississippi, 2002), 243; John E. Laughton Jr., "The Sharpshooters of Mahone's Brigade," *Southern Historical Society Papers* 22 (1894): 99; I. G. Bradwell, "Fort Steadman and Subsequent Events," *Confederate Veteran* 23 (1915): 21; "Lane's Corps of Sharpshooters," *Southern Historical Society Papers* 28 (1900): 1; General Orders No. 10, Headquarters Battle's Brigade, April 30, 1864, Orders, Rodes's and Battle's Brigades, Army of Northern Virginia, 1861–65, chap. 2, vol. 66, RG1–9, NARA; Winter, "Battalion of Sharpshooters," 89; Ray, *Shock Troops,* 95; Dunlop, *Lee's Sharpshooters,* 17–18, 361; Anthony Wood, ed., *Reminiscences of the 35th Ga. Regt. as Seen by a Sharpshooter at the Front* (Conyers, GA: THP, n.d.), 40; John D. Young, "A Campaign with Sharpshooters," *Philadelphia Weekly Times,* January 26, 1878.

35. Dunlop, *Lee's Sharpshooters,* 486; Winter, "Battalion of Sharpshooters," 89; Casey, *Infantry Tactics,* 1: 11. I have found documentation for the existence of sharpshooter battalions in only twelve of the thirty-four brigades of Lee's army in May 1864. Half of those brigades belonged to Wilcox's or Rodes's divisions; nine of the brigades were in Hill's 3rd Corps, and the rest in Ewell's 2nd Corps. Dunlop contends that every brigade had one, and that may well be true, but there is a frustrating lack of evidence to prove the point. The twelve brigades were Evans, Battle, and Daniel in the 2nd Corps; Harris, Mahone, Lane, McGowan, Thomas, Stone, Kirkland, Scales, and Perrin in the 3rd Corps.

36. Ray, "Shock Troops of the South," 38; Bradwell, "Fort Steadman and Subsequent Events," 20–21; Laughton, "Sharpshooters of Mahone's Brigade," 99; Dunlop, *Lee's Sharpshooters,* 22; Ray, *Shock Troops,* 272.

37. I. G. Bradwell, "Second Day's Battle of the Wilderness," *Confederate Veteran* 28 (1920): 20; Robert F. Ward to William S. Dunlop, December 30, 1898, in Dunlop, *Lee's Sharpshooters,* 362–363; Marion Hill Fitzpatrick to Amanda, March 21, 1864, in Jeffrey C. Lowe and Sam Hodges, eds., *Letters to Amanda: The Civil War Letters of Marion Hill Fitzpatrick, Army of Northern Virginia* (Macon, GA: Mercer University Press, 1998), 128; Jerome B. Yates to Ma, April 20, 1864, in Evans, *16th Mississippi,* 243; Laughton, "Sharpshooters of Mahone's Brigade," 100.

38. Robert F. Ward to William S. Dunlop, December 30, 1898, in Dunlop, *Lee's Sharpshooters,* 362; Elias Davis to Mrs. G. A. Davis, August 3, 1864, Elias Davis Papers, SHC-UNC.

39. "Lane's Corps of Sharpshooters," 1; Marion Hill Fitzpatrick to Amanda, March 21, 1864, in Lowe and Hodges, *Letters to Amanda,* 128–129; Dunlop, *Lee's Sharpshooters,* 19, 22–23, 361;

Young, "A Campaign with Sharpshooters"; Wood, *Reminiscences of the 35th Ga.*, 54; "Sharpshooting in Lee's Army," 98.

40. Dunlop, *Lee's Sharpshooters*, 19–20, 364–366; Wood, *Reminiscences of the 35th Ga.*, 40; Laughton, "Sharpshooters of Mahone's Brigade," 100; Ray, *Shock Troops*, 94.

41. Ray, *Shock Troops*, 344–345; Laughton, "Sharpshooters of Mahone's Brigade," 100; Leon, *Diary*, 59; Wood, *Reminiscences of the 35th Ga.*, 40–41; Dunlop, *Lee's Sharpshooters*, 20–21.

42. "Sharpshooting in Lee's Army," 98; Marion Hill Fitzpatrick to Amanda, March 21, 1864, in Lowe and Hodges, *Letters to Amanda*, 128–129; Ray, *Shock Troops*, 285; Hess, *Lee's Tar Heels*, 237–238.

43. Wood, *Reminiscences of the 35th Ga.*, 44; Young, "A Campaign with Sharpshooters"; Winter, "Battalion of Sharpshooters," 95.

Chapter Six. Skirmishing in Battle

1. William H. Merritt to John M. Schofield, n.d.; John A. Logan to John A. McClernand, November 11, 1861, *OR*, vol. 3, 81, 288.

2. Charles C. Nott, *Sketches of the War: A Series of Letters to the North Moore Street School of New York*, 2nd ed. (New York: Anson D. F. Randolph, 1865), 25; Record of Events, Company I, 66th Illinois, February 13–15, 1862, in *Supplement to the Official Records of the Union and Confederate Armies*, 100 vols. (Wilmington, NC: Broadfoot, 1993–2000), pt. 2, vol. 12, 727; letter to not stated, February 18, 1862, in Mildred Throne, ed., "Civil War Letters of Abner Dunham, 12th Iowa Infantry," *Iowa Journal of History* 53, 4 (1955): 309; entry of February 14, 1862, in Mary Ann Andersen, ed., *The Civil War Diary of Allen Morgan Geer, Twentieth Regiment, Illinois Volunteers* (New York: Cosmos, 1977), 18.

3. Entry of February 14, 1862, in Jill Knight Garrett, ed., *The Civil War Diary of Andrew Jackson Campbell* (Columbia, TN: Jill Knight Garrett, 1965), 15; Ferrell, *Holding the Line*, 17–18.

4. David A. Enyart to S. T. Corn, April 9, 1862; Charles S. Hanson to S. T. Corn, April 9, 1862; Edward H. Hobson to Jeremiah T. Boyle, April 10, 1862; Samuel Beatty to Boyle, April 9, 1862; W. K. Patterson to S. A. M. Wood, April 9, 1862, *OR*, vol. 10, pt. 1, 350–351, 352–353, 361, 363, 598.

5. General Orders No. 8, Headquarters, Army of the Ohio, April 15, 1862, ibid., vol. 52, pt. 1, 238; Daniel D. T. Cowen to J. A. Mallory, October 9, 1862, ibid., vol. 16, pt. 1, 1085.

6. Ferrell, *Holding the Line*, 82, 85–86.

7. Hagerman Tripp to P. P. Baldwin, January 4, 1863; Charles F. Manderson to W. H. H. Sheets, January 6, 1863; Bernard F. Mullen to Samuel W. Price, January 5, 1863; A. J. Vaughan to John Ingram, January 9, 1863, *OR*, vol. 20, pt. 1, 339, 593, 610, 745.

8. William W. Berry to William Mangan, January 8, 1863, ibid., 342.

9. James J. Dollins to F. Whitehead, May 4, 1863, ibid., vol. 24, pt. 1, 654; Ephraim McD. Anderson, *Memoirs: Historical and Personal; Including the Campaigns of the First Missouri Confederate Brigade* (Dayton, OH: Morningside Bookshop, 1972), 328, 333; entry of June 5, 1863, in Mark Grimsley and Todd D. Miller, eds., *The Union Must Stand: The Civil War Diary of John Quincy Adams Campbell, Fifth Iowa Volunteer Infantry* (Knoxville: University of Tennessee Press, 2000), 105; William Henry Harrison Clayton to father and mother, June 18, 1863, in Donald C. Elder III, ed., *A Damned Iowa Greyhound: The Civil War Letters of William Henry Harrison Clayton* (Iowa City: University of Iowa Press, 1998), 74; Samuel E. Snure to Sir, June 21,

1863, in William A. Russ Jr., "The Vicksburg Campaign as Viewed by an Indiana Soldier," *Journal of Mississippi History* 19, 4 (1957): 268–269; Edmund Newsome, *Experience in the War of the Great Rebellion* (Carbondale, IL: E. Newsome, 1879), 32; Stewart Bennett and Barbara Tillery, eds., *The Struggle for the Life of the Republic: A Civil War Narrative by Brevet Major Charles Dana Miller, 76th Ohio Volunteer Infantry* (Kent, OH: Kent State University Press, 2004), 96, 98; Edward O. C. Ord to Charles P. Stone, September 26, 1863, *OR,* vol. 26, pt. 1, 738.

10. Lewis Guion log, May 19, 20, 21, 1863, and Jared Sanders diary, May 20, 21, June 4, 1863, in Allan C. Richard Jr. and Mary Margaret Higginbotham Richard, *The Defense of Vicksburg: A Louisiana Chronicle* (College Station: Texas A & M University Press, 2004), 154, 156, 180; Stephen D. Lee to J. J. Reeve, July 25, 1863; John C. Moore to S. Croom, July 8, 1863; Ashbel Smith to J. M. Loughborough, July 10, 1863; William E. Baldwin to J. G. Devereux, July 10, 1863; Francis A. Shoup to Devereux, July 8, 1863, *OR,* vol. 24, pt. 2, 351, 382, 390–391, 403, 408.

11. John W. DeForest, "Port Hudson," in James H. Croushore, ed., *A Volunteer's Adventures: A Union Captain's Record of the Civil War* (New Haven, CT: Yale University Press, 1946), 111.

12. Andrew Haughton, *Training, Tactics and Leadership in the Confederate Army of Tennessee: Seeds of Failure* (Portland, OR: Frank Cass, 2000), 115, 126; Richard H. Whiteley to S. A. Moreno, October 4, 1863; William Green to [J. W. Harris], September 30, 1863; Daniel Coleman to O. S. Palmer, October 6, 1863; J. E. Austin to H. H. Bein, September 26, 1863; W. C. Richards to Walker Anderson, October 5, 1863, *OR,* vol. 30, pt. 2, 90, 115–116, 172–173, 227, 328; Samuel F. Gray to Carl Schmitt, September 26, 1863, *OR,* vol. 30, pt. 1, 551.

13. Joshua K. Callaway to Dulcinea B. Callaway, September 24, 1863, in Judith Lee Hallock, ed., *The Civil War Letters of Joshua K. Callaway* (Athens: University of Georgia Press, 1997), 136.

14. Orlando M. Poe to Richard Delafield, October 8, 1865, *OR,* vol. 44, 61.

15. Henry L. Abbott to Mamma, April 20, 1862, in Robert Garth Scott, ed., *Fallen Leaves: The Civil War Letters of Major Henry Livermore Abbott* (Kent, OH: Kent State University Press, 1991), 110; William Blaisdell to Joseph Hibbert Jr., n.d.; L. Q. C. Lamar to W. A. Harris, May 13, 1862, *OR,* vol. 11, pt. 1, 476, 598; Carnot Posey to Thomas S. Mills, May 12, 1863, *OR,* vol. 25, pt. 1, 871; Eugene Blackford to Mary, May 21, 1863, Gordon-Blackford Papers, Mary-HS; Joseph B. Carr to Charles Hamlin, August 1, 1863, *OR,* vol. 27, pt. 1, 543; Samuel W. Hankins, *Simple Story of a Soldier: Life and Service in the 2d Mississippi Infantry* (Tuscaloosa: University of Alabama Press, 2004), 42–43.

16. Elwood W. Christ, *The Struggle for the Bliss Farm at Gettysburg, July 2nd and 3rd, 1863,* 2nd ed. (Baltimore: Butternut and Blue, 1994), 4–6, 97–100, 104, 106, 108.

17. Eugene Blackford memoir, in Noah Andre Trudeau, ed., "Taking Aim at Cemetery Hill," *America's Civil War* 14, 1 (2001): 50–51; Fred L. Ray, *Shock Troops of the Confederacy: The Sharpshooter Battalions of the Army of Northern Virginia* (Asheville, NC: CFS, 2006), 69–70.

18. Edmund Rice to William R. Driver, October 17, 1863; George Sykes to M. T. McMahon, November 11, 1863; Kenner Garrard to Fred T. Locke, November 10, 1863; Joseph J. Bartlett to Locke, November 10, 1863; Joseph Hayes to C. B. Mervine, November 10, 1863; Joshua L. Chamberlain to Mervine, November 10, 1863; David A. Russell to McMahon, November 16, 1863; George Fuller to C. H. Hurd, November 10, 1863, *OR,* vol. 29, pt. 1, 285, 576–580, 582, 588, 599.

19. W. H. Chamberlin, "The Skirmish Line in the Atlanta Campaign," in *Sketches of War History, 1861–1865: Papers Prepared for the Ohio Commandery of the Military Order of the Loyal Legion of the United States,* vol. 3. (Cincinnati: Robert Clarke, 1890), 182; Samuel W. Price, "The Skirmish Line in the Atlanta Campaign," in *War Papers: Being Papers Read before the Comman-*

dery of the District of Columbia, Military Order of the Loyal Legion of the United States, vol. 3. (Wilmington, NC: Broadfoot, 1993), 97.

20. Adolphus Bushbeck to Thomas H. Elliott, May 14, 1864; Lewis D. Warner to G. W. Mindil, September 8, 1864, *OR,* vol. 38, pt. 2, 203, 246; Kenneth A. Hafendorfer, ed., *Civil War Journal of William L. Trask: Confederate Sailor and Soldier* (Louisville, KY: KH, 2003), 141.

21. Entry of May 28, 30, 1864, in William C. Davis, ed. *Diary of a Confederate Soldier: John S. Jackman of the Orphan Brigade* (Columbia: University of South Carolina Press, 1990), 132, 134; A. D. Kirwan, ed., *Johnny Green of the Orphan Brigade: The Journal of a Confederate Soldier* (Lexington: University of Kentucky Press, 1956), 134.

22. Entry of June 15, 19, July 3, 1864, in Arnold Gates, ed., *The Rough Side of War: The Civil War Journal of Chesley A. Mosman, 1st Lieutenant, Company D, 59th Illinois Volunteer Infantry Regiment* (Garden City, NY: Basin, 1987), 215, 220, 234.

23. General Field Orders, Headquarters, Hardee's Corps, May 31, 1864, *OR,* vol. 38, pt. 4, 751.

24. Alpheus Baker to Johnston, July 4, 1874, Joseph E. Johnston Papers, SC-CWM; [Thomas M. Jack] to Samuel G. French, James Cantey, Winfield S. Featherston, June 17, 1864, Special Orders from May 9th 1864 to June 19th 1864, Head Quarters, Army of the Mississippi, chap. 2, no. 221 1/2, RG109, NARA.

25. Andrew Jackson Neal to Ra, June 20, 1864, Andrew Jackson Neal Papers, SC-EU.

26. Henry Stone, "From the Oostenaula to the Chattahoochee," in *The Mississippi Valley, Tennessee, Georgia, Alabama, 1861–1864: Papers of the Military Historical Society of Massachusetts,* vol. 8. (Boston: Cadet Armory, 1910), 413; Oliver Otis Howard, *Autobiography of Oliver Otis Howard,* 2 vols. (New York: Baker & Taylor, 1907), 1: 548–549.

27. Chamberlin, "Skirmish Line," 192–193.

28. Charles A. Partridge, ed., *History of the Ninety-sixth Regiment Illinois Volunteer Infantry* (Chicago: Brown, Pettibone, 1887), 387; William B. Westervelt diary, August 14, 1864, in George S. Maharay, ed., *Lights and Shadows of Army Life: From Bull Run to Bentonville* (Shippensburg, PA: Burd Street, 1998), 204; entry of August 9, 1864, in Janet Correll Ellison, ed., *On to Atlanta: The Civil War Diaries of John Hill Ferguson, Illinois Tenth Regiment of Volunteers* (Lincoln: University of Nebraska Press, 2001), 73; entry of July 31, 1864, in Gates, *Rough Side of War,* 252; William B. Hazen, *A Narrative of Military Service* (Boston: Ticknor, 1885), 419–420; J. B. Jordan statement, August 14, 1864, *OR,* vol. 38, pt. 5, 495.

29. Stephen D. Lee to A. P. Mason, January 30, 1865; Patton Anderson to J. W. Ratchford, February 9, 1865, *OR,* vol. 38, pt. 3, 763, 770.

30. Arthur M. Manigault memoir, in R. Lockwood Tower, ed., *A Carolinian Goes to War: The Civil War Narrative of Arthur Middleton Manigault, Brigadier General, C.S.A.* (Columbia: University of South Carolina Press, 1988), 253–254.

31. Ibid., 239–240.

32. Samuel G. French, *Two Wars: An Autobiography of Gen. Samuel G. French* (Nashville, TN: Confederate Veteran, 1901), 220–222; Samuel G. French to William D. Gale, December 6, 1864; William H. Young to D. W. Sanders, September 17, 1864, *OR,* vol. 38, pt. 3, 905, 907, 911.

33. Entry of December 9, 1864, in M. A. De Wolfe Howe, ed., *Marching with Sherman: Passages from the Letters and Campaign Diaries of Henry Hitchcock, Major and Assistant Adjutant General of Volunteers, November 1864–May 1865* (Lincoln: University of Nebraska Press, 1995), 165; Price, "Skirmish Line in the Atlanta Campaign," 2; J. C. Thompson to A. P. Stewart, December 8, 1867, Joseph E. Johnston Papers, SC-CWM; David S. Stanley to J. S. Fullerton,

September 1864; Absalom Baird to A. C. McClurg, September 7, 1864, *OR,* vol. 38, pt. 1, 226, 747; Benjamin Harrison to John Speed, September 14, 1864, *OR,* vol. 38, pt. 2, 349; John M. Corse to J. W. Barnes, September 8, 1864, *OR,* vol. 38, pt. 3, 411.

34. Robert F. Ward to Dunlop, December 30, 1898, in W. S. Dunlop, *Lee's Sharpshooters; Or, The Forefront of Battle* (Dayton, OH: Morningside Bookshop, 1982), 368–370.

35. Ibid., 371–373.

36. Ibid., 381; John E. Laughton Jr., "The Sharpshooters of Mahone's Brigade," *Southern Historical Society Papers* 22 (1894): 101; George J. Winter, "A Battalion of Sharpshooters," *Transactions of the Huguenot Society of South Carolina* 79 (1974): 94; James Fleming to Martin Binney, n.d., *OR,* vol. 36, pt. 1, 388.

37. John D. Young, "A Campaign with Sharpshooters," *Philadelphia Weekly Times,* January 26, 1878; Laughton, "Sharpshooters of Mahone's Brigade," 100–101; Ray, *Shock Troops,* 120; Berry Benson, "How I Lifted the Colonel's Mare," *Confederate Veteran* 27 (1919): 20–21.

38. John R. Brooke to Assistant Adjutant General, 1st Division, November 1, 1865, *OR,* vol. 36, pt. 1, 409.

39. Dunlop, *Lee's Sharpshooters,* 70–71.

40. Ward, "Mississippi Sharpshooters," in Dunlop, *Lee's Sharpshooters,* 427, 431–432.

41. Laughton, "Sharpshooters of Mahone's Brigade," 103; Frank Wilkeson, *Recollections of a Private Soldier in the Army of the Potomac* (New York: G. P. Putnam's Sons, 1893), 120–121; Ray, *Shock Troops,* 145.

42. George B. Osborn to Lysander Cutler, June 1, 1864; George J. Clarke to Gouverneur K. Warren, June 2, 1864, *OR,* vol. 36, pt. 3, 453, 487; entry of June 5, 1864, in W. Springer Menge and J. August Shimrak, eds., *The Civil War Notebook of Daniel Chisholm: A Chronicle of Daily Life in the Union Army, 1864–1865* (New York: Ballantine, 1989), 21; Ed Malles, ed., *Bridge Building in Wartime: Colonel Wesley Brainerd's Memoir of the 50th New York Volunteer Engineers* (Knoxville: University of Tennessee Press, 1997), 236.

43. Thomas D. Cockrell and Michael B. Ballard, eds., *A Mississippi Rebel in the Army of Northern Virginia: The Civil War Memoirs of Private David Holt* (Baton Rouge: Louisiana State University Press, 1995), 278–279; August Forsberg memoirs, 27, SC-WLU; Robert Stiles, *Four Years under Marse Robert* (New York: Neale, 1904), 302.

44. Unidentified member of Manly's Battery to editor, June 6, 1864, *Daily Confederate,* Raleigh, NC, June 10, 1864; Stiles, *Fours Years under Marse Robert,* 302; Louis H. Manarin, ed., *North Carolina Troops, 1861–1865: A Roster,* 15 vols. (Raleigh, NC: Division of Archives and History, 1966–2003), 1: 41–50.

45. J. F. J. Caldwell, *The History of a Brigade of South Carolinians, Known First as "Gregg's," and Subsequently as "McGowan's Brigade"* (Philadelphia: King & Baird, 1866), 159–160.

46. Entry of July 1, 1864, in George A. Bowen, ed., "The Diary of Captain George D. Bowen, 12th Regiment New Jersey Volunteers," *Valley Forge Journal* 2 (1985): 200; James Mitchell to David, July 17, 1864, PNB; Clarion Miltmore to mother, June 26, 1864, Ira Miltmore Collection, CHS; Amos M. Judson, *History of the Eighty-third Regiment Pennsylvania Volunteers* (Alexandria, VA: Stonewall House, 1985), 104.

47. Bushrod R. Johnson to Maj. Duncan, December 2, 1864, *OR,* vol. 42, pt. 1, 918; John Forrest Robinson memoirs, 74–75, *Confederate Veteran* Papers, SCL-DU; Caldwell, *History of a Brigade of South Carolinians,* 206.

48. Joab Goodson to Nannie, June 27, 1864, in W. Stanley Hoole, ed., "The Letters of Captain Joab Goodson, 1862–1864 (Part II)," *Alabama Review* 10, 3 (1957): 2, 224–225.

49. Laughton, "Sharpshooters of Mahone's Brigade," 104; W. R. S., "The Sharpshooters of Mahone's Old Brigade at the Crater," *Southern Historical Society Papers* 28 (1900): 307–308.

50. Dunlop, *Lee's Sharpshooters,* 190–195; Francis A. Walker, *History of the Second Army Corps in the Army of the Potomac* (New York: Charles Scribner's Sons, 1887), 589, 592–593; John Horn, *The Destruction of the Weldon Railroad: Deep Bottom, Globe Tavern, and Reams Station, August 14–25, 1864* (Lynchburg, VA: H. E. Howard, 1991), 133. Dunlop's brigade commander reported that the sharpshooters fired on average 100 rounds per man at Reams Station, with some men firing more than 150 cartridges. Samuel McGowan to Joseph A. Engelhard, September 1, 1864, *Supplement to the Official Records,* pt. 1, vol. 7, 509.

51. Gilbert P. Robinson to C. J. Mills, August 20, 1864; Napoleon B. McLaughlen to John D. Bertolette, October 16, 1864, *OR,* vol. 42, pt. 1, 564, 575; Orlando B. Willcox to P. M. Lydig, April 2, 1865, *OR,* vol. 46, pt. 1, 323; R. E. McBride, *In the Ranks: From the Wilderness to Appomattox Court-House* (Cincinnati: Walden & Stowe, 1881), 97–98.

52. "Lane's Corps of Sharpshooters," *Southern Historical Society Papers* 28 (1900): 4; Dunlop, *Lee's Sharpshooters,* 224.

53. Walker, *History of the Second Army Corps,* 607; William P. Wilson to Gershom Mott, September 4, 1862; Special Orders No. 77, 1st Brigade, 3rd Division, 2nd Corps, September 9, 1864; Circular, Headquarters, 1st Brigade, 3rd Division, 2nd Corps, September 9, 1864; Hancock to Orlando B. Willcox, September 9, 1864; Hancock to Humphreys, September 10, 1864; Gershom Mott to Wilson, September 10, 1864, *OR,* vol. 42, pt. 2, 687, 761–762, 764, 774, 777; John Haley journal, September 10, 1864, in Ruth L. Silliker, ed., *The Rebel Yell and the Yankee Hurrah: The Civil War Journal of a Maine Volunteer* (Camden, ME: Down East, 1985), 198; John Horn, *The Petersburg Campaign, June 1864–April 1865* (Conshohocken, PA: Combined, 2000), 196.

54. A. Wilson Greene, *Breaking the Backbone of the Rebellion: The Final Battles of the Petersburg Campaign* (Mason City, IA: Savas, 2000), 201–205; Dunlop, *Lee's Sharpshooters,* 248–255; Susan Williams Benson, ed., *Berry Benson's Civil War Book: Memoirs of a Confederate Scout and Sharpshooter* (Athens: University of Georgia Press, 1992), 179–180; Ray, *Shock Troops,* 254–256.

55. Benson, *Berry Benson's Civil War Book,* 180, 185, 189.

56. Ray, *Shock Troops,* 143, 222, 309, 329; Frederick L. Ray, "Shock Troops of the South," *America's Civil War* 15, 3 (2002): 35, 39.

57. Young, "A Campaign with Sharpshooters."

58. Dunlop, *Lee's Sharpshooters,* 10, 75–76, 155, 157; Winter, "Battalion of Sharpshooters," 98–99; Laughton, "Sharpshooters of Mahone's Brigade," 98.

59. Stan C. Harley letter to editor, n.d., *Confederate Veteran* 7 (1899): 307; Haughton, *Training, Tactics and Leadership,* 86, 126.

60. Ray, *Shock Troops,* 310–313.

61. William T. Sherman, *Memoirs of General William T. Sherman,* 2 vols. (New York: D. Appleton, 1875), 2: 384; Silas Casey, *Infantry Tactics, for the Instruction, Exercise, and Manoeuvres of the Soldier, a Company, Line of Skirmishers, Battalion, Brigade, or Corps d'Armee,* 3 vols. (New York: D. Van Nostrand, 1862), vol. 1, Edwin M. Stanton's preface, August 11, 1862, not paginated.

62. Paddy Griffith, *Battle Tactics of the Civil War* (New Haven, CT: Yale University Press, 1989), 155; Grady McWhiney and Perry D. Jamieson, *Attack and Die: Civil War Military Tactics and the Southern Heritage* (University: University of Alabama Press, 1982), 100–101;

Haughton, *Training, Tactics and Leadership,* 159; Thomas Vernon Moseley, "Evolution of the American Civil War Infantry Tactics" (Ph.D. diss., University of North Carolina, Chapel Hill, 1967), 369.

63. Sherman, *Memoirs,* 2: 394; Arthur L. Wagner, *Organization and Tactics* (New York: B. Westermann, 1895), 93; Moseley, "Evolution of the American Civil War Infantry Tactics," 335.

64. Wagner, *Organization and Tactics,* 144–148.

Chapter Seven. Sniping

1. Martin Pegler, *Out of Nowhere: A History of the Military Sniper* (Oxford: Osprey, 2004), 17, 59–61; Brent Nosworthy, *The Bloody Crucible of Courage: Fighting Methods and Combat Experience of the Civil War* (New York: Carroll & Graf, 2003), 190; Horace William Shaler Cleveland, "Rifle-Clubs," *Atlantic Monthly,* September 1862, 307.

2. Andrew Haughton, *Training, Tactics and Leadership in the Confederate Army of Tennessee: Seeds of Failure* (Portland, OR: Frank Cass, 2000), 85.

3. John Anderson Morrow, *The Confederate Whitworth Sharpshooters* (N.p., 2002), 7; Pegler, *Out of Nowhere,* 16.

4. Pegler, *Out of Nowhere,* 19–29, 144, 164–167.

5. C. A. Stevens, *Berdan's United States Sharpshooters in the Army of the Potomac, 1861–1865* (Dayton, OH: Morningside Bookshop, 1972), 39, 49.

6. Lew Wallace, "The Capture of Fort Donelson," in Robert Underwood Johnson and Clarence Clough Buel, eds., *Battles and Leaders of the Civil War* (New York: Thomas Yoseloff, 1956), 1: 407.

7. William Y. W. Ripley, *A History of Company F, 1st United States Sharpshooters* (Rutland, VT: Tuttle, 1883), 179.

8. C. C. Marsh to I. P. Rumsey, February 17, 1862; William T. Shaw to J. G. Lauman, February 19, 1862, *OR,* vol. 7, 201, 232.

9. Flavel C. Barber journal, February 13–15, 1862, Ferrell, *Holding the Line,* 17, 20–23.

10. Samuel Evans to father, May 1, 1862, in Robert F. Engs and Corey M. Brooks, eds., *Their Patriotic Duty: The Civil War Letters of the Evans Family of Brown County, Ohio* (New York: Fordham University Press, 2007), 21–22.

11. Stevens, *Berdan's United States Sharpshooters,* 38–39, 49–50; Edgeworth Bird to Sallie, April 28, 1862, in John Rozier, ed., *The Granite Farm Letters: The Civil War Correspondence of Edgeworth and Sallie Bird* (Athens: University of Georgia Press, 1988), 84; Charles F. Bryan Jr. and Nelson D. Lankford, eds., *Eye of the Storm: A Civil War Odyssey* (New York: Free Press, 2000), 54.

12. Bryan and Lankford, *Eye of the Storm,* 54.

13. Stevens, *Berdan's United States Sharpshooters,* 51.

14. Ibid., 55–56.

15. Ibid., 63.

16. Winfield S. Hancock to J. H. Taylor, September 29, 1862, *OR,* vol. 19, pt. 1, 280.

17. Jeffrey L. Patrick, ed., *Three Years With Wallace's Zouaves: The Civil War Memoirs of Thomas Wise Durham* (Macon: Mercer University Press, 2003), 129; James Harrison Wilson, *Under the Old Flag,* 2 vols. (New York: D. Appleton, 1912), 1: 219.

18. Wilson, *Under the Old Flag,* 1: 219.

19. Andrew Hickenlooper, "The Vicksburg Mine," in Robert Underwood Johnson and Clarence Clough Buel, eds., *Battles and Leaders of the Civil War* (New York: Thomas Yoseloff, 1956), 3: 541n.

20. Thomas Kilby Smith to W. D. Green, May 24, 1863; William E. Baldwin to J. C. Devereux, July 10, 1863, *OR,* vol. 24, pt. 2, 268, 403.

21. John W. DeForest, "Port Hudson," in James H. Croushore, ed., *A Volunteer's Adventures: A Union Captain's Record of the Civil War* (New Haven, CT: Yale University Press, 1946), 118.

22. Ibid., 117, 144.

23. T. B. Brooks to Quincy A. Gillmore, August 2, 1863, and endorsement by Gillmore and Brooks, *OR,* vol. 28, pt. 1, 323–324.

24. S. A. Ashe, "Life at Fort Wagner," *Confederate Veteran* 35 (1927): 254–255.

25. Fred L. Ray, *Shock Troops of the Confederacy: The Sharpshooter Battalions of the Army of Northern Virginia* (Asheville, NC: CFS, 2006), 275; John Anderson Morrow, *The Confederate Whitworth Sharpshooters* (N.p., 2002), 17; Stan C. Harley letter, *Confederate Veteran* 7 (1899): 307; "Attention, Whitworth Sharpshooters," *Confederate Veteran* 1 (1893): 117.

26. Morrow, *Confederate Whitworth Sharpshooters,* 27, 46; Stan C. Harley letter, 307; "Attention, Whitworth Sharpshooters," 117.

27. "Attention, Whitworth Sharpshooters," 117.

28. Stan C. Harley letter, 307; Morrow, *Confederate Whitworth Sharpshooters,* 26–27; "Attention, Whitworth Sharpshooters," 117; Patrick R. Cleburne to Archer Anderson, August 3, 1863, *OR,* vol. 23, pt. 1, 587.

29. Emerson Opdycke to wife, September 28, 1863, in Glenn V. Longacre and John E. Haas, eds., *To Battle for God and the Right: The Civil War Letterbooks of Emerson Opdycke* (Urbana: University of Illinois Press, 2003), 101; John M. Palmer to P. P. Oldershaw, September 30, 1863, *OR,* vol. 30, pt. 1, 715.

30. J. W. Minnich, "Famous Rifles," *Confederate Veteran* 30 (1922): 247–248.

31. Ed Porter Thompson, *History of the Orphan Brigade* (Louisville, KY: Lewis N. Thompson, 1898), 240–241; Ray, *Shock Troops,* 278; A. D. Kirwan, ed., *Johnny Green of the Orphan Brigade: The Journal of a Confederate Soldier* (Lexington: University of Kentucky Press, 1956), 127–128. The Confederates believed the Whitworth was better than the Kerr. See Irving A. Buck, *Cleburne and His Command* (Jackson, TN: McCowat-Mercer, 1959), 201.

32. Thompson, *History of the Orphan Brigade,* 241, 243.

33. Sam R. Watkins, *"Co. Aytch": A Side Show of the Big Show* (New York: Collier, 1962), 135–136; Buck, *Cleburne and His Command,* 201; James I. Hall, "History of Company C," in James R. Fleming, *Band of Brothers: Company C, 9th Tennessee Infantry* (Shippensburg, PA: White Mane, 1996), 64.

34. Buck, *Cleburne and His Command,* 202; Morrow, *Confederate Whitworth Sharpshooters,* 50; Stan C. Harley letter, 307.

35. Isaac N. Shannon, "Sharpshooters with Hood's Army," *Confederate Veteran* 15 (1907): 126.

36. Thompson, *History of the Orphan Brigade,* 242.

37. Ellison Capers to B. B. Smith, September 10, 1864, *OR,* vol. 38, pt. 3, 716; [Henry O. Dwight], "How We Fight at Atlanta," *Harper's New Monthly Magazine* 29 (1864): 666. Lt. Col. James W. Langley of the 125th Illinois might have been referring to the practice of placing mirrors on the butts of guns when he reported that his sharpshooters "did excellent service" at short range by "using an invention called the refracting sight." Langley to Theodore Wiseman, September 9, 1864, *OR,* vol. 38, pt. 1, 711.

38. Leonidas Polk to W. W. Mackall, June 1, 1864, Special Orders from May 9th 1864 to June 19th 1864, Head Quarters, Army of the Mississippi, chap. 2, no. 221 1/2 , RG109, NARA; Larry J. Daniel, *Soldiering in the Army of Tennessee: A Portrait of Life in a Confederate Army* (Chapel Hill: University of North Carolina Press, 1991), 156.

39. John T. Bell, *Tramps and Triumphs of the Second Iowa Infantry* (Omaha: Gibson, Miller & Richardson, 1886), 21–22; *Roster and Record of Iowa Soldiers in the War of the Rebellion, Together with Historical Sketches of Volunteer Organizations, 1861–1866,* 6 vols. (Des Moines: Emory H. English, 1908–1911), 1: 180, 262, 495.

40. William H. Heath to Henry Hoover, July 21, 1864, *OR,* vol. 39, pt. 1, 275; Shannon, "Sharpshooters with Hood's Army," 124–125; William B. Bate to James D. Porter, January 25, 1865, *OR,* vol. 45, pt. 1, 751.

41. Thomas G. Jones to John W. Daniel, July 3, 1904, in *Supplement to the Official Records of the Union and Confederate Armies,* 100 vols. (Wilmington, NC: Broadfoot, 1993–2000), pt. 1, vol. 6, 672; "Sharpshooting in Lee's Army," *Confederate Veteran* 3 (1895): 98.

42. Gordon C. Rhea, *The Battles for Spotsylvania Court House and the Road to Yellow Tavern, May 7–12, 1864* (Baton Rouge: Louisiana State University Press, 1997), 94, 96; Berry Benson, "How General Sedgwick Was Killed," *Confederate Veteran* 26 (1918): 115–116; B. M. Powell to wife, November 21, 1907, FSNMP; Berry Benson, "How I Lifted the Colonel's Mare," *Confederate Veteran* 27 (1919): 20; Ray, *Shock Troops,* 118–119.

43. Stevens, *Berdan's United States Sharpshooters,* 417.

44. Ibid., 435, 440.

45. Raymond J. Herek, *These Men Have Seen Hard Service: The First Michigan Sharpshooters in the Civil War* (Detroit: Wayne State University Press, 1998), 58, 169–171.

46. Theodore G. Ellis to H. J. Morse, August 9, 1864; Otho H. Binkley to John A. Gump, September 7, 1864, *OR,* vol. 36, pt. 1, 458, 744; Francis A. Walker to David B. Birney, June 4, 1864, ibid., vol. 36, pt. 3, 575; George M. Edgar to Colonel Johnston, July 24, 1902, George M. Edgar Collection, SHC-UNC.

47. George G. Meade to Ulysses S. Grant, July 5, 1864, in John Y. Simon, ed., *The Papers of Ulysses S. Grant,* 24 vols. (Carbondale: Southern Illinois University Press, 1967–2000), 11: 175n; Zenas R. Bliss reminiscences, Bliss Papers, USAMHI, 146.

48. Abbott Spear et al., eds., *The Civil War Recollections of General Ellis Spear* (Orono: University of Maine Press, 1997), 128–129.

49. Circular, July 9, 1864, 2nd Brigade, 1st Division, 9th Corps, Special Orders and Circulars, June–September, 1864 (2nd Brig., 1st Div., 9th AC), RG393, NARA.

50. Augustus Buell, *"The Cannoneer:" Recollections of Service in the Army of the Potomac* (Washington, DC: National Tribune, 1890), 236; Edmund J. Cleveland diary, entry of July 19, 1864, in Edmund J. Cleveland Jr., ed., "The Siege of Petersburg," [pt. 1], *Proceedings of the New Jersey Historical Society* 66 (1948): 93; Gouverneur K. Warren to Winfield S. Hancock, September 10, 1864; Hancock to Warren, September 10, 1864; J. H. Lockwood to Regis De Trobriand, September 12, 1864, *OR,* vol. 42, pt. 2, 779.

51. Edgeworth Bird to Sallie Bird, July 17, 1864, in Rozier, *Granite Farm Letters,* 176.

52. Stevens, *Berdan's United States Sharpshooters,* 235–236, 451–452, 460–461, 480.

53. Thomas D. Cockrell and Michael B. Ballard, eds., *A Mississippi Rebel in the Army of Northern Virginia: The Civil War Memoirs of Private David Holt* (Baton Rouge: Louisiana State University Press, 1995), 292–293; I. G. Bradwell, "Fort Steadman and Subsequent Events," *Confederate Veteran* 23 (1915): 20.

54. Benson, "How General Sedgwick Was Killed," 115; F. S. Harris, "Fine Shots in the Virginia Army," *Confederate Veteran* 4 (1896): 73–74.

55. Daniel W. Sawtelle reminiscences, and Sawtelle to sister, July 2, [1864], in Peter H. Buckingham, ed., *All's for the Best: The Civil War Reminiscences and Letters of Daniel W. Sawtelle, Eighth Maine Volunteer Infantry* (Knoxville: University of Tennessee Press, 2001), 90, 115, 283–284, 286.

56. Sawtelle reminiscences, in ibid., 90.

57. Sawtelle reminiscences, and Sawtelle to sister, July 10, 15, 1864, in ibid., 115–116, 133–134, 153, 287, 291.

58. Sawtelle reminiscences, and Sawtelle to sister, August 15, 21, 1864, in ibid., 116, 119–120, 128, 299.

59. Sawtelle reminiscences, in ibid., 112, 118.

60. Sawtelle reminiscences, and Sawtelle to sister, October 12, 1864, in ibid., 130, 134, 144, 307.

61. Sawtelle reminiscences, in ibid., 183.

62. Morrow, *Confederate Whitworth Sharpshooters,* 7; Shannon, "Sharpshooters with Hood's Army," 126–127.

63. Richard Delafield to Edwin M. Stanton, October 7, 1864, in Simon, *Papers of Ulysses S. Grant,* 12: 468n; General Orders no. 286, Adjutant General's Office, War Department, November 22, 1864, *OR,* vol. 45, pt. 1, 984.

64. Article by unidentified veteran of Pickett's Division in *Baltimore Herald,* November 3, 1886, in Stevens, *Berdan's United States Sharpshooters,* 462–463.

65. I counted fifty-eight divisions in the Army of the Potomac, the Army of the James, Sherman's army group, the Army of Tennessee, and the Army of Northern Virginia in 1864–1865. These units were engaged in the most intense and prolonged action during the period when sniping reached its peak in the war. It is not known how widespread was the practice of organizing sniper squads in units that mostly engaged in mobile operations during the last year of the war, because they either rarely or only occasionally engaged the enemy. However, the commander of the 14th Maine in the 19th Corps "permanently detailed" three of his men as sharpshooters during the Shenandoah Valley campaign. See Thomas W. Porter to Joseph Hibbert, Jr., October 23, 1864, *Supplement to the Official Records,* pt. 1, vol. 7, 555. Information on total enlistments during the Civil War comes from James M. McPherson, *Ordeal by Fire: The Civil War and Reconstruction,* 3rd ed. (New York: McGraw-Hill, 2001), 202.

Chapter Eight. The Rifle's Impact on Civil War Combat

1. Grady McWhiney and Perry D. Jamieson, *Attack and Die: Civil War Military Tactics and the Southern Heritage* (University: University of Alabama Press, 1982), 13–24.

2. Ibid., 19–21.

3. John A. Lynn, *The Wars of Louis XIV, 1667–1714* (London: Longman, 1999), 293, 320, 331–332, 334; Steven Ross, *From Flintlock to Rifle: Infantry Tactics, 1740–1866* (Rutherford, NJ: Fairleigh Dickinson University Press, 1979), 31–32, 58, 107–108, 110–111; Gunther E. Rothenburg, *The Art of Warfare in the Age of Napoleon* (Bloomington: Indiana University Press, 1980), 81.

4. Lynn, *Wars of Louis XIV,* 331–332, 334; Ross, *Flintlock to Rifle,* 31; B. P. Hughes, *Firepower: Weapons Effectiveness on the Battlefield, 1630–1850* (New York: Sarpedon, 1997), 165.

5. Ross, *Flintlock to Rifle,* 82; Paddy Griffith, *The Art of War of Revolutionary France, 1789–1802* (London: Greenhill, 1998), 233; Christopher Duffy, *Austerlitz, 1805* (Hamden, CT: Archon, 1977), 156; Rothenburg, *Art of Warfare,* 81.

6. David Eggenberger, *Encyclopedia of Battles: Accounts of Over 1,560 Battles from 1479 B.C. to the Present* (New York: Dover, 1985), 146; Rory Muir, *Tactics and the Experience of Battle in the Age of Napoleon* (New Haven, CT: Yale University Press, 1998), 85, 95, 99; Rothenburg, *Art of Warfare,* 81.

7. Paddy Griffith, *Battle Tactics of the Civil War* (New Haven, CT: Yale University Press, 1989), 19–20.

8. Craig L. Symonds, *A Battlefield Atlas of the American Revolution* (Baltimore: Nautical & Aviation, 1991), 19; Donald E. Graves, *Field of Glory: The Battle of Crysler's Farm, 1813* (Toronto: Robin Brass Studio, 2000), 268, 270; Graves, *Where Right and Glory Lead! The Battle of Lundy's Lane, 1814* (Toronto: Robin Brass Studio, 1999), 195–196; K. Jack Bauer, *The Mexican War, 1846–1848* (New York: Macmillan, 1974), 217; Justin H. Smith, *The War with Mexico,* 2 vols. (New York: Macmillan, 1919), 1: 386, 396.

9. Antulio J. Echevarria II, *After Clausewitz: German Military Thinkers before the Great War* (Lawrence: University Press of Kansas, 2000), 18; John A. English, *A Perspective on Infantry* (New York: Praeger, 1981), 118, 217.

10. John Lazenby, *Hannibal's War: A Military History of the Second Punic War* (Warminster, UK: Aris & Phillips, 1978), 84; Eggenberger, *Encyclopedia of Battles,* 75.

11. Robert A. Doughty and Ira D. Gruber, *American Military History and the Evolution of Warfare in the Western World* (Lexington, MA: D. C. Heath, 1996), 163; William B. Greene to mother, February 5, 1862, in William H. Hastings, ed., *Letters from a Sharpshooter: The Civil War Letters of Private William B. Greene, Co. G, 2nd United States Sharpshooters (Berdan's), Army of the Potomac, 1861–1865* (Belleville, WI: Historic Publications, 1993), 70.

12. Russell F. Weigley, *A Great Civil War: A Military and Political History, 1861–1865* (Bloomington: Indiana University Press, 2000), 32; Doughty and Gruber, *American Military History,* 163.

13. Francis J. Lippitt, *A Treatise on the Tactical Use of the Three Arms: Infantry, Artillery, and Cavalry* (Harrah, OK: Brandy Station Bookshelf, 1994), 7, 128, 134; H. Wager Halleck, *Elements of Military Art and Science; Or, Course of Instruction in Strategy, Fortification, Tactics of Battles, & C, Embracing the Duties of Staff, Infantry, Cavalry, Artillery, and Engineers, Adapted to the Use of Volunteers and Militia,* 3rd ed. (New York: D. Appleton, 1862), 125.

14. William T. Sherman, *Memoirs of General William T. Sherman,* 2 vols. (New York: D. Appleton, 1875), 2: 394; Gouverneur K. Warren to Daniel Butterfield, May 12, 1863, *OR,* vol. 25, pt. 1, 193.

15. Griffith, *Battle Tactics of the Civil War,* 197–198.

16. Doughty and Gruber, *American Military History,* 163; Griffith, *Battle Tactics of the Civil War,* 190–192.

17. Russell F. Weigley, *The Age of Battles: The Quest for Decisive Warfare from Breitenfeld to Waterloo* (Bloomington: Indiana University Press, 1991), xiii; Lynn, *Wars of Louis XIV,* 367–370.

18. Weigley, *Great Civil War,* 34.

19. Graves, *Field of Glory,* 268, 270; Graves, *Where Right and Glory Lead!* 182.

20. Note the title of Geoffrey Perret's book, *There's a War to Be Won: The United States Army in World War II* (New York: Random House, 1991).

21. Duffy, *Austerlitz,* 5–6, 50–51, 54–55, 72, 77, 87, 92–93, 100, 104–108, 113, 120–121, 130, 140, 142, 149, 152–154, 156–157, 162–163.

22. Edward Hagerman, *The American Civil War and the Origins of Modern Warfare: Ideas, Organization, and Field Command* (Bloomington: Indiana University Press, 1988), xii, 34; McWhiney and Jamieson, *Attack and Die,* 60, 112–123; Larry J. Daniel, *Cannoneers in Gray: The Field Artillery of the Army of Tennessee, 1861–1865* (Tuscaloosa: University of Alabama Press, 1984), 64–66; Earl J. Hess, *Pickett's Charge—The Last Attack at Gettysburg* (Chapel Hill: University of North Carolina Press, 2001), 163–164.

23. Kenneth W. Noe, ed., *A Southern Boy in Blue: The Memoirs of Marcus Woodcock, 9th Kentucky Infantry (U.S.A.)* (Knoxville: University of Tennessee Press, 1996), 207; J. Monroe, *The Company Drill of the Infantry of the Line, Together with the Skirmishing Drill of the Company and Battalion, after the Method of Gen. LeLouteril, Bayonet Fencing, with a Supplement on the Handling and Service of Light Infantry* (New York: D. Van Nostrand, 1863), 146.

24. Brent Nosworthy, *The Anatomy of Victory: Battle Tactics, 1689–1763* (New York: Hippocrene, 1992), 314.

25. Gary W. Gallagher, ed., *Fighting for the Confederacy: The Personal Recollections of General Edward Porter Alexander* (Chapel Hill: University of North Carolina Press, 1989), 261–262.

26. Daniel, *Cannoneers in Gray,* 38; Griffith, *Battle Tactics of the Civil War,* 170–171, 173–174, 176–177.

27. Hagerman, *American Civil War,* xii; McWhiney and Jamieson, *Attack and Die,* 127, 132.

28. S. B. Barron, *The Lone Star Defenders: A Chronicle of the Third Texas Cavalry, Ross' Brigade* (New York: Neale, 1908), 194; Brent Nosworthy, *The Bloody Crucible of Courage: Fighting Methods and Combat Experience of the Civil War* (New York: Carroll & Graf, 2003), 301.

29. Arthur L. Wagner, *Organization and Tactics* (New York: B. Westermann, 1895), 165–167; Nosworthy, *Bloody Crucible,* 303, 305.

30. Nosworthy, *Bloody Crucible,* 282; C. M. Wilcox, *Rifles and Rifle Practice: An Elementary Treatise upon the Theory of Rifle Firing* (New York: D. Van Nostrand, 1859), 246.

31. Nosworthy, *Anatomy of Victory,* 303–304, 307.

32. Griffith, *Battle Tactics of the Civil War,* 179–180; Geoffrey Wawro, *The Franco-Prussian War: The German Conquest of France in 1870–1871* (Cambridge: Cambridge University Press, 2003), 132–133.

33. Hagerman, *American Civil War,* 198; McWhiney and Jamieson, *Attack and Die,* 75; Wilcox, *Rifles and Rifle Practice,* 249, 251–252; Wagner, *Organization and Tactics,* 93; Andrew Haughton, *Training, Tactics and Leadership in the Confederate Army of Tennessee: Seeds of Failure* (Portland, OR: Frank Cass, 2000), 152–153; Horace William Shaler Cleveland, "Rifle-Clubs," *Atlantic Monthly,* September 1862, 307; Francis W. Palfrey, "The Period Which Elapsed between the Fall of Yorktown and the Beginning of the Seven Days' Battles," in *Campaigns in Virginia, 1861–1862: Papers of the Military Historical Society of Massachusetts,* vol. 1 (Boston: Houghton, Mifflin, 1895), 212.

34. William B. Hazen, *A Narrative of Military Service* (Boston: Ticknor, 1885), 381–383; William B. Hazen to R. R. Townes, September 10, 1864, *OR,* vol. 38, pt. 3, 184.

35. Griffith, *Battle Tactics of the Civil War,* 127, 129, 133–135, 189.

36. Earl J. Hess, *Field Armies and Fortifications in the Civil War: The Eastern Campaigns, 1861–1864* (Chapel Hill: University of North Carolina Press, 2005), 308–314.

37. Wawro, *Franco-Prussian War,* 107–112, 124–132, 154–160, 166–184.

38. Patton Anderson to J. W. Ratchford, February 9, 1865, *OR,* vol. 38, pt. 3, 770–771; William B. Hazen to R. R. Townes, September 10, 1864, *OR,* vol. 38, pt. 3, 184.

Chapter Nine. After the Rifle Musket

1. Francis J. Lippitt, *Reminiscences* (Providence: Preston & Rounds, 1902), 5, 28–32, 34, 37, 46, 50, 53, 58, 108–110.

2. Emory Upton, *A New System of Infantry Tactics, Double and Single Rank, Adapted to American Topography and Improved Fire-arms* (New York: D. Appleton, 1867); Paddy Griffith, *Battle Tactics of the Civil War* (New Haven, CT: Yale University Press, 1989), 104.

3. Upton, *Infantry Tactics,* 43, 45–47, 131–133.

4. Stephen Vincent Benet to William T. Sherman, January 29, 1883, in John C. Tidball, "Rifle Target Practice in the Army," *Ordnance Notes,* January 30, 1883, 3; Perry D. Jamieson, *Crossing the Deadly Ground: United States Army Tactics, 1865–1899* (Tuscaloosa: University of Alabama Press, 1994), 54–56, 59; Russell Gilmore, "'The New Courage': Rifles and Soldier Individualism, 1876–1918," *Military Affairs* 40, 3 (1976): 97–98.

5. Robert A. Doughty and Ira D. Gruber, *American Military History and the Evolution of Warfare in the Western World* (Lexington, MA: D. C. Heath, 1996), 233; Martin Pegler, *Out of Nowhere: A History of the Military Sniper* (Oxford: Osprey, 2004), 73.

6. Fred L. Ray, *Shock Troops of the Confederacy: The Sharpshooter Battalions of the Army of Northern Virginia* (Asheville, NC: CFS, 2006), 315–316.

7. Allan R. Millett and Peter Maslowski, *For the Common Defense: A Military History of the United States of America* (New York: Free Press, 1984), 238, 315; James M. Morris, *America's Armed Forces: A History,* 2nd ed. (Upper Saddle River, NJ: Prentice Hall, 1996), 146; Sidney B. Brinckerhoff and Pierce Chamberlin, "The Army's Search for a Repeating Rifle: 1873–1903," *Military Affairs* 32 (Spring 1968): 25–26, 28–29; Jamieson, *Crossing the Deadly Ground,* 110.

8. Gilmore, "'New Courage,'" 99.

9. Arthur L. Wagner, *Organization and Tactics* (New York: B. Westermann, 1895), 46–47, 111–113.

10. Ibid., 114–115, 118–119.

11. Steven Ross, *From Flintlock to Rifle: Infantry Tactics, 1740–1866* (Rutherford, NJ: Fairleigh Dickinson University Press, 1979), 177; Geoffrey Wawro, *The Franco-Prussian War: The German Conquest of France in 1870–1871* (Cambridge: Cambridge University Press, 2003), 51.

12. Ross, *Flintlock to Rifle,* 190; Wawro, *Franco-Prussian War,* 52.

13. Wawro, *Franco-Prussian War,* 95–107, 124–132, 143–144, 154, 168–169, 174.

14. Albrecht von Boguslawski, *Tactical Deductions from the War of 1870–71* (Minneapolis: Absinthe, 1996), 150–151.

15. Bill Nasson, *The South African War, 1899–1902* (London: Arnold, 1999), 57–58, 74, 161; M. A. Ramsay, *Command and Cohesion: The Citizen Soldier and Minor Tactics in the British Army, 1870–1918* (Westport, CT: Praeger, 2002), 87.

16. Gilmore, "'New Courage,'" 100; Antulio J. Echevarria II, *After Clausewitz: German Military Thinkers before the Great War* (Lawrence: University Press of Kansas, 2000), 164–165.

17. Gilmore, "'New Courage,'" 100; Kenneth Finlayson, *An Uncertain Trumpet: The Evolu-*

tion of U.S. Army Infantry Doctrine, 1919–1941 (Westport, CT; Greenwood, 2001), xv–xvi, 30–37, 106, 154, 160.

18. Thomas Christianson, "M-1 Rifle," in John Whiteclay Chambers, ed., *The Oxford Companion to American Military History* (New York: Oxford University Press, 1999) and Christianson, "M-16 Rifle," in ibid., 405; Pegler, *Out of Nowhere,* 294.

19. Griffith, *Battle Tactics of the Civil War,* 148.

20. Francis J. Lippitt, *A Treatise on the Tactical Use of the Three Arms: Infantry, Artillery, and Cavalry* (Harrah, OK: Brandy Station Bookshelf, 1994), 54–55.

21. Upton, *Infantry Tactics,* 98.

22. Wagner, *Organization and Tactics,* 109.

23. Pegler, *Out of Nowhere,* 82, 85–87, 150–151.

24. Ibid., 120–123, 143–145, 148, 151, 154.

25. Ibid., 19–29.

26. Ibid., 164–167, 169–170, 173, 176–178, 184, 199, 214; John Anderson Morrow, *The Confederate Whitworth Sharpshooters* (N.p., 2002), 8.

27. Pegler, *Out of Nowhere,* 294, 312–314.

28. Ibid., 298–299, 314, 317, 320–327, 329.

Bibliography

Manuscripts

Chicago History Museum
Ira Miltmore Collection

College of William and Mary, Special Collections, Williamsburg, VA
Joseph E. Johnston Papers

Duke University, Special Collections Library, Durham, NC
John Gross Barnard Papers
Confederate Veteran Papers
Charles Powell Reminiscences

Emory University, Special Collections, Atlanta, GA
Andrew Jackson Neal Papers

Fredericksburg-Spotsylvania National Military Park, Fredericksburg, VA
B. M. Powell Letter

Library of Congress, Manuscripts Division, Washington, DC
William Franklin Patterson Papers
Hazard Stevens Papers

Maryland Historical Society, Baltimore
Gordon-Blackford Papers

National Archives and Records Administration, Washington, DC
Compiled Service Records of Volunteer Union Soldiers Who Served in Organizations from the State of Missouri (M405)
14th Army Corps, Letters Sent (3rd Brigade, 2nd Division), vol. 42/109, part 2, vol. 43/112, part 2
Orders, Rodes's and Battle's Brigades, Army of Northern Virginia, 1861–65, chap. 2, vol. 66
Records of United States Continental Commands, Polyonymous Succession of Commands, 1861–1870 (RG393)
Records of Volunteer Union Soldiers Who Served during the Civil War (RG94)
17th Army Corps, Letters Sent, vol. 20/79, part 2
Special Orders and Circulars, June–September, 1864 (2nd Brigade 1st Division, 9th Corps)

Special Orders from May 9th 1864 to June 19th 1864, Head Quarters, Army of the Mississippi, chap. 2, no. 221 & 1/2
War Department Collection of Confederate Records (RG109)

New York Public Library, Rare Books and Manuscripts, New York
United States Sanitary Commission Records

Petersburg National Battlefield, Petersburg, VA
James Mitchell Letter

United States Army Military History Institute, Carlisle, PA
Zenas R. Bliss Papers

University of North Carolina, Southern Historical Collection, Chapel Hill
Elias Davis Papers

University of Tennessee, Special Collections, Knoxville
Edwin H. Rennolds Diary

Washington and Lee University, Special Collections, Lexington, VA
Augustus Forsberg Memoir

Newspapers

Aberdeen Examiner, Aberdeen, MS
Augusta Telegraph, Augusta, GA
Daily Confederate, Raleigh, NC
Daily Missouri Democrat, St. Louis, MO

Articles, Books, and Dissertations

Acken, J. Gregory, ed. *Inside the Army of the Potomac: The Civil War Experience of Captain Francis Adams Donaldson.* Mechanicsburg, PA: Stackpole, 1998.

Alderson, William T., ed. "The Civil War Reminiscences of John Johnston, 1861–1865." *Tennessee Historical Quarterly* 13 (1954): 65–82, 156–178.

Aldrich, C. Knight, ed. *Quest for a Star: The Civil War Letters and Diaries of Colonel Francis T. Sherman of the 88th Illinois.* Knoxville: University of Tennessee Press, 1999.

Allin, E. S. *Rules for the Management and Cleaning of the Rifle Musket, Model 1855.* Washington, DC: Government Printing Office, 1862.

Ambrose, D. Leib. *From Shiloh to Savannah: The Seventh Illinois Infantry in the Civil War.* DeKalb: Northern Illinois University Press, 2003.

Andersen, Mary Ann, ed. *The Civil War Diary of Allen Morgan Geer, Twentieth Regiment, Illinois Volunteers.* New York: Cosmos, 1977.

Anderson, Ephraim McD. *Memoirs: Historical and Personal; Including the Campaigns of the First Missouri Confederate Brigade.* Dayton, OH: Morningside Bookshop, 1972.

Angle, Paul M., ed. *Three Years in the Army of the Cumberland: The Letters and Diary of Major James A. Connolly.* Bloomington: Indiana University Press, 1959.

Ashe, S. A. "Life at Fort Wagner." *Confederate Veteran* 35 (1927): 254–256.

"Attention, Whitworth Sharpshooters." *Confederate Veteran* 1 (1893): 117.

Austerman, Wayne. "Abhorrent to Civilization." *Civil War Times Illustrated* 24, 5 (1985): 36–40.

Bailey, Hugh C., ed. "An Alabamian at Shiloh: The Diary of Liberty Independence Nixon." *Alabama Review* 11 (1958): 144–155.

Bailey, Ronald H. *Forward to Richmond.* Alexandria, VA: Time-Life, 1983.

Barron, S. B. *The Lone Star Defenders: A Chronicle of the Third Texas Cavalry, Ross' Brigade.* New York: Neale, 1908.

"The Battle at Fort Gregg." *Southern Historical Society Papers* 28 (1900): 265–267.

Bauer, K. Jack. *The Mexican War, 1846–1848.* New York: Macmillan, 1974.

———, ed. *Soldiering: The Civil War Diary of Rice C. Bull, 123rd New York Volunteer Infantry.* San Rafael, CA: Presidio, 1978.

Baumann, Ken. *Arming the Suckers, 1861–1865: A Compilation of Illinois Civil War Weapons.* Dayton, OH: Morningside Bookshop, 1989.

Baumgart, Winfried. *The Crimean War, 1853–1856.* London: Arnold, 1999.

Baumgartner, Richard A., ed. *Blood and Sacrifice: The Civil War Journal of a Confederate Soldier.* Huntington, WV: Blue Acorn, 1997.

Beals, Thomas P. "In a Charge Near Fort Hell, Petersburg, April 2, 1865." In *War Papers Read before the Commandery of the State of Maine, Military Order of the Loyal Legion of the United States,* vol. 2, 105–115. Portland, ME: Lefavor-Tower, 1902.

Bell, John T. *Tramps and Triumphs of the Second Iowa Infantry.* Omaha: Gibson, Miller & Richardson, 1886.

Bellesiles, Michael A. *Arming America: The Origins of a National Gun Culture.* New York: Alfred A. Knopf, 2000.

Bennett, Stewart, and Barbara Tillery, eds. *The Struggle for the Life of the Republic: A Civil War Narrative by Brevet Major Charles Dana Miller, 76th Ohio Volunteer Infantry.* Kent, OH: Kent State University Press, 2004.

Benson, Berry. "How General Sedgwick Was Killed." *Confederate Veteran* 26 (1918): 115–116.

———. "How I Lifted the Colonel's Mare." *Confederate Veteran* 27 (1919): 20–23.

Benson, Susan Williams, ed. *Berry Benson's Civil War Book: Memoirs of a Confederate Scout and Sharpshooter.* Athens: University of Georgia Press, 1992.

Benton, J. G. *Ordnance and Gunnery.* New York: D. Van Nostrand, 1867.

Bergeron, Arthur W. Jr., ed. *The Civil War Reminiscences of Major Silas T. Grisamore, C.S.A.* Baton Rouge: Louisiana State University Press, 1993.

Best, Isaac O. *History of the 121st New York State Infantry.* Chicago: W. B. Conkey, 1921.

Biel, John G., ed. "The Evacuation of Corinth: From the Diary and a Letter of Joseph Dimmit Thompson." *Journal of Mississippi History* 24, 1 (1962): 40–56.

Bilby, Joseph G. *Civil War Firearms: Their Historical Background and Tactical Use.* Conshohocken, PA: Combined, 1996.

Binney, Capt. "Muskets and Musketry." *Aide Memoir to the Military Sciences* 2 (1860): 442–467.

Bircher, William. *A Drummer-Boy's Diary: Comprising Four Years of Service with the Second Regiment Minnesota Veteran Volunteers, 1861 to 1865.* St. Paul, MN: St. Paul Book and Stationery, 1889.

Black, John D. "Reminiscences of the Bloody Angle." In *Glimpses of the Nation's Struggle, Fourth Series: Papers Read before the Minnesota Commandery of the Military Order of the Loyal Legion of the United States, 1892–1897,* 420–436. St. Paul, MN: H. L. Collins, 1898.

Blomquist, Ann K., and Robert A. Taylor, eds. *This Cruel War: The Civil War Letters of Grant and Malinda Taylor, 1862–1865.* Macon, GA: Mercer University Press, 2000.

Boguslawski, Albrecht von. *Tactical Deductions from the War of 1870–71.* Minneapolis: Absinthe, 1996.

Bohrnstedt, Jennifer Cain, ed. *Soldiering with Sherman: Civil War Letters of George F. Cram.* DeKalb: Northern Illinois University Press, 2000.

Bond, Otto F., ed. *Under the Flag of the Nation: Diaries and Letters of a Yankee Volunteer in the Civil War.* Columbus: Ohio State University Press, 1961.

Bowen, George A., ed. "The Diary of Captain George D. Bowen, 12th Regiment New Jersey Volunteers." *Valley Forge Journal* 2 (1984): 116–145, and 2 (1985): 176–231.

Bowen, James L. "General Edwards's Brigade at the Bloody Angle." In Robert Underwood Johnson and Clarence Clough Buel, eds., *Battles and Leaders of the Civil War,* 4: 177. New York: Thomas Yoseloff, 1956.

———. *History of the Thirty-seventh Regiment Mass. Volunteers, in the Civil War of 1861–1865.* Holyoke, MA: Clark W. Bryan, 1884.

Bradwell, I. G. "Fort Steadman and Subsequent Events." *Confederate Veteran* 23 (1915): 20–23.

———. "Second Day's Battle of the Wilderness." *Confederate Veteran* 28 (1920): 20–22.

———. "Soldier Life in the Confederate Army." *Confederate Veteran* 24 (1916): 20–25.

Brinckerhoff, Sidney B., and Pierce Chamberlin. "The Army's Search for a Repeating Rifle: 1873–1903." *Military Affairs* 32 (Spring 1968): 20–30.

Brown, George. "Explosive Bullets." *Confederate Veteran* 24 (1916): 95.

Brown, Russell K. *"Our Connection with Savannah": History of the First Battalion Georgia Sharpshooters, 1861–1865.* Macon, GA: Mercer University Press, 2004.

Brown, Varina D. *A Colonel at Gettysburg and Spotsylvania.* Columbia, SC: State Company, 1931.

Bruce, Robert V. *Lincoln and the Tools of War.* Indianapolis: Bobbs-Merrill, 1956.

Bryan, Charles F. Jr., and Nelson D. Lankford, eds. *Eye of the Storm: A Civil War Odyssey.* New York: Free Press, 2000.

Buck, Irving A. *Cleburne and His Command.* Jackson, TN: McCowat-Mercer, 1959.

Buckingham, Peter H., ed. *All's for the Best: The Civil War Reminiscences and Letters of Daniel W. Sawtelle, Eighth Maine Volunteer Infantry.* Knoxville: University of Tennessee Press, 2001.

Buell, Augustus. *"The Cannoneer": Recollections of Service in the Army of the Potomac.* Washington, DC: National Tribune, 1890.

Burdette, Robert J. *The Drums of the 47th.* Urbana: University of Illinois Press, 2000.

Busk, Hans. *The Rifle: And How to Use It.* 8th ed. London: Routledge, Warne, & Routledge, 1862.

Byrne, Frank L., ed. *Uncommon Soldiers: Harvey Reid and the 22nd Wisconsin March with Sherman.* Knoxville: University of Tennessee Press, 2001.

Byrne, Frank L., and Jean Powers Soman, eds. *Your True Marcus: The Civil War Letters of a Jewish Colonel.* Kent, OH: Kent State University Press, 1985.

Caldwell, J. F. J. *The History of a Brigade of South Carolinians, Known First as "Gregg's," and Subsequently as "McGowan's Brigade."* Philadelphia: King & Baird, 1866.

Calhoun, W. L. *History of the 42d Regiment, Georgia Volunteers, Confederate States Army, Infantry.* University, AL: Confederate, 1977.

Campbell, Eric A. "The Aftermath and Recovery of Gettysburg." Pt. 2. *Gettysburg Magazine* 12 (January 1995): 97–110.

Carr, Pat, ed. *In Fine Spirits: The Civil War Letters of Ras Stirman.* Fayetteville, AR: Washington County Historical Society, 1986.

Carson, J. P. "Fort Steadman's Fall." *Confederate Veteran* 22 (1914): 460–462.

Casey, Silas. *Infantry Tactics, for the Instruction, Exercise, and Manoeuvres of the Soldier, a Company, Line of Skirmishers, Battalion, Brigade, or Corps d'Armee.* 3 vols. New York: D. Van Nostrand, 1862.

Chamberlin, W. H. "The Skirmish Line in the Atlanta Campaign." In *Sketches of War History, 1861–1865: Papers Prepared for the Ohio Commandery of the Military Order of the Loyal Legion of the United States,* 3: 182–196. Cincinnati: Robert Clarke, 1890.

Chance, Joseph E. *The Second Texas Infantry: From Shiloh to Vicksburg.* Austin, TX: Eakin, 1984.

Chapman, Sarah Bahnson, ed. *Bright and Gloomy Days: The Civil War Correspondence of Captain Charles Frederic Bahnson, a Moravian Confederate.* Knoxville: University of Tennessee Press, 2003.

Christ, Elwood W. *The Struggle for the Bliss Farm at Gettysburg, July 2nd and 3rd, 1863.* 2nd ed. Baltimore: Butternut & Blue, 1994.

Christ, Mark K., ed. *Getting Used to Being Shot At: The Spence Family Civil War Letters.* Fayetteville: University of Arkansas Press, 2002.

Christianson, Thomas. "M-1 Rifle." In John Whiteclay Chambers II, ed., *The Oxford Companion to American Military History,* 405. New York: Oxford University Press, 1999.

———. "M-16 Rifle." In John Whiteclay Chambers II, ed., *The Oxford Companion to American Military History,* 405. New York: Oxford University Press, 1999.

Clark, Charles T. *Opdycke Tigers, 125th O.V.I.: A History of the Regiment and of the Campaigns and Battles of the Army of the Cumberland.* Columbus, OH: Spahr & Glenn, 1895.

Cleveland, Edmund J. Jr., ed. "The Siege of Petersburg." [Pt. 1.] *Proceedings of the New Jersey Historical Society* 66 (1948): 76–95.

Cleveland, Horace William Shaler. *Hints to Riflemen.* New York: D. Appleton, 1864.

———. "Rifle-Clubs." *Atlantic Monthly,* September 1862, 303–310.

———. "The Use of the Rifle." *Atlantic Monthly,* March 1862, 300–306.

Cockrell, Thomas D., and Michael B. Ballard, eds. *A Mississippi Rebel in the Army of Northern Virginia: The Civil War Memoirs of Private David Holt.* Baton Rouge: Louisiana State University Press, 1995.

Coggins, Jack. *Arms and Equipment of the Civil War.* Garden City, NY: Doubleday, 1962.

———. "The Engineers Played a Key Role in Both Armies." *Civil War Times Illustrated* 3, 9 (1965): 40–44.

"Col. Erasmus I. Stirman." *Confederate Veteran* 22 (1914): 226.

Collins, Donald R. "How Bullets Were Made." *Civil War Times Illustrated* 4, 9 (1966): 22–25.

———. "More on Bullets." *Civil War Times Illustrated* 5, 4 (1966): 36–39.

Comey, Lyman Richard, ed. *A Legacy of Valor: The Memoirs and Letters of Captain Henry Newton Comey, 2nd Massachusetts Infantry.* Knoxville: University of Tennessee Press, 2004.

Cook, Henry H. "Explosive Bullets." *Confederate Veteran* 7 (1899): 27.

Cox, Jacob D. *Atlanta.* New York: Charles Scribner's Sons, 1882.

Cozzens, Peter. *The Darkest Days of the War: The Battles of Iuka and Corinth.* Chapel Hill: University of North Carolina Press, 1997.

———. *The Shipwreck of Their Hopes: The Battles for Chattanooga.* Urbana: University of Illinois Press, 1994.

———. *This Terrible Sound: The Battle of Chickamauga.* Urbana: University of Illinois Press, 1992.

Crist, Lynda Lasswell, ed. *The Papers of Jefferson Davis.* 10 vols. Baton Rouge: Louisiana State University Press, 1971–1999.

Croushore, James H., ed. *A Volunteer's Adventures: A Union Captain's Record of the Civil War.* New Haven, CT: Yale University Press, 1946.

Crute, Joseph J. Jr. *Units of the Confederate States Army.* Midlothian, VA: Derwent, 1987.

Curtis, Finley P. "The Black Shadow of the Sixties." *Confederate Veteran* 24 (1916): 353–357.

Daniel, Larry J. *Cannoneers in Gray: The Field Artillery of the Army of Tennessee, 1861–1865.* Tuscaloosa: University of Alabama Press, 1984.

———. *Shiloh: The Battle That Changed the Civil War.* New York: Simon & Schuster, 1997.

———. *Soldiering in the Army of Tennessee: A Portrait of Life in a Confederate Army.* Chapel Hill: University of North Carolina Press, 1991.

Davis, Carl L. *Arming the Union: Small Arms in the Civil War.* Port Washington, NY: Kennikat, 1973.

Davis, William C., ed. *Diary of a Confederate Soldier: John S. Jackman of the Orphan Brigade.* Columbia: University of South Carolina Press, 1990.

Dederer, John Morgan. "Side Arms, Standard Infantry." In John Whiteclay Chambers II, ed., *The Oxford Companion to American Military History,* 659–661. New York: Oxford University Press, 1999.

"Diary of Colonel William Camm, 1861 to 1865." *Journal of the Illinois State Historical Society* 18 (1925–1926): 793–969.

Dornbusch, Charles E. *Military Bibliography of the Civil War.* 3 vols. New York: New York Public Library, 1994.

Douchy, George K. "The Battle of Ream's Station." In *Military Essays and Recollections: Papers Read before the Commandery of the State of Illinois, Military Order of the Loyal Legion of the United States,* 3: 125–140. Chicago: Dial, 1890.

Doughty, Robert A., and Ira D. Gruber. *American Military History and the Evolution of Warfare in the Western World.* Lexington, MA: D. C. Heath, 1996.

Duane, William. *The American Military Library; Or, Compendium of the Modern Tactics.* 2 vols. Philadelphia: William Duane, 1809.

Duffy, Christopher. *The Army of Frederick the Great.* London: David & Charles, 1974.

———. *Austerlitz, 1805.* Hamden, CT: Archon, 1977.

Dunlop, W. S. *Lee's Sharpshooters; Or, The Forefront of Battle.* Dayton, OH: Morningside Bookshop, 1982.

Durham, Roger S., ed. *The Blues in Gray: The Civil War Journal of William Daniel Dixon and the Republican Blues Daybook.* Knoxville: University of Tennessee Press, 2000.

[Dwight, Henry O.] "How We Fight at Atlanta." *Harper's New Monthly Magazine* 29 (1864): 663–666.

Echevarria, Antulio J. II. *After Clausewitz: German Military Thinkers before the Great War.* Lawrence: University Press of Kansas, 2000.

Eggenberger, David. *Encyclopedia of Battles: Accounts of Over 1,560 Battles from 1479 B.C. to the Present.* New York: Dover, 1985.

Eisenschiml, Otto, ed. *Vermont General: The Unusual War Experiences of Edward Hastings Ripley, 1861–1865.* New York: Devin-Adair, 1960.

Elder, Donald C. III, ed. *A Damned Iowa Greyhound: The Civil War Letters of William Henry Harrison Clayton.* Iowa City: University of Iowa Press, 1998.

———. *Love amid the Turmoil: The Civil War Letters of William and Mary Vermilion.* Iowa City: University of Iowa Press, 2003.

Elliott, Charles G. "Kirkland's Brigade, Hoke's Division, 1864–65." *Southern Historical Society Papers* 23 (1895): 166–174.

Elliott, Isaac H. *History of the Thirty-third Regiment Illinois Veteran Volunteer Infantry in the Civil War, 22nd August, 1861, to 7th December, 1865.* Gibson City, IL: Gibson Courier, 1902.

Ellison, Janet Correll, ed. *On to Atlanta: The Civil War Diaries of John Hill Ferguson, Illinois Tenth Regiment of Volunteers.* Lincoln: University of Nebraska Press, 2001.

Ellsworth, E. E. *Manual of Arms for Light Infantry, Adapted to the Rifled Musket, with, or without, the Priming Attachment, Arranged for the U.S. Zouave Cadets, Governors Guard of Illinois.* N.p., 1859.

English, John A. *A Perspective on Infantry.* New York: Praeger, 1981.

Engs, Robert F., and Corey M. Brooks, eds. *Their Patriotic Duty: The Civil War Letters of the Evans Family of Brown County, Ohio.* New York: Fordham University Press, 2007.

Evans, Robert G., ed. *The 16th Mississippi Infantry: Civil War Letters and Reminiscences.* Jackson: University Press of Mississippi, 2002.

Ferrell, Robert H., ed. *Holding the Line: The Third Tennessee Infantry, 1861–1864.* Kent, OH: Kent State University Press, 1994.

Finlayson, Kenneth. *An Uncertain Trumpet: The Evolution of U.S. Army Infantry Doctrine, 1919–1941.* Westport, CT: Greenwood, 2001.

Fleet, Betsy, and John D. P. Fuller, eds. *Green Mount: A Virginia Plantation Family during the Civil War: Being the Journal of Benjamin Robert Fleet and Letters of His Family.* Lexington: University of Kentucky Press, 1962.

Fleming, James R. *Band of Brothers: Company C, 9th Tennessee Infantry.* Shippensburg, PA: White Mane, 1996.

Floyd, David Bittle. *History of the Seventy-fifth Regiment Indiana Infantry Volunteers.* Philadelphia: Lutheran Publication Society, 1893.

Floyd, Fred C. *History of the Fortieth (Mozart) Regiment New York Volunteers.* Boston: F. H. Gilson, 1909.

Folmar, John Kent, ed. *From That Terrible Field: Civil War Letters of James M. Williams, Twenty-first Alabama Infantry Volunteers.* Tuscaloosa: University of Alabama Press, 1981.

Fratt, Steve. "American Civil War Tactics: The Theory of W. J. Hardee and the Experience of E. C. Bennett." *Indiana Military History Journal* 10, 1 (1985): 4–15.

French, Samuel G. *Two Wars: An Autobiography of Gen. Samuel G. French.* Nashville, TN: Confederate Veteran, 1901.

Gallagher, Gary W., ed. *Fighting for the Confederacy: The Personal Recollections of General Edward Porter Alexander.* Chapel Hill: University of North Carolina Press, 1989.

Gallagher, John J. *The Battle of Brooklyn, 1776.* New York: Sarpedon, 1995.

Galloway, G. Norton. "Hand-to-Hand Fighting at Spotsylvania." In Robert Underwood Johnson and Clarence Clough Buel, eds., *Battles and Leaders of the Civil War,* 4: 170–174. New York: Thomas Yoseloff, 1956.

Garavaglia, Louis A., and Charles G. Worman. "Arms and the Man." *North and South* 4, 6 (2001): 44–56.

Garrett, Jill Knight, ed. *The Civil War Diary of Andrew Jackson Campbell.* Columbia, TN: Jill Knight Garrett, 1965.

Gaskill, J. W. *Footprints through Dixie: Everyday Life of the Man under a Musket on the Firing Line and in the Trenches, 1862–1865.* Alliance, OH: Bradshaw, 1919.

Gates, Arnold, ed. *The Rough Side of War: The Civil War Journal of Chesley A. Mosman, 1st Lieutenant, Company D, 59th Illinois Volunteer Infantry Regiment.* Garden City, NY: Basin, 1987.

Gibson, J. Catlett, and William W. Smith. "The Battle of Spotsylvania Courthouse, May 12, 1864." *Southern Historical Society Papers* 32 (1904): 200–215.

Gilmore, Russell. "'The New Courage': Rifles and Soldier Individualism, 1876–1918." *Military Affairs* 40, 3 (1976): 97–102.

Gist, W. W. "The Battle of Franklin." *Tennessee Historical Magazine* 6, 3 (1920): 213–265.

Gorgas, Josiah. "Contributions to the History of the Confederate Ordnance Department." *Southern Historical Society Papers* 12 (1884): 66–94.

Gould, David, and James B. Kennedy, eds. *Memoirs of a Dutch Mudsill: The "War Memories" of John Henry Otto, Captain, Company D, 21st Regiment Wisconsin Volunteer Infantry.* Kent, OH: Kent State University Press, 2004.

Grant, Ulysses S. *Personal Memoirs.* 2 vols. New York: Viking, 1990.

———. "The Vicksburg Campaign." In Robert Underwood Johnson and Clarence Clough Buel, eds., *Battles and Leaders of the Civil War,* 3: 493–539. New York: Thomas Yoseloff, 1956.

Graves, Donald E. *Field of Glory: The Battle of Crysler's Farm, 1813.* Toronto: Robin Brass Studio, 2000.

———. *Where Right and Glory Lead! The Battle of Lundy's Lane, 1814.* Toronto: Robin Brass Studio, 1999.

Greene, A. Wilson. *Breaking the Backbone of the Rebellion: The Final Battles of the Petersburg Campaign.* Mason City, IA: Savas, 2000.

Griffith, Paddy. *The Art of War of Revolutionary France, 1789–1802.* London: Greenhill, 1998.

———. *Battle Tactics of the Civil War.* New Haven, CT: Yale University Press, 1989.

Grimsley, Mark. "Review Essay: The Continuing Battle of Gettysburg." *Civil War History* 49, 2 (2003): 181–187.

———. "Surviving Military Revolution: The U.S. Civil War." In Macgregor Knox and Williamson Murray, eds., *The Dynamics of Military Revolution, 1300–2050,* 74–91. Cambridge: Cambridge University Press, 2001.

Grimsley, Mark, and Todd D. Miller, eds. *The Union Must Stand: The Civil War Diary of John Quincy Adams Campbell, Fifth Iowa Volunteer Infantry.* Knoxville: University of Tennessee Press, 2000.

Hafendorfer, Kenneth A., ed. *Civil War Journal of William L. Trask: Confederate Sailor and Soldier.* Louisville, KY: KH, 2003.

Hagerman, Edward. *The American Civil War and the Origins of Modern Warfare: Ideas, Organization, and Field Command.* Bloomington: Indiana University Press, 1988.

Hall, Winchester. *The Story of the 26th Louisiana Infantry, in the Service of the Confederate States.* N.p., n.d.

Halleck, H. Wager. *Elements of Military Art and Science; Or, Course of Instruction in Strategy, Fortification, Tactics of Battles, & C, Embracing the Duties of Staff, Infantry, Cavalry, Artillery, and Engineers, Adapted to the Use of Volunteers and Militia.* 3rd ed. New York: D. Appleton, 1862.

Hallock, Judith Lee, ed. *The Civil War Letters of Joshua K. Callaway.* Athens: University of Georgia Press, 1997.

Hancock, Richard R. *Hancock's Diary; Or, A History of the Second Tennessee Confederate Cavalry, with Sketches of First and Seventh Battalions.* 2 vols. Nashville: Brandon, 1887.

Hankins, Samuel W. *Simple Story of a Soldier: Life and Service in the 2d Mississippi Infantry.* Tuscaloosa: University of Alabama Press, 2004.

Hardee, William J. *Rifle and Light Infantry Tactics, for the Instruction, Exercises and Manoeuvres of Riflemen and Light Infantry, Including School of the Soldier and School of the Company.* New York: J. O. Kane, 1862.

Hardy, Robert R. "Explosive Bullets." *Civil War Times Illustrated* 5, 6 (1966): 43–45.

Harley, Stan C. Letter. *Confederate Veteran* 7 (1899): 307.

Harrington, Sion H. III, and John Hairr, eds. *Eyewitnesses to Averasboro.* Erwin, NC: Averasboro Press, 2001.

Harris, F. S. "Fine Shots in the Virginia Army." *Confederate Veteran* 4 (1896): 73–74.

Harris, N. H. "Defence of Battery Gregg." *Southern Historical Society Papers* 8 (1880): 475–488.

Harris, Robert F., and John Niflot, eds. *Dear Sister: The Civil War Letters of the Brothers Gould.* Westport, CT: Praeger, 1998.

Harwell, Richard, and Philip N. Racine, eds. *The Fiery Trail: A Union Officer's Account of Sherman's Last Campaigns.* Knoxville: University of Tennessee Press, 1986.

Hastings, William H., ed. *Letters from a Sharpshooter: The Civil War Letters of Private William B. Greene, Co. G, 2nd United States Sharpshooters (Berdan's), Army of the Potomac, 1861–1865.* Belleville, WI: Historic Publications, 1993.

Hatch, Carl E., ed. *Dearest Susie: A Civil War Infantryman's Letters to His Sweetheart.* Jericho, NY: Exposition, 1971.

Hattaway, Herman. "The Changing Face of Battle." *North and South* 4, 6 (2001): 34–43.

Haughton, Andrew. *Training, Tactics and Leadership in the Confederate Army of Tennessee: Seeds of Failure.* Portland, OR: Frank Cass, 2000.

Hayden, Horace Edwin. "Explosive and Poisoned Bullets." *Confederate Veteran* 7 (1899): 156–158.

———. "Explosive or Poisoned Musket or Rifle Balls." *Southern Historical Society Papers* 8 (1880): 18–28.

Hazen, William B. *A Narrative of Military Service.* Boston: Ticknor, 1885.

Hedley, F. Y. *Marching through Georgia: Pen-Pictures of Every-Day Life in General Sherman's Army, from the Beginning of the Atlanta Campaign until the Close of the War.* Chicago: Donohue, Henneberry, 1890.

Herek, Raymond J. *These Men Have Seen Hard Service: The First Michigan Sharpshooters in the Civil War.* Detroit: Wayne State University Press, 1998.

Herring, Marcus D. "General Rodes at Winchester." *Confederate Veteran* 28 (1920): 184.

Hess, Earl J. *Field Armies and Fortifications in the Civil War: The Eastern Campaigns, 1861–1864.* Chapel Hill: University of North Carolina Press, 2005.

———. *Lee's Tar Heels: The Pettigrew-Kirkland-MacRae Brigade.* Chapel Hill: University of North Carolina Press, 2002.

———. *Pickett's Charge—The Last Attack at Gettysburg.* Chapel Hill: University of North Carolina Press, 2001.

———. "Tactics, Trenches, and Men in the Civil War." In Stig Forster and Jorg Nagler, eds., *On the Road to Total War: The American Civil War and the German Wars of Unification, 1861–1871,* 481–496. New York: Cambridge University Press, 1997.

———. *Trench Warfare under Grant and Lee: Field Fortifications in the Overland Campaign.* Chapel Hill: University of North Carolina Press, 2007.

———. *The Union Soldier in Battle: Enduring the Ordeal of Combat.* Lawrence: University Press of Kansas, 1997.

Hewett, Janet B., ed. *The Roster of Confederate Soldiers, 1861–1865.* 16 vols. Wilmington, NC: Broadfoot, 1995–1996.

Hickenlooper, Andrew. "The Vicksburg Mine." In Robert Underwood Johnson and Clarence Clough Buel, eds., *Battles and Leaders of the Civil War,* 3: 539–542. New York: Thomas Yoseloff, 1956.

Higginson, Thomas Wentworth. *Army Life in a Black Regiment.* Lansing: Michigan State University Press, 1960.

Hill, Daniel H. "Chickamauga—The Great Battle of the West." In Robert Underwood Johnson and Clarence Clough Buel, eds., *Battles and Leaders of the Civil War,* 3: 638–662. New York: Thomas Yoseloff, 1956.

Hoole, W. Stanley, ed. "The Letters of Captain Joab Goodson, 1862–1864 (Part II)." *Alabama Review* 10, 3 (1957): 215–231.

Horn, John. *The Destruction of the Weldon Railroad: Deep Bottom, Globe Tavern, and Reams Station, August 14–25, 1864.* Lynchburg, VA: H. E. Howard, 1991.

———. *The Petersburg Campaign, June 1864–April 1865.* Conshohocken, PA: Combined, 2000.

Horner, Thomas A. "Killers, Fillers, and Fodder." *Parameters* 12, 3 (1982): 27–34.

Howard, Oliver Otis. *Autobiography of Oliver Otis Howard.* 2 vols. New York: Baker & Taylor, 1907.

Howe, M. A. De Wolfe, ed. *Marching with Sherman: Passages from the Letters and Campaign Diaries of Henry Hitchcock, Major and Assistant Adjutant General of Volunteers, November 1864–May 1865.* Lincoln: University of Nebraska Press, 1995.

———. *Touched with Fire: Civil War Letters and Diary of Oliver Wendell Holmes, Jr., 1861–1864.* Cambridge, MA: Harvard University Press, 1946.

Hughes, B. P. *Firepower: Weapons Effectiveness on the Battlefield, 1630–1850.* New York: Sarpedon, 1997.

Hughes, Nathaniel C., ed. *Liddell's Record: St. John Richardson Liddell, Brigadier General, C.S.A.* Dayton, OH: Morningside Bookshop, 1985.

———. *Sir Henry Morton Stanley, Confederate.* Baton Rouge: Louisiana State University Press, 2000.

Hyde, Thomas W. *Following the Greek Cross; Or, Memories of the Sixth Army Corps.* Columbia: University of South Carolina Press, 2005.

Instruction for Field Artillery, Prepared by a Board of Artillery Officers. Philadelphia: J. B. Lippincott, 1864.

Jamieson, Perry D. "Background to Bloodshed: The Tactics of the U.S.-Mexican War and the 1850s." *North and South* 4, 6 (2001): 24–31.

———. *Crossing the Deadly Ground: United States Army Tactics, 1865–1899.* Tuscaloosa: University of Alabama Press, 1994.

———. Review of *Field Armies and Fortifications,* by Earl J. Hess. *America's Civil War* 18, 5 (2005): 66, 68.

Jarman, R. A. "History of Company K, 27th Mississippi." *Aberdeen Examiner,* March 14, 1890.

Johnson, Charles Beneulyn. *Muskets and Medicine; Or, Army Life in the Sixties.* Philadelphia: F. A. Davis, 1917.

Johnson, Richard W. *A Soldier's Reminiscences in Peace and War.* Philadelphia: J. B. Lippincott, 1886.

Jones, A. K. "The Battle of Fort Gregg." *Southern Historical Society Papers* 31 (1903): 56–60.

Jones, Cadwallader. "Tree Cut Down by Bullets." *Confederate Veteran* 34 (1926): 8.

Jordan, Thomas. "Notes of a Confederate Staff-Officer at Shiloh." In Robert Underwood Johnson and Clarence Clough Buel, eds., *Battles and Leaders of the Civil War,* 1: 594–603. New York: Thomas Yoseloff, 1956.

Joyce, Marion D. "Tactical Lessons of the War." *Civil War Times Illustrated* 2, 10 (1964): 42–47.

Judson, Amos M. *History of the Eighty-third Regiment Pennsylvania Volunteers.* Alexandria, VA: Stonewall House, 1985.

Kendall, John Smith. "Recollections of a Confederate Officer." *Louisiana Historical Quarterly* 29 (1946): 1041–1228.

Kennett, Lee, and James LaVerne Anderson. *The Gun in America: The Origins of a National Dilemma.* Westport, CT: Greenwood, 1975.

Kepler, Virginia. "'My God, We Thought You Had a Division Here!'" *Civil War Times Illustrated* 5, 9 (1967): 4–11, 47–48.

Kern, Albert. "Bullets Used in the Civil War." *Confederate Veteran* 24 (1916): 310–311.

Kirwan, A. D., ed. *Johnny Green of the Orphan Brigade: The Journal of a Confederate Soldier.* Lexington: University of Kentucky Press, 1956.

Klein, Frederic Shriver. "The Civil War as a Pitchman's Paradise." *Civil War Times Illustrated* 2, 3 (1963): 30–33.

Ladd, David L., and Audrey J. Ladd, eds. *The Bachelder Papers: Gettysburg in Their Own Words.* 3 vols. Dayton, OH: Morningside Bookshop, 1994–1995.

Laidley, T. T. S. "Breech-Loading Musket." *United States Service Magazine* 3 (January 1865): 67–70.

Laine, J. Gary, and Morris M. Penny. *Law's Alabama Brigade in the War between the Union and the Confederacy.* Shippensburg, PA: White Mane, 1996.

"Lane's Corps of Sharpshooters." *Southern Historical Society Papers* 28 (1900): 1–6.

Laughton, John E. Jr. "The Sharpshooters of Mahone's Brigade." *Southern Historical Society Papers* 22 (1894): 98–105.

Lazenby, John. *Hannibal's War: A Military History of the Second Punic War.* Warminster, UK: Aris & Phillips, 1978.

Leon, Louis. *Diary of a Tar Heel Confederate Soldier.* Charlotte, NC: Stone, 1913.

Lippitt, Francis J. *Reminiscences.* Providence: Preston & Rounds, 1902.

———. *A Treatise on the Tactical Use of the Three Arms: Infantry, Artillery, and Cavalry.* Harrah, OK: Brandy Station Bookshelf, 1994.

Lockett, Samuel H. "The Defense of Vicksburg." In Robert Underwood Johnson and Clarence Clough Buel, eds., *Battles and Leaders of the Civil War,* 3: 482–492. New York: Thomas Yoseloff, 1956.

Longacre, Glenn V., and John E. Haas, eds. *To Battle for God and the Right: The Civil War Letterbooks of Emerson Opdycke.* Urbana: University of Illinois Press, 2003.

Looby, Christopher, ed. *The Complete Civil War Journal and Selected Letters of Thomas Wentworth Higginson.* Chicago: University of Chicago Press, 2000.

Lord, Francis A. "Disposal of Post-war Surplus." *Civil War Times Illustrated* 6, 10 (1968): 35–40.

———. "Springfield Armory Produced 793,000 Muskets." *Civil War Times Illustrated* 2, 7 (1963): 18–19.

Lossing, Benson J. *Pictorial Field Book of the Civil War: Journeys through the Battlefields in the Wake of Conflict.* 3 vols. Baltimore: Johns Hopkins University Press, 1997.

Lowe, Jeffrey C., and Sam Hodges, eds. *Letters to Amanda: The Civil War Letters of Marion Hill Fitzpatrick, Army of Northern Virginia.* Macon, GA: Mercer University Press, 1998.

Lowe, Richard, ed. *A Texas Cavalry Officer's Civil War: The Diary and Letters of James C. Bates.* Baton Rouge: Louisiana State University Press, 1999.

Lynn, John A. *Giant of the Grand Siecle: The French Army, 1610–1715.* New York: Cambridge University Press, 1997.

———. *The Wars of Louis XIV, 1667–1714.* London: Longman, 1999.

Maharay, George S., ed. *Lights and Shadows of Army Life: From Bull Run to Bentonville.* Shippensburg, PA: Burd Street, 1998.

Mahon, John K. "Civil War Infantry Assault Tactics." In *Military Analysis of the Civil War: An Anthology by the Editors of Military Affairs.* Millwood, NY: KTO, 1977.

Malles, Ed, ed. *Bridge Building in Wartime: Colonel Wesley Brainerd's Memoir of the 50th New York Volunteer Engineers.* Knoxville: University of Tennessee Press, 1997.

Manarin, Louis H., ed. *North Carolina Troops, 1861–1865: A Roster.* 15 vols. Raleigh, NC: Division of Archives and History, 1966–2003.

Marraro, Howard R. "Unpublished Letters of Colonel Nicola, Revolutionary Soldier." *Pennsylvania History* 12, 4 (1946): 275–282.

Martin, James Kirby, ed. *Ordinary Courage: The Revolutionary War Adventures of Joseph Plumb Martin.* St. James, NY: Brandywine, 1993.

Mason, Edwin C. "Through the Wilderness to the Bloody Angle at Spotsylvania Court House." In *Glimpses of the Nation's Struggle, Fourth Series: Papers Read before the Minnesota Commandery of the Military Order of the Loyal Legion of the United States, 1892–1897,* 281–312. St. Paul: H. L. Collins, 1898.

McBride, R. E. *In the Ranks: From the Wilderness to Appomattox Court-House.* Cincinnati: Walden & Stowe, 1881.

McGee, B. F. *History of the 72d Indiana Volunteer Infantry of the Mounted Lightning Brigade.* Lafayette, IN: S. Vater, 1882.

McPherson, James M. *Battle Cry of Freedom: The Civil War Era.* New York: Oxford University Press, 1988.

———. *Ordeal by Fire: The Civil War and Reconstruction.* 3rd ed. New York: McGraw-Hill, 2001.

McWhiney, Grady, and Perry D. Jamieson. *Attack and Die: Civil War Military Tactics and the Southern Heritage.* University: University of Alabama Press, 1982.

The Medical and Surgical History of the Civil War. 12 vols. Wilmington, NC: Broadfoot, 1991.

Menge, W. Springer, and J. August Shimrak, eds. *The Civil War Notebook of Daniel Chisholm: A Chronicle of Daily Life in the Union Army, 1864–1865.* New York: Ballantine, 1989.

Military Service Records: A Select Catalog of National Archives Microfilm Publications. Washington, DC: National Archives and Service Administration, 1985.

Miller, Robert Royal, ed. *The Mexican War Journal and Letters of Ralph W. Kirkham.* College Station: Texas A & M University Press, 1991.

Millett, Allan R., and Peter Maslowski. *For the Common Defense: A Military History of the United States of America.* New York: Free Press, 1984.

Minnich, J. W. "Famous Rifles." *Confederate Veteran* 30 (1922): 247–248.

Mitchell, Enoch L., ed. "The Civil War Letters of Thomas Jefferson Newberry." *Journal of Mississippi History* 10, 1 (1948): 44–80.

Monroe, J. *The Company Drill of the Infantry of the Line, Together with the Skirmishing Drill of the Company and Battalion, after the Method of Gen. LeLouteril, Bayonet Fencing, with a Supplement on the Handling and Service of Light Infantry.* New York: D. Van Nostrand, 1863.

Moore, John C. "Battle of Lookout Mountain." *Confederate Veteran* 6 (1898): 426–429.

Morris, James M. *America's Armed Forces: A History.* 2nd ed. Upper Saddle River, NJ: Prentice Hall, 1996.

Morris, William S. *History 31st Regiment Volunteers, Organized by John A. Logan.* Evansville, IN: Keller, 1902.

Morrow, John Anderson. *The Confederate Whitworth Sharpshooters.* N.p., 2002.

Moseley, Thomas Vernon. "Evolution of the American Civil War Infantry Tactics." Ph.D. diss., University of North Carolina, Chapel Hill, 1967.

Muir, Rory. *Tactics and the Experience of Battle in the Age of Napoleon.* New Haven, CT: Yale University Press, 1998.

Mullins, Michael A. *The Fremont Rifles: A History of the 37th Illinois Veteran Volunteer Infantry.* Wilmington, NC: Broadfoot, 1990.

Murphy, Kevin C., ed. *The Civil War Letters of Joseph K. Taylor of the Thirty-seventh Massachusetts Volunteer Infantry.* Lewiston, NY: Edwin Mellen, 1998.

Nasson, Bill. *The South African War, 1899–1902.* London: Arnold, 1999.

Neal, W. A., ed. *An Illustrated History of the Missouri Engineer and the Twenty-fifth Infantry Regiments.* Chicago: Donohue & Henneberry, 1889.

Newsome, Edmund. *Experience in the War of the Great Rebellion.* Carbondale, IL: E. Newsome, 1879.

Nichols, G. W. *A Soldier's Story of His Regiment.* Kennesaw, GA: Continental, 1961.

Nicola, Lewis. *A Treatise of Military Exercise, Calculated for the Use of the Americans.* Philadelphia: Stymer & Cist, 1776.

Noe, Kenneth W., ed. *A Southern Boy in Blue: The Memoirs of Marcus Woodcock, 9th Kentucky Infantry (U.S.A.).* Knoxville: University of Tennessee Press, 1996.

Nosworthy, Brent. *The Anatomy of Victory: Battle Tactics, 1689–1763.* New York: Hippocrene, 1992.

———. *The Bloody Crucible of Courage: Fighting Methods and Combat Experience of the Civil War.* New York: Carroll & Graf, 2003.

Nott, Charles C. *Sketches of the War: A Series of Letters to the North Moore Street School of New York.* 2nd ed. New York: Anson D. F. Randolph, 1865.

Osterhoudt, Henry Jerry. "The Evolution of U.S. Army Assault Tactics, 1778–1919: The Search for Sound Doctrine." Ph.D. diss., Duke University, 1986.

Painter, John S., ed. "Bullets, Hardtack and Mud: A Soldier's View of the Vicksburg Campaign." *Journal of the West* 4, 2 (1965): 129–168.

Palfrey, Francis W. "The Period Which Elapsed between the Fall of Yorktown and the Beginning of the Seven Days' Battles." In *Campaigns in Virginia, 1861–1862: Papers of the Military Historical Society of Massachusetts,* 1: 153–215. Boston: Houghton, Mifflin, 1895.

Parker, John L. *Henry Wilson's Regiment: History of the Twenty-second Massachusetts Infantry, the Second Company Sharpshooters, and the Third Light Battery, in the War of the Rebellion.* Baltimore: Butternut & Blue, 1997.

Parker, Thomas H. *History of the 51st Regiment of P.V. and V.V.* Philadelphia: King & Baird, 1869.

Partridge, Charles A., ed. *History of the Ninety-sixth Regiment Illinois Volunteer Infantry.* Chicago: Brown, Pettibone, 1887.

Patrick, Jeffrey L., ed. *Three Years with Wallace's Zouaves: The Civil War Memoirs of Thomas Wise Durham.* Macon, GA: Mercer University Press, 2003.

Patrick, Jeffrey L., and Robert J. Willey, eds. *Fighting for Liberty and Right: The Civil War Diary of William Bluffton Miller, First Sergeant, Company K, Seventy-fifth Indiana Volunteer Infantry.* Knoxville: University of Tennessee Press, 2005.

Pegler, Martin. *Out of Nowhere: A History of the Military Sniper.* Oxford: Osprey, 2004.

Perret, Geoffrey. *There's a War to Be Won: The United States Army in World War II.* New York: Random House, 1991.

Pfanz, Harry W. *Gettysburg—The First Day.* Chapel Hill: University of North Carolina Press, 2001.

Polley, J. B. "J. B. Polley to 'Charming Nellie.'" *Confederate Veteran* 5 (1897): 425–426.

———. *A Soldier's Letters to Charming Nellie.* New York: Neale, 1908.

———. "Texas in the Battle of the Wilderness." *Confederate Veteran* 5 (1897): 290–293.

Price, Samuel W. "The Skirmish Line in the Atlanta Campaign." In *War Papers: Being Papers Read before the Commandery of the District of Columbia, Military Order of the Loyal Legion of the United States,* 3: 95–106. Wilmington, NC: Broadfoot, 1993.

Prokopowicz, Gerald J. *All for the Regiment: The Army of the Ohio, 1861–1862.* Chapel Hill: University of North Carolina Press, 2001.

Ramsay, M. A. *Command and Cohesion: The Citizen Soldier and Minor Tactics in the British Army, 1870–1918.* Westport, CT: Praeger, 2002.

Ray, Fred L. *Shock Troops of the Confederacy: The Sharpshooter Battalions of the Army of Northern Virginia.* Asheville, NC: CFS, 2006.

———. "Shock Troops of the South." *America's Civil War* 15, 3 (2002): 34–40.

Reid-Green, Marcia, ed. *Letters Home: Henry Matrau of the Iron Brigade.* Lincoln: University of Nebraska Press, 1993.

Reinhart, Joseph R. *A History of the 6th Kentucky Volunteer Infantry, U.S.: The Boys Who Feared No Noise.* Louisville, KY: Beargrass, 2000.

Review of *Field Tactics for Infantry,* by William H. Morris. *North American Review* 99 (July 1864): 287–292.

Rhea, Gordon C. *The Battles for Spotsylvania Court House and the Road to Yellow Tavern, May 7–12, 1864.* Baton Rouge: Louisiana State University Press, 1997.

———. *Cold Harbor: Grant and Lee, May 26–June 3, 1864.* Baton Rouge: Louisiana State University Press, 2002.

Rhodes, Robert Hunt, ed. *All for the Union: The Civil War Diary and Letters of Elisha Hunt Rhodes.* New York: Orion, 1985.

Richard, Allan C. Jr., and Mary Margaret Higginbotham Richard. *The Defense of Vicksburg: A Louisiana Chronicle.* College Station: Texas A & M University Press, 2004.

Ripley, William Y. W. *A History of Company F, 1st United States Sharpshooters.* Rutland, VT: Tuttle, 1883.

Roback, Henry. *The Veteran Volunteers of Herkimer and Otsego Counties in the War of the Rebellion: Being a History of the 152nd N.Y.V.* Utica, NY: L. C. Childs, 1888.

Robertson, James I. Jr., ed. *The Civil War Letters of General Robert McAllister.* New Brunswick, NJ: Rutgers University Press, 1965.

Robins, Colin, ed. *Captain Dunscombe's Diary.* Bowdon, UK: Withycut House, 2003.

Rogers, Robert M. *The 125th Regiment Illinois Volunteer Infantry.* Champaign, IL: Gazette Steam, 1882.

Rosenblatt, Emil, and Ruth Rosenblatt, eds. *Hard Marching Every Day: The Civil War Letters of Private Wilbur Fisk, 1861–1865.* Lawrence: University Press of Kansas, 1992.

Ross, Steven. *From Flintlock to Rifle: Infantry Tactics, 1740–1866.* Rutherford, NJ: Fairleigh Dickinson University Press, 1979.

Roster and Record of Iowa Soldiers in the War of the Rebellion Together with Historical Sketches of Volunteer Organizations, 1861–1866. 6 vols. Des Moines: Emory H. English, 1908–1911.

Rothenburg, Gunther E. *The Art of Warfare in the Age of Napoleon.* Bloomington: Indiana University Press, 1980.

Rozier, John, ed. *The Granite Farm Letters: The Civil War Correspondence of Edgeworth and Sallie Bird.* Athens: University of Georgia Press, 1988.

Rules for the Management and Cleaning of the Rifle Musket, Model 1863. Washington, DC: Government Printing Office, 1863.

Russ, William A. Jr. "The Vicksburg Campaign as Viewed by an Indiana Soldier." *Journal of Mississippi History* 19, 4 (1957): 263–269.

Samito, Christian G., ed. *"Fear Was Not in Him": The Civil War Letters of Major General Francis C. Barlow, U.S.A.* New York: Fordham University Press, 2004.

Schofield, John M. *Forty-six Years in the Army.* New York: Century, 1897.

Scott, Robert Garth, ed. *Fallen Leaves: The Civil War Letters of Major Henry Livermore Abbott.* Kent, OH: Kent State University Press, 1991.

Sears, Stephen W. *Landscape Turned Red: The Battle of Antietam.* New Haven, CT: Ticknor & Fields, 1983.

Shannon, Isaac N. "Sharpshooters with Hood's Army." *Confederate Veteran* 15 (1907): 123–127.

"Sharpshooting in Lee's Army." *Confederate Veteran* 3 (1895): 98.

Shaver, Lewellyn A. *A History of the Sixtieth Alabama Regiment: Gracie's Alabama Brigade.* Montgomery, AL: Barrett & Brown, 1867.

Sherman, William T. *Memoirs of General William T. Sherman.* 2 vols. New York: D. Appleton, 1875.

Sifakis, Stewart. *Compendium of the Confederate Armies.* 10 vols. New York: Facts on File, 1991.

Silliker, Ruth L., ed. *The Rebel Yell and the Yankee Hurrah: The Civil War Journal of a Maine Volunteer.* Camden, ME: Down East, 1985.

Simon, John Y., ed. *The Papers of Ulysses S. Grant.* 24 vols. Carbondale: Southern Illinois University Press, 1967–2000.

Simpson, Brooks D., and Jean V. Berlin, eds. *Sherman's Civil War: Selected Correspondence of William T. Sherman, 1860–1865.* Chapel Hill: University of North Carolina Press, 1999.

Simpson, Harold B., ed. *The Bugle Softly Blows: The Confederate Diary of Benjamin M. Seaton.* Waco, TX: Texian, 1965.

Smith, Justin H. *The War with Mexico.* 2 vols. New York: Macmillan, 1919.

Smith, W. B. *On Wheels and How I Came There.* New York: Hunt & Eaton, 1892.

Spear, Abbott, et al., eds. *The Civil War Recollections of General Ellis Spear.* Orono: University of Maine Press, 1997.

Sperry, A. F. *History of the 33d Iowa Infantry Volunteer Regiment, 1863–6.* Fayetteville: University of Arkansas Press, 1999.

Spruill, Matt, ed. *Guide to the Battle of Chickamauga.* Lawrence: University Press of Kansas, 1993.

"Steel Breast Plates." *Southern Historical Society Papers* 32 (1904): 221–222.

Steuben, Baron von. *Revolutionary War Drill Manual.* New York: Dover, 1985.

Stevens, C. A. *Berdan's United States Sharpshooters in the Army of the Potomac, 1861–1865.* Dayton, OH: Morningside Bookshop, 1972.

Stevens, Michael E., ed. *As if It Were Glory: Robert Beecham's Civil War from the Iron Brigade to the Black Regiments.* Madison, WI: Madison House, 1998.

Stiles, Robert. *Four Years under Marse Robert.* New York: Neale, 1904.

Stillwell, Leander. *The Story of a Common Soldier of Army Life in the Civil War, 1861–1865.* 2nd ed. Kansas City, MO: Franklin Hudson, 1920.

Stone, Henry. "From the Oostenaula to the Chattahoochee." In *The Mississippi Valley, Tennessee, Georgia, Alabama, 1861–1864: Papers of the Military Historical Society of Massachusetts,* 8: 397–427. Boston: Cadet Armory, 1910.

The Story of the Fifty-fifth Regiment Illinois Volunteer Infantry in the Civil War, 1861–1865. Clinton, MA: W. J. Coulter, 1887.

Supplement to the Official Records of the Union and Confederate Armies. 100 vols. Wilmington, NC: Broadfoot, 1993–2000.

Switzer, Charles I., ed. *Ohio Volunteer: The Childhood and Civil War Memoirs of Captain John Calvin Hartzell, OVI.* Athens: Ohio University Press, 2005.

Symonds, Craig L. *A Battlefield Atlas of the American Revolution.* Baltimore: Nautical & Aviation, 1991.

Tancig, W. J., ed. *Confederate Military Land Units, 1861–1865.* New York: Thomas Yoseloff, 1967.

Tapert, Annette, ed. *The Brothers' War: Civil War Letters to Their Loved Ones from the Blue and Gray.* New York: Vintage, 1988.

Taylor, Michael W. *To Drive the Enemy from Southern Soil: The Letters of Col. Francis Marion Parker and the History of the 30th Regiment North Carolina Troops.* Dayton, OH: Morningside Bookshop, 1998.

Thompson, Ed Porter. *History of the Orphan Brigade.* Louisville, KY: Lewis N. Thompson, 1898.

Throne, Mildred, ed. *The Civil War Diary of Cyrus F. Boyd, Fifteenth Iowa Infantry, 1861–1863.* Baton Rouge: Louisiana State University Press, 1998.

———. "Civil War Letters of Abner Dunham, 12th Iowa Infantry." *Iowa Journal of History* 53, 4 (1955): 303–340.

Tidball, John C. "Rifle Target Practice in the Army." *Ordnance Notes,* January 30, 1883.

Tilley, Nannie M., ed. *Federals on the Frontier: The Diary of Benjamin F. McIntyre, 1862–1864.* Austin: University of Texas Press, 1963.

Tower, R. Lockwood, ed. *A Carolinian Goes to War: The Civil War Narrative of Arthur Middleton Manigault, Brigadier General, C.S.A.* Columbia: University of South Carolina Press, 1988.

Trudeau, Noah Andre. *The Last Citadel: Petersburg, Virginia, June 1864–April 1865.* Baton Rouge: Louisiana State University Press, 1991.

———, ed. "Taking Aim at Cemetery Hill." *America's Civil War* 14, 1 (2001): 46–53.

Turner, Nat S. III, ed. *A Southern Soldier's Letters Home: The Civil War Letters of Samuel A. Burney, Cobb's Georgia Legion, Army of Northern Virginia.* Macon, GA: Mercer University Press, 2002.

Tyler, Mason Whiting. *Recollections of the Civil War, with Many Original Diary Entries and Let-*

ters Written from the Seat of War, and with Annotated References. New York: G. P. Putnam's Sons, 1912.

Upton, Emory. *A New System of Infantry Tactics, Double and Single Rank, Adapted to American Topography and Improved Fire-arms.* New York: D. Appleton, 1867.

Wagner, Arthur L. *Organization and Tactics.* New York: B. Westermann, 1895.

Wagner, Margaret E., Gary W. Gallagher, and Paul Finkelman, eds. *Civil War Desk Reference.* New York: Simon & Schuster, 2002.

Walker, Arthur. *The Rifle: Its Theory and Practice.* Westminster: J. B. Nichols and Sons, 1864.

Walker, Charles N., and Rosemary Walker, eds. "Diary of the War, by Robt. S. Robertson." Pt. 4. *Old Fort News* 28 (1965): 175–232.

Walker, Edward L. *Shiloh to Vicksburg: Dear Eliza—An Eyewitness Account in the Civil War Letters of Major Virgil H. Moats.* Pebble Beach, CA: Hedgehog, [1984].

Walker, Francis A. *History of the Second Army Corps in the Army of the Potomac.* New York: Charles Scribner's Sons, 1887.

Walker, James A. "Gordon's Assault on Fort Stedman." *Southern Historical Society Papers* 31 (1903): 19–31.

Wallace, Lew. "The Capture of Fort Donelson." In Robert Underwood Johnson and Clarence Clough Buel, eds., *Battles and Leaders of the Civil War,* 1: 398–428. New York: Thomas Yoseloff, 1956.

Ware, E. F. *The Lyon Campaign in Missouri: Being a History of the First Iowa Infantry.* Iowa City: Press of the Camp Pope Bookshop, 1991.

The War of the Rebellion: A Compilation of the Official Records of the Union and Confederate Armies. 70 vols. Washington, DC: Government Printing Office, 1880–1901.

Watkins, Sam R. *"Co. Aytch": A Side Show of the Big Show.* New York: Collier, 1962.

Watson, William. *Life in the Confederate Army.* New York: Scribner & Welford, 1888.

Wawro, Geoffrey. *The Franco-Prussian War: The German Conquest of France in 1870–1871.* Cambridge: Cambridge University Press, 2003.

Weigley, Russell F. *The Age of Battles: The Quest for Decisive Warfare from Breitenfeld to Waterloo.* Bloomington: Indiana University Press, 1991.

———. *A Great Civil War: A Military and Political History, 1861–1865.* Bloomington: Indiana University Press, 2000.

———. *History of the United States Army.* Rev. ed. Bloomington: Indiana University Press, 1984.

White, Russell C., ed. *The Civil War Diary of Wyman S. White.* Baltimore: Butternut & Blue, 1991.

Wilcox, C. M. *Rifles and Rifle Practice: An Elementary Treatise upon the Theory of Rifle Firing.* New York: D. Van Nostrand, 1859.

Wiley, Bell Irvin. *The Life of Billy Yank: The Common Soldier of the Union.* Baton Rouge: Louisiana State University Press, 1983.

———, ed. *"This Infernal War": The Confederate Letters of Sgt. Edwin H. Fay.* Austin: University of Texas Press, 1958.

Wilkeson, Frank. *Recollections of a Private Soldier in the Army of the Potomac.* New York: G. P. Putnam's Sons, 1893.

"William J. Rogers' Memorandum Book." *West Tennessee Historical Society Papers* 9 (1955): 59–92.

Williams, Frederick D., ed. *The Wild Life of the Army: Civil War Letters of James A. Garfield.* Lansing: Michigan State University Press, 1964.

Williams, Samuel C. *General John T. Wilder: Commander of the Lightning Brigade.* Bloomington: Indiana University Press, 1936.

Wills, Charles W. *Army Life of an Illinois Soldier.* Washington, DC: Globe, 1906.

Wilson, George S. "Wilder's Brigade of Mounted Infantry in the Tullahoma-Chickamauga Campaigns." In *War Talks in Kansas: A Series of Papers Read before the Kansas Commandery of the Military Order of the Loyal Legion of the United States,* 45–76. Kansas City, MO: Franklin Hudson, 1906.

Wilson, James Harrison. *Under the Old Flag.* 2 vols. New York: D. Appleton, 1912.

Winschel, Terrence J., ed. *The Civil War Diary of a Common Soldier: William Wiley of the 77th Illinois Infantry.* Baton Rouge: Louisiana State University Press, 2001.

Winter, George J. "A Battalion of Sharpshooters." *Transactions of the Huguenot Society of South Carolina* 79 (1974): 89–101.

Winther, Oscar Osburn, ed. *With Sherman to the Sea: The Civil War Letters, Diaries and Reminiscences of Theodore F. Upson.* Bloomington: Indiana University Press, 1958.

Wood, Anthony, ed. *Reminiscences of the 35th Ga. Regt. as Seen by a Sharpshooter at the Front.* Conyers, GA: THP, n.d.

Wright, James W. A. "An Eyewitness Account of General Bragg's Chattanooga Campaign." In William Stanley Hoole, *A Historical Sketch of the Thirty-sixth Alabama Infantry Regiment, 1862–1865,* 8–39. University, AL: Confederate, [1986].

Wright, John W. The Corps of Light Infantry in the Continental Army." *American Historical Review* 31 (1925–1926): 454–461.

———. "The Rifle in the American Revolution." *American Historical Review* 29 (1923–1924): 293–299.

Wright, Robert Kenneth Jr. "Organization and Doctrine in the Continental Army, 1774 to 1784." Ph.D. diss., College of William and Mary, 1980.

W. R. S. "The Sharpshooters of Mahone's Old Brigade at the Crater." *Southern Historical Society Papers* 28 (1900): 307–308.

Young, John D. "A Campaign with Sharpshooters." *Philadelphia Weekly Times,* January 26, 1878.

Index